D1499270

THE SHELL PROCESS CONTROL WORKSHOP

David M. Prett
Manfred Morari

WITHDRAWN

Butterworths

Boston   London   Durban   Singapore   Sydney   Toronto   Wellington

The fonts in this book are Computer Modern, set using Leslie Lamport's LaTeX document preparation facility, with help from Janet M. Fox and David B. Garrison.

TeX is a trademark of the American Mathematical Society.

Library of Congress Cataloging-in-Publication Data

Shell Process Control Workshop (1986 : Houston, Tex.)
   The Shell Process Control Workshop.

   Bibliography: p.
   Includes index.
   1.   Chemical process control—Congresses. I. Prett, David M.
II. Morari, Manfred. III. Title.
TP155.75.S49   1986   660.2'81   87-15822
ISBN 0-409-90136-9

Butterworth Publishers
80 Montvale Avenue
Stoneham, MA 01280

10 9 8 7 6 5 4 3 2 1

Printed in the United States of America

Contents

Acknowledgment by D. M. Prett v
Acknowledgment by M. Morari vi
From Sindermann's *The Joy of Science* vii
Objectives viii
Participants ix
Author Index xi
Schedule of Events xii

1 Workshop Papers   **1**
  1.1   Design Methodology Based on the Fundamental Control Problem Formulation, by C. E. Garcia and D. M. Prett 3
  1.2   Control Of Autoclave Processing of Polymeric Composites, by B. R. Holt 27
  1.3   Expert Systems in Process Control and Optimization: A Perspective, by D. B. Garrison, D. M. Prett, and P. E. Steacy . . 37
  1.4   An Artificial Intelligence Perspective in the Design of Control Systems for Complete Chemical Processes, by G. Stephanopoulos, J. Johnston, and R. Lakshmanan 49
  1.5   Process Identification – Past, Present, Future, by D. M. Prett, T. A. Skrovanek, and J. F. Pollard 79
  1.6   Adaptive Control Software for Processes with Significant Deadtime, by C. Brosilow, W. Belias, and C. Cheng 105
  1.7   A Systems Engineering Approach to Process Modeling, by O. A. Asbjørnsen 139
  1.8   Modeling and Control of Dispersed Phase Systems, by J. B. Rawlings 183
  1.9   Robust Control, An Overview and Some New Directions, by V. Manousiouthakis 213
  1.10 An Overview of Nonlinear Geometrical Methods for Process Control, by J. C. Kantor 225
  1.11 Recent Advances in the use of the Internal Model Control Structure for the Synthesis of Robust Multivariable Controllers, by E. Zafiriou 251
  1.12 Characterization of Distillation Nonlinearity for Control System Design and Analysis, by K. A. McDonald 279

iv

1.13 Selecting Sensor Location and Type for Multivariable Processes, by C. Moore, J. Hackney, and D. Canter 291

1.14 Three Critiques of Process Control Revisited a Decade Later, by M. Morari 309

2 Discussion Sessions   **323**

3 Shell Control Problem   **349**

3.1 Introduction 351

3.2 Tutorial Sessions Summary 351

3.3 Problem Description 355

3.4 Schedule of Events 361

4 Summary   **363**

ACKNOWLEDGMENT

The President's Science Advisor has stated that "... we realize that the government has to modify the way it supports university research; we can no longer rely exclusively on narrow, project-by-project mechanisms of support for individual researchers ... Working in a partnership with both academia and industry, government is finding ways for university faculty and students to work on the forefront of industrial problems – which these days happen to include some of the most challenging scientific and technical problems as well." As is already widely known, Shell Development Company has been a strong and active supporter of this position. Over the years we have been involved in many research interactions with academia in order to promote forefront research on industrial problems. In this continuing spirit we invited the workshop participants and the future readers of this final report to embark with us in developing an improved understanding of the process control challenges facing us.

The problems we faced are challenging and include not only the purely technical issues. Not the least of these problems was that associated with the difficulty in facilitating communication between disparate groups. These included both academic and industrial researchers, as well as members of the engineering community who must manufacture the end use products using the insights of the research community. Also included were the more general problems of synergistically assimilating advanced technological artifacts such as computers with the remainder of the world around us. For this reason our workshop included attention to the emerging fields of Expert Systems and Artificial Intelligence.

In bringing this illustrious group together for in-depth discussion we hoped to raise many of these problems for open debate. From this debate may not have come solutions, but at least we have begun the solution process. The process itself will of course greatly facilitate communication between us all in the future. This is of itself of great value.

Finally, let us congratulate the efforts of the workshop participants who gave of their limited time to the endeavor. Their leadership in the field is recognized and their interest in the objectives of the workshop is appreciated.

D. M. Prett

ACKNOWLEDGMENT

A review of the history of control reveals that all the great advances were made by people with a solid fundamental background <u>and</u> an intimate familiarity with real practical problems. Recent examples are the Structured Singular Value introduced by Doyle to deal with model uncertainty in the context of feedback design for multivariable systems and Dynamic Matrix Control invented at Shell to cope with the large numbers of constraints encountered in the real time operation of chemical processes.

During the last ten years we have observed a resurgence of interest in process control, both in industry and academia, stimulated in part by the documented successes of Dynamic Matrix Control. Much new talent is entering the field, and industry can play an important role in channeling the theoretical ideas of these researchers into the right practical directions. The Shell Process Control Workshop is a pioneering step in this direction. In the name of the academic participants I would like to thank Shell Development Company for providing the forum for this fruitful and productive interaction.

Manfred Morari

Science is an art form. Just as with painting or sculpture there are those who develop its commercial aspects, using "formulas" to produce a salable product. Some professionals stumble on a formula early and grind out successive papers which elaborate on a single theme. Then there are those who search for elegant portrayals of concepts—of visions—and who become bored quickly with the routine and the mundane. Such individuals may spend an entire career searching for but never finding the right experiment or never making the unique synthesis. It is from the ranks of those searchers, though, that brilliant new insights can be expected.[†]

Sindermann
The Joy of Science

[†]Carl J. Sindermann, *The Joy of Science: Excellence and its Rewards.* (New York: Plenum Press, 1985), p. 76.

OBJECTIVES

The objective of this process control research workshop is to improve the communication process between academic researchers, industrial researchers, and the engineering community in the field of process control. From this communication process it is intended for us all to improve our understanding of the nature of the control problems facing us. Where does the theory apply and where does it not apply? What is the optimal role for academic researchers, and what is the correspondingly appropriate role for industrial researchers. Are these roles different? In what ways must we modify our respective approaches to problem definition and solution so as to improve the complementary aspect of our efforts? These are some of the issues we explore in this workshop.

ACADEMIC COORDINATOR ...... MANFRED MORARI
INDUSTRIAL COORDINATOR .... DAVID M. PRETT

PARTICIPANTS:

ACADEMIA

O. Asbjørnsen	University of Maryland
C. Brosilow	Case Western Reserve University
B. Holt	University of Washington
J. Johnston	Massachusetts Institute of Technology
J. Kantor	University of Notre Dame
R. Lakshmanan	Massachusetts Institute of Technology
V. Manousiouthakis	University of California at L. A.
K. McDonald	University of California at Davis
C. Moore	University of Tennessee
M. Morari	California Institute of Technology
J. Rawlings	University of Texas
E. Zafiriou	University of Maryland

SHELL PERSONNEL

R. M. Carroll	Shell Development Co., Houston, TX
A. N. Eken	Shell Internationale Petroleum Mij. B. V., The Hague, The Netherlands
J. M. Fox	Shell Development Co., Houston, TX
C. E. Garcia	Shell Development Co., Houston, TX
D. B. Garrison	Shell Development Co., Houston, TX
R. D. Gerard	Shell Development Co., Houston, TX
J. J. Haydel	Shell Development Co., Houston, TX
R. P. Jensen	Shell Development Co., Houston, TX
J. M. Keeton	Shell Development Co., Houston, TX
J. W. Kovach, III	Shell Development Co., Houston, TX
H. K. Lau	Shell Development Co., Houston, TX
J. R. Leis	Shell Development Co., Houston, TX
R. E. Linn	Shell Development Co., Houston, TX
P. Marquis	Shell Recherche S. A. – Centre de Recherche, Grand Couronne, France
S. G. Metchis	Shell Development Co., Houston, TX
J. F. Pollard	Shell Development Co., Houston, TX
D. M. Prett	Shell Development Co., Houston, TX

B. L. Ramaker	Shell Oil Co., Houston, TX
W. J. Schmidt, II	Shell Development Co., Houston, TX
P. E. Steacy	Shell Development Co., Houston, TX
T. A. Skrovanek	Shell Oil Co., Houston, TX
P. Sommelet	Shell Recherche S. A. – Centre de Recherche, Grand Couronne, France
K. Turner	Shell Internationale Petroleum Mij. B. V., The Hague, The Netherlands

AUTHOR INDEX

Asbjørnsen, O. A.	139-182, 336-337
Belias, W.	105-138
Brosilow, C. B.	105-138, 334-335
Canter, D.	291-308
Cheng, C.	105-138
Garcia, C. E.	3-26, 325-327
Garrison, D. B.	37-48, 330-332
Hackney, J.	291-308
Holt, B. R.	27-36, 327-329
Johnston, J.	49-78, 332-333
Kantor, J. C.	225-250, 340-341
Lakshmanan, R.	49-78, 332-333
Manousiouthakis, V.	213-224, 339
McDonald, K. A.	279-290, 342
Moore, C.	291-308, 343-344
Morari, M.	309-322, 344-347
Pollard, J. F.	79-104, 333-334
Prett, D. M.	3-26, 37-48, 79-104
Rawlings, J. B.	183-212, 337-338
Skrovanek, T. A.	79-104
Steacy, P. E.	37-48
Stephanopoulos, G.	49-78
Zafiriou, E.	251-274, 341

SCHEDULE OF EVENTS

MONDAY, DECEMBER 15, 1986
Crowne Plaza Holiday Inn, 14703 Park Row, Houston, TX

7:30 – 8:00	Breakfast – *Fairbanks 1,2*
8:00 – 8:15	**Welcome/Introduction**...R. D. Gerard, General Manager, Westhollow Research Center
8:15 – 9:00	**Opening Business**...D. M. Prett

Coordinator, Morning Session – J. W. Kovach III

9:00 – 9:45	**Opening Presentation**...C. Garcia, *Design Methodology Based on the Fundamental Control Problem Formulation*
9:45 –10:15	Discussion
10:15 –11:00	Research Presentation...B. Holt, *Control of Autoclave Processing of Polymeric Composites*
11:00 –11:30	Discussion
11:30 – 1:30	Lunch/Break
1:30 – 1:45	Meet in Lobby for Transport
2:00 – 5:00	Workshop on Control Design–I *Shell Process Control Training Center Suite 300, 10565 Katy Fwy., Houston, TX*
5:00 – 6:30	Dinner/Break

Coordinator, Evening Session – J. R. Leis

6:30 – 7:15	Research Presentation...D. Garrison, *Expert Systems in Process Control and Optimization: A Perspective*
7:15 – 7:45	Discussion
7:45 – 8:30	Research Presentation...J. Johnston/R. Lakshmanan, *An Artificial Intelligence Perspective in the Design of Control Systems for Complete Chemical Processes*
8:30 – 9:00	Discussion
9:00 – 9:30	General Discussion and Review

TUESDAY, DECEMBER 16, 1986
Crowne Plaza Holiday Inn, 14703 Park Row, Houston, TX

8:00 – 8:30   Breakfast – *Fairbanks 1,2*

Coordinator, Morning Session – J. M. Keeton

8:30 – 9:15   Research Presentation...J. Pollard,
Process Identification – Past, Present, Future

9:15 – 9:45   Discussion

9:45 –10:30   Research Presentation...C. Brosilow,
Adaptive Control Software for Processes with Significant Deadtime

10:30 –11:00   Discussion

11:00 –12:45   Lunch/Break

12:45 – 1:00   Meet in Lobby for Transport

1:15 – 5:00   Workshop on Control Design – II
Shell Process Control Training Center Suite 300, 10565 Katy Fwy., Houston, TX

5:00 – 6:30   Dinner/Break

Coordinator, Evening Session – W. J. Schmidt, II

6:30 – 7:15   Research Presentation...O. Asbjørnsen,
A Systems Engineering Approach to Process Modeling

7:15 – 7:45   Discussion

7:45 – 8:30   Research Presentation...J. Rawlings,
Modeling and Control of Dispersed Phase Systems

8:30 – 9:00   Discussion

9:00 – 9:30   General Discussion and Review

WEDNESDAY, DECEMBER 17, 1986
Crowne Plaza Holiday Inn, 14703 Park Row, Houston, TX

8:00 – 8:30   Breakfast – *Fairbanks 1,2*

Coordinator, Morning Session – C. E. Garcia

8:30 – 9:15   Research Presentation...V. Manousiouthakis,
Robust Control, An Overview and Some New

	Directions
9:15 – 9:45	Discussion
9:45 –10:30	Research Presentation...J. Kantor, *An Overview of Nonlinear Geometrical Methods for Process Control*
10:30 –11:00	Discussion
11:00 – 1:00	Lunch/Break

Coordinator, Afternoon Session – H. K. Lau

1:00 – 1:45	Research Presentation...E. Zafiriou, *Recent Advances in the use of the Internal Model Control Structure for the Synthesis of Robust Multivariable Controllers*
1:45 – 2:15	Discussion
2:15 – 3:00	Research Presentation...K. McDonald, *Characterization of Distillation Nonlinearity for Control System Design and Analysis*
3:00 – 3:30	Discussion
3:30 – 4:00	Break
4:00 – 4:45	Research Presentation...C. Moore, *Selecting Sensor Location and Type for Multivariable Processes*
4:45 – 5:15	Discussion
5:15 – 5:45	General Discussion and Review
6:00 – 6:30	Social Time
6:30 –10:30	Banquet

THURSDAY, DECEMBER 18, 1986
Crowne Plaza Holiday Inn, 14703 Park Row, Houston, TX

7:30 – 8:00	Breakfast – *Fairbanks 1,2*
8:00 – 8:45	**Closing Presentation**...M. Morari, *Three Critiques of Process Control Revisited a Decade Later*
8:45 – 9:15	Discussion
9:15 –11:00	Summary of Workshop *Research Directions*
	• *Adaptive Control*
	• *Process Identification*
	• *Expert Systems in Process Control*

- *Process Representation (Process Modeling)*
- *Other Areas*

11:00   Optional Tour of Westhollow Research Center

Chapter 1

Workshop Papers

Design Methodology Based On the Fundamental Control Problem Formulation

Carlos E. Garcia and David M. Prett

December 15-18, 1986

Abstract

After more than ten years of advanced process control developments at Shell, we have come to the realization that a Unified Approach to control design is needed. Two main reasons for this need exist. On the one hand, the costs associated with treating each control problem as a separate and unique case study are becoming increasingly higher. On the other hand, with the increase in the number of control applications, it is becoming prohibitively costly to have expert manpower maintaining loops. Our own experience dictates that a unification of approach is only possible after there is a complete understanding of the Fundamental Control Problem. This involves a recognition that every control system attempts to meet certain Performance Criteria given a Process Representation. Therefore, a design methodology that recognizes all elements of this Fundamental Control Problem up front and allows for intelligent introduction of necessary assumptions and compromises by the designer should provide the desired unification. In this paper we outline a design methodology based on this Fundamental Control Problem definition indicating the future research efforts needed in order to realize it.

1   Summary

The status of our industry dictates the need to competitively respond to rapidly changing marketplace conditions in order to maintain and enhance profitability. This requires synergism between the many system technologies that are involved in the decision making process: measurement, control, optimization, and logistics. Control technology thus encompasses not only the traditional regulatory aspects but must also take into account all these other technologies in an integrated manner.

In order to be able to ensure an optimal response of the system to changing conditions, all of our manufacturing facilities must be involved in the automated decision making process and therefore, there is a need to find a systematic way of designing an increasingly larger number of control systems. This dictates the need for a Unified Approach to control so that the costs associated with design and maintenance of loops are minimized. This unification

3

can only be achieved by employing the same methodology for every control loop. An understanding of the Fundamental Control Problem provides the desired unification.

In this paper we give a definition of the Fundamental Control Problem which is the result of years of process control experience at Shell. It is shown that every control problem attempts to meet certain Performance Criteria given a particular Process Representation or model. It is demonstrated how every control methodology known to date arrives at a mathematical formulation for the performance criteria and the process representation via assumptions and compromises. Therefore, the usual discussions about which technique is better or worse than another become meaningless since each technique solves the problem for which it was designed. In many cases the real problem to be solved and the problem actually posed for the solution are quite different. The degree to which they are different very often determines the success in implementation.

We propose how the Fundamental Control Problem formulation can yield a sound procedure for control design by making the designer formulate the true performance criteria first. Given the best available process representation, the control problem can be solved allowing the designer to trade off the performance criteria in the face of model errors. Then, the control hardware restrictions are imposed but only after a realization that the performance criteria could indeed be achieved if unlimited computing power were available. This design procedure may dictate that a simple Proportional - Integral - Derivative (PID) controller is all that is needed. Conversely, the required design might involve an on-line nonlinear optimization. In any case, the designer will make a decision after considering all the aspects of the control problem, without making a biased selection of a technique. The designer will have a tool for evaluating the performance criteria and the trade-offs imposed by the uncertainties in the process representation and therefore will be able to make intelligent compromises and assumptions. Major advances in the fields of optimization, artificial intelligence and super computing will be required in order to achieve the realization of this design procedure.

2   Introduction

In [1] we presented our position on the subject of Industrial Model Predictive Control. In this paper we expand on those ideas by proposing a design procedure for process control.

The petro-chemical industry is characterized by having very dynamic and unpredictable marketplace conditions. In the course of 15 years we have witnessed an enormous variation in crude and product prices that forces our processes to operate over a wide range of conditions *(Figure 1)*. In the past, typical responses of our industry to such changes have involved redesign of our plants to meet the new conditions. However, due to the nature and frequency of such changes which have reduced dramatically our industry's profitability, it is becoming less profitable to embark upon major plant redesigns where the payback cannot be guaranteed. As a result, there is a need to extract the most profit out of existing processes while responding to those changes.

The most effective way of responding to marketplace changes with minimal capital investments is provided by the integration of Systems Technologies. This involves all aspects of automation of the decision making process. It is only through this automation that it is possible to achieve rapid and effective response.

The Systems Technologies integrated in the decision making process are the following *(Figure 2)*:

- **Measurement**: The gathering and monitoring of process variables via instrumentation. It includes important process variables such as temperature, pressure, flow, level, composition, etc.

- **Control**: The manipulation of process degrees of freedom for the satisfaction of operating criteria. This typically involves two layers of implementation: the single loop control which is performed via analog controllers or rapid sampling digital controllers; and the control performed using mainframe computers with large CPU capabilities.

- **Optimization**: The manipulation of process degrees of freedom for the satisfaction of plant economic objectives. It is usually implemented at a rate such that the controlled plant is assumed to be at steady-state. Therefore, the distinction between control and optimization is primarily due to a difference in implementation frequencies.

- **Logistics**: The allocation of raw materials and scheduling of operating plants for the maximization of profits and the realization of the company's program. It is implemented to respond to external market changes and involves the integration of all manufacturing locations.

Each one of these technologies plays a unique and complementary role in automating the decision making process and allowing the company to react rapidly to changes. Therefore, one technology cannot be effective without the others. In addition, the effectiveness of the whole approach is only possible when all manufacturing plants are integrated into the system.

In the process control area (now including both control and optimization) due to this necessary increase in the number of applications, there is a need to minimize both design and maintenance costs. Since each petro-chemical process is unique we cannot exploit the population factor as done in other industries (e.g. aerospace). That is, we cannot afford extreme expenses in designing a control system that we know will not work in another process and therefore its cost cannot be spread over a large number of applications. However, we also know that it is not cost effective to treat each problem as a case study. Therefore, a Unified Approach to control is needed allowing treatment of each problem within the same framework resulting in a significant reduction in design costs.

This Unified Approach is also required for more effective maintenance of control systems. We can no longer afford the utilization of expert manpower to maintain each system, but rather desire to have only a small core of expertise maintaining all loops. This is only achievable by implementing techniques designed via a Unified Approach.

In this report, we first propose what we believe to be the Fundamental Control Problem. The understanding of this problem allows us to be able to propose a Unified Approach to control. This unification is justified by the realization that all existing control methodologies to date fit the proposed problem formulation. Finally, a design methodology is outlined based on the solution of this problem and new tools required for its realization are discussed.

3   The Fundamental Control Problem

Control systems in our industry are characterized by constantly changing performance criteria, primarily due to the dynamics of the marketplace that demand changes in operating strategies. Also, our processes are highly nonlinear and not well modeled. In the face of these unique characteristics, the control problem to be solved is:

On-line update the manipulated variables to satisfy multiple, changing performance criteria based on a process representation which includes a description of the uncertainties.

The whole spectrum of process control methodologies in use today is faced with the solution of this problem. The difference between these methodologies lies in the particular assumptions and compromises made to simplify the mathematical problem so that its solution fits the existing hardware capabilities. These assumptions and compromises have generally prevented the solution of the real problems in the past.

As was exposed in [1] the issues of performance criteria formulation and uncertainties pose an interesting challenge for the designer. In the following, we will briefly discuss each issue for completeness before describing the details of the Fundamental Control Problem.

3.1   Performance Criteria for Process Control

The control system will move the process manipulated variables in order to satisfy one or more of the following practical performance criteria:

- **Economic**: These can be associated with either maintaining process variables at the targets dictated by the optimization phase or dynamically minimizing an operating cost function.

- **Safety and Environmental**: Some process variables must not violate specified bounds for reasons of personnel or equipment safety, or because of environmental regulations.

- **Equipment**: The control system must not drive the process outside the physical limitations of the equipment.

- **Product Quality**: Consumer specifications on products must be satisfied.

- **Human Preferences**: There exist excessive levels of variable oscillations or jaggedness that the operator will not tolerate. There can also be preferred modes of operation.

In order for the controller to solve for the manipulated variables, these criteria must be translated to mathematical expressions. It will be assumed that every variable needed to evaluate the practical criteria stated above is measurable or can be inferred by secondary measurements. Then it follows that any of the above practical criteria can be stated as being one of two types of mathematical criteria:

- **Objectives**: Functions of variables to be optimized dynamically where optimal means best satisfaction of the criterion.

- **Constraints**: Functions of variables to be kept within bounds which in turn can be of two kinds:

  - **Hard constraints**: No dynamic violations of the bounds are allowed at any time.
  - **Soft constraints**: Violations of bounds can be allowed temporarily for satisfaction of other criteria.

For example, maintaining a variable at its target can be stated as a minimization of the sum of squared deviations of the measured variable and the target, and therefore is an objective criterion. Keeping a valve within its fully open and closed limits is considered as a satisfaction of a hard constraint.

For almost all problems, the translation of practical performance criteria to mathematical criteria necessarily involves some sort of compromise or assumption by the designer. For instance, in our experience we have found that one of the most crucial compromises made in formulating performance criteria is to ignore constraint criteria. This is motivated by the fact that techniques that handle constraints require significantly more computational power than the standard recursive algorithms [1]. This compromise removes the capability of the control system to automatically reject process disturbances and keep the process close to constraints without violations. Such inefficiency necessarily forces the control system to operate within some distance of the constraint limits therefore preventing the achievement of profits.

In addition, formulation of control performance criteria is made more difficult by the fact that even the practical performance criteria are sometimes not clearly stated and are of an inherent qualitative nature. The field of Artificial Intelligence (AI) has an important role to play in facilitating the formulation of practical criteria for a given process and in the translation of practical criteria to mathematical criteria. In the framework of the proposed design methodology based on the Fundamental Control Problem solution, AI technology will be superimposed upon state-of-the-art control technology to facilitate its implementation[2].

3.2   Process Representation for Process Control

Petro-chemical processes are invariably described by nontrivial, nonlinear thermodynamic and kinetic models. These models are hard to formulate primarily because the feedstocks to the units are not well characterized. Therefore, these models are not only nonlinear but also the model parameters are uncertain.

Since uncertainties impose restrictions on the satisfaction of the control system performance criteria [1], the design procedure for process control must necessarily involve an evaluation of trade-offs between performance criteria satisfaction and model accuracy. For instance, for a given model error, satisfaction of a hard constraint might be impossible to achieve. The designer must then decide whether his/her constraints are too restrictive and therefore should be relaxed, or whether a more detailed (and/or accurate) model of the process is required.

In order to be able to perform this "negotiation" off-line in a design environment, the process representation used must contain not only the model but also the uncertainty description. It is well known that the more detailed (or structured) this uncertainty description is, the less conservative the design will be. Given the difficulty in formulating this uncertainty description one could argue that there may be no advantage in using an uncertainty description at all. However, we believe that inclusion of uncertainties of any structure allows the designer to analyze and design control systems by performing "what if" cases off-line. These studies would be done using quantitative approaches rather than ad-hoc sensitivity arguments. As a result, more robust designs would be produced.

This uncertainty description could be obtained in several ways. We believe that at some stage in the modeling process some sort of on-line experimentation will be required to not only determine more accurate model parameters but also to obtain ranges of parameter uncertainty to be used explicitly for control design. This added requirement on the process representation dictates a need for a change in emphasis in the field of process identification[3].

Of course, the choice of a process representation influences the complexity of the controller. It is clear that much simpler algorithms are obtained if a linear process representation is used. Should the inaccuracies introduced by the linearization still allow the satisfaction of the stipulated performance criteria, then the linear model should sufficient. For more stringent criteria, a nonlinear model might be required.

3.3   Elements of the Fundamental Control Problem and the Unified Approach

Given the performance criteria and the process representation in a quantifiable form, the Fundamental Control Problem stated above can be solved. This problem contains five elements as shown in *Figure 3*. The performance criteria consists of objectives to be optimized subject to inequality and equality constraints. The process representation consists of the model equations plus an uncertainty description of its parameters. Also, each element can be fixed or allowed to change during the life of the control system.

Depending on the mathematical (quantitative) formulation of each element the problem to be solved becomes more or less difficult. For instance, in the absence of inequality constraints and uncertainty description, and for a fixed linear model and objective, the controller can be formulated as a recursive controller with trivial computational load. However, as indicated above, if assumptions and compromises must be made to reach a simpler algorithm, these must be made being aware of the trade-offs.

We must stress the fact that we are not necessarily proposing the on-line solution of a nonlinear multiobjective optimization in the face of uncertainties. A design procedure based on the Fundamental Control Problem formulation is a framework for designing controllers and evaluating trade-offs between performance criteria and model accuracy off-line. The result of this design procedure might very well be a simple PID controller. However, if all the steps are followed, the decision to use such a simple controller would have been arrived at after having considered more advanced designs.

In the following section, the most widely used control methodologies are overviewed in the light of this Fundamental Control Problem. As we shall demonstrate, all of them are the result of specific assumptions and compromises in defining each of the elements.

4   Survey of Existing Control Methodologies

The formulation of the Fundamental Control Problem allows us to now compare the different control methodologies in a unified form. As a result of this comparison we have arrived at the following observations:

- Any control problem to be solved consists of the following five elements:

  - Performance Criteria
    * Objectives
    * Inequality Constraints
    * Equality Constraints

       – Process Representation

          * Model Equations

          * Uncertainty Description

- Any of these elements are specified at various degrees of quantification. Therefore, all existing control methodologies are the result of compromises and assumptions in formulating these elements, which in many cases are made outside awareness.

- All elements can change during the life of the control system.

Let us review the most commonly known control methodologies in the light of the elements of the Fundamental Control Problem. In order to clarify the discussion, figures showing the Fundamental Control Problem elements and the corresponding assumptions are included. The compromised elements are shown in dashed boxes.

4.1   Single Input - Single Output (SISO) PID

An SISO PID controller *(Figure 4)* can be obtained simply by minimizing a single, fixed, quadratic objective function that penalizes deviations of the controlled variable from an exponential trajectory [4]. The model equation is at most of second order, linear and fixed. Note that no constraints nor an explicit uncertainty description are used (dashed boxes). Since these elements are fixed, the control algorithm can be solved off-line to obtain a recursive control law which can be implemented in the most elementary piece of hardware.

It would be unfair to say that uncertainties are totally ignored in the implementation of PID (or other) controllers. In fact, we are always aware of model errors but these are handled implicitly by on-line detuning the controller. This can be interpreted as an on-line compromising of a performance criterion (speed of tracking) in the face of uncertainties.

4.2   Linear Quadratic Optimal Control (LQC)

This classical control algorithm minimizes an objective function that lumps a set of objectives by weighting them together *(Figure 5)*. The objective is quadratic and includes setpoint tracking and input penalties for smoothing the manipulated variables [5]. The model used is linear, and multivariable. No constraints are handled by this technique. Since these elements are fixed, the off-line solution of the optimization problem produces a multivariable recursive controller (e.g. the gain matrix obtained via a solution of a Riccati equation). Depending on the formulation, some on-line tuning might be performed for implicit handling of model errors.

4.3   Self Tuning Regulator (STR)

One of the most popular forms of adaptive control *(Figure 6)* actually solves for exactly the same objective as the Linear Quadratic Optimal Controller with a linear MIMO model [6]. The difference is that the model is no longer fixed but its parameters are updated on-line through an identification procedure. A simple solution of the controller in terms of the model parameters allows the on-line updating of the controller with minimal computational effort. No constraints are handled by this technique.

Because the model is updated on-line, adaptive controllers are heralded to be the solution to the uncertainty problem. However, identification experiments have inherent limitations that do not allow exact identification of all process modes or of all disturbance characteristics. Therefore, the model updates might not be good enough to allow the control system to satisfy the specified objective and therefore, a detuning procedure is inevitable.

4.4   Structured Singular Value Synthesis (SSV)

New developments in the area of robust control have produced a technique where for the first time model uncertainty is explicitly handled [7]. In this technique the same objectives and models as in the Linear Quadratic Optimal Control problem are used *(Figure 7)*. However, the whole approach is formulated in the frequency domain. This allows the designer to specify performance as a function of frequency which is used to weight the frequency response of the setpoint errors. Also, the dynamic characteristics of the inputs to the loop (that is, setpoints and disturbances) are specified as frequency functions.

Structured uncertainties are handled for the "worst" case. This means that, given an uncertainty region for the model parameters, a controller is found that minimizes the worst performance over that uncertainty region.

Since all these elements are fixed, the controller is obtained off-line yielding a multivariable frequency response of the controller. Realization of such a controller produces a transfer function readily implementable with standard hardware. The frequency response formulation used in this methodology immediately rules out consideration of inequality constraints.

4.5   Quadratic Dynamic Matrix Control (QDMC)

This is the state-of-the-art multivariable control technique in use at Shell [8]. Compared to the Linear Quadratic Optimal Control problem, the same objective and models are used *(Figure 8)*. The difference lies in the fact that it can explicitly deal with constraints in process variables. This is possible by solving a quadratic problem on-line and therefore, a computational price is paid for improved capabilities. The on-line solution of the problem also allows for the modification of all elements of the control problem during the operation without redesigning the controller. For example, soft constraints can be added or removed as desired on-line by modifying the objective function. The model can also be updated on-line if desired as done by the Self Tuning Regulator.

However, with all its advantages, the uncertainty issue is not handled by QDMC any differently that by any other controller. Since no uncertainty description is used, one has to rely on on-line detuning to achieve robustness. Again, the objective performance criterion of speed of response is traded-off for robustness.

These comparisons are summarized in *Figure 9*. Note that by the realization that all methodologies are the result of compromises and assumptions in defining the Fundamental Control Problem elements, we can better judge the suitability of a particular scheme for the job. For example, if the given application requires constraint handling but LQC is used, we expect the control system to require extensive supervision for its successful operation. This explains why QDMC has been so successful as a control methodology. Since for improved profitability it is necessary for the process to operate close to process constraints, QDMC provides the designer with a technique for smooth transfers of control action to different output variables as they become constrained. There is no need in QDMC to design different control structures a priori and switch between them on-line using ad-hoc procedures as would be required if any of the other "fixed structure" techniques are used. Such ad-hoc procedure invariably demands increased supervision and maintenance by expert personnel. Conversely, in the absence of constraints, a "fixed structure" controller might be the best choice because of its small on-line computational requirements. In the next section a design procedure utilizing this formulation is outlined.

5   A Design Procedure for Process Control

The formulation of the Fundamental Control Problem naturally yields a design procedure for robust control systems. This design procedure cannot be fully realized with the technology at hand. However, our intention in presenting it is to generate discussion and to stimulate researchers in the control field to look into the different supporting technologies that need to be investigated for its realization. In the discussion that follows, we will assume that unlimited off-line computing capability is available including both number crunching and logical processing. This is by no means restrictive in the light of current advances in the area of Super Computing.

The following design steps are suggested by the formulation described above:

1. Provide the most accurate model and uncertainty description available for the process. A process representation is provided by the user which can be of any kind, depending on the availability of a model. A description of the uncertainties is required. A model building facility is

envisioned that would combine both empirical and fundamental (first principle) models to formulate the process representation required to start the design. This will demand the use of advanced graphics facilities and AI technology superimposed on chemical engineering technology.

2. Formulate the true performance criteria. The designer specifies all the requirements of the control system without restricting his/her judgments to the available control hardware. An advanced graphics terminal inter-facing an Expert System will facilitate the formulation of criteria.

3. Translate to mathematical criteria. The designer will interactively translate the list of performance criteria into mathematical criteria with the help of an Expert System. The type and mathematical form will be chosen. Note that at this point all available measurements will be considered to be able to formulate these mathematical expressions of performance. Also, no attempt will be made at this point to restrict the set of manipulated variables used. Rather, all possible degrees of freedom will be considered. Selection of the final structure of the controller will be done later in the design stage.

4. Solve the Fundamental Control Problem. The resulting optimization problem is solved to produce a solution. Should the problem have no solution, then the designer must either go to Step 1 or Step 2 (or 3) to either get a better model or relax the performance criteria. Successful solution of the problem means that the criteria can be met with the given process representation, however, the implementation might require solution of the problem on-line. If this can be done with the available hardware, then the design is complete. If there are hardware restrictions, then proceed to next step.

5. Specify controller equations. The designer now must pick a control scheme that fits the available hardware. This control scheme consists of not only the algorithm but also of the structure (that is, the sets of controlled and manipulated variables). Then the equations that govern the chosen control system are added as constraints to the optimization problem for solution. A particular formulation of the optimization problem could be used that produces the structure as a direct solution. However, we prefer the approach where, if the original optimization problem can be solved efficiently, the designer can analyze several candidate strategies during one session, allowing the introduction of his/her judgment and expertise. Of course, the inexperienced user will be provided with "expert" guidance via AI technology.

6. Solve the Fundamental Control Problem with controller equations. The controller equations basically determine the manipulated variable moves.

If these equations are substituted into the optimization problem, the controller tuning parameters become now the independent variables. Successful solution of the problem will indicate to the designer that the chosen controller will meet the performance criteria for the given uncertainty description. Moreover, an initial set of tuning parameters will be provided. If no solution exists, then the controller chosen is not appropriate. Several alternatives are available to the designer at this point. Either a different controller must chosen (back to Step 5) or the performance criteria must be relaxed.

6   Areas of Research for the Realization of the Fundamental Control Problem

In order to realize the solution of this problem, several area of research are currently being pursued.

6.1   Time Domain Optimization Under Uncertainties

In order to allow naturally for the inclusion of inequality constraints, the formulation of the control problem should be in the time domain. Therefore, there is a need to find techniques that solve the optimal control problem in the face of uncertainties in the time domain. The issue of conservativeness of time domain uncertainty representations would need to be addressed. Also, the solution method should be very efficient and robust to allow the realization of the proposed design facility.

6.2   Multiobjective Optimization Techniques

All the control methodologies described above handle the satisfaction of multiple objectives by weighting together all objectives into one. The relative importance of each objective is influenced by the weight selected. There is a need to develop optimization techniques that consider multiple objectives and allow a transparent specification of priorities. Expert System technology might help in this area.

6.3   Nonlinear Process Representations

Process control has traditionally relied on linear models, primarily because of their simplicity and the availability of design tools. However, we believe that there is enough wealth of modeling technology in chemical engineering to allow the control engineer to improve on the performance of control systems by utilizing nonlinear models. We need to develop generic models for common processes and methods to validate their parameters with on-line experiments. These models would be made available to the control engineer through sophisticated model building facilities.

6.4 Uncertainty Descriptions

Model parameters and disturbances are never precisely known. The techniques currently available for describing uncertainty rely on a linear model description and therefore, the majority of the problems related to unstructured representations are due to nonlinearities. The use of nonlinear models should provide parameters with more physical significance and therefore, the designer will be able to assess uncertainties based on physical grounds. However, in case of limited hardware capability, linear models might be required. We need to develop techniques to convert from one process representation to another as required by the designer: use a nonlinear model with uncertainties or a linear model with uncertainties.

6.5 Artificial Intelligence

As it was mentioned above, Expert Systems have a definite role to play in the realization of the design methodology under the Fundamental Control Problem. In the context that we have discussed, Expert Systems will never be used to substitute state-of-the-art technology but rather will be superimposed over advanced control technology. In general, these systems will help in facilitating the synergism between qualitative and quantitative knowledge bases. We have already discussed one implementation consisting of facilitating the formulation of practical criteria and their translation into mathematical criteria. This could be done not only at the design stage but also in an on-line environment in order to automate the redefinition of performance criteria in the event of unexpected changes.

6.6 Super Computing

Perhaps the most important of the areas of research, current advances in the field of Super Computing indicate that there is presently the capability to execute the type of optimization problem that we are proposing, not only off-line but on-line as well. Moreover, the availability of high power computing facilities will allow us to employ any model rigor required for achieving the desired performance.

7 References

1. Garcia, C. E. and D. M. Prett, "Advances in Industrial Model Predictive Control," *Chemical Process Control Conference - III*, Asilomar, CA, January 1986.

2. Garrison, D. B., D. M. Prett and P. E. Steacy, "Expert Systems in Process Control and Optimization: A Perspective," *Shell Process Control Workshop*, Houston, TX, December 1986.

3. Prett, D. M., T. A. Skrovanek and J. F. Pollard, "Process Identification - Past, Present and Future," *Shell Process Control Workshop*, Houston, TX, December 1986.

4. Morari, M., S. Skogestad, and D. Rivera, "Implications of Internal Model Control for PID Controllers," *American Control Conference*, San Diego, CA, June 6-8, 1984.

5. Kwakernaak H. and R. Sivan, *Linear Optimal Control Systems*, Wiley Interscience, New York, 1972.

6. Borison, U., "Self-Tuning Regulators for a Class of Multivariable Systems," *Automatica*, **15**, 209-215, 1979.

7. Doyle, J., "Analysis of Feedback Systems with Structure Uncertainties," *IEE Proc.*, **129**, 242-250, 1982.

8. Garcia, C. E. and A. M. Morshedi, "Quadratic Programming Solution of Dynamic Matrix Control (QDMC)," *Chemical Engineering Communications*, **46**, 73-87, 1986.

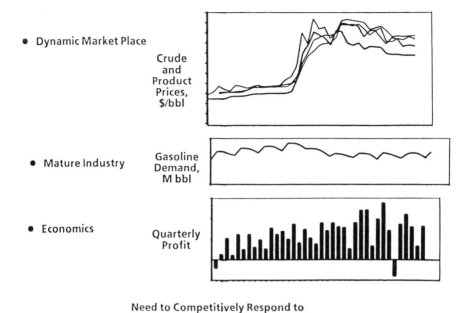

- Dynamic Market Place

  Crude
  and
  Product
  Prices,
  $/bbl

- Mature Industry

  Gasoline
  Demand,
  M bbl

- Economics

  Quarterly
  Profit

Need to Competitively Respond to
Rapidly Changing Marketplace Conditions

Figure 1: Industry Status – Characteristics of the Petrochemical Industry.

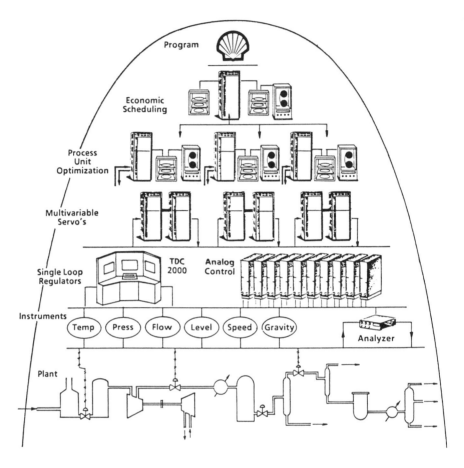

Figure 2.

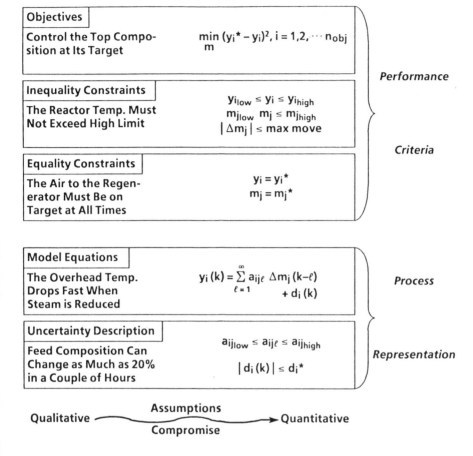

Figure 3: The Fundamental Control Problem.

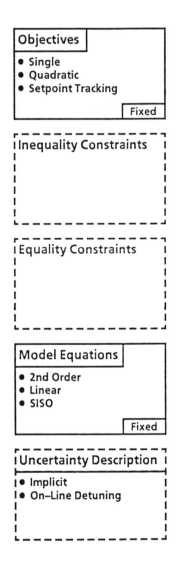

Figure 4: SISO PID.

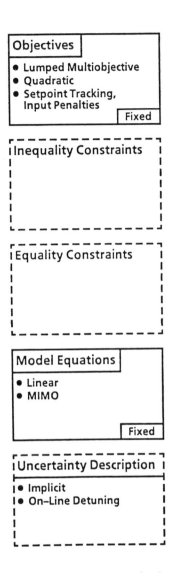

Figure 5: Linear Quadratic Optimal Control.

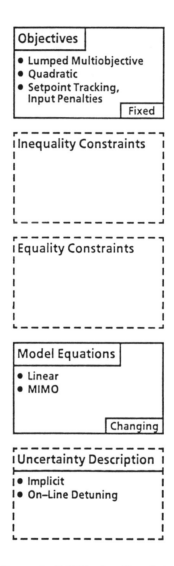

Figure 6: Self-Tuning Regulator.

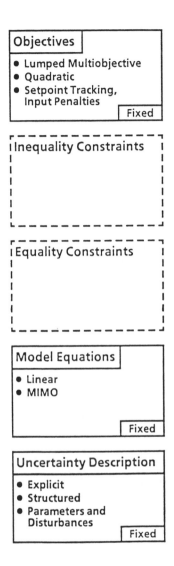

Figure 7: Structured Singular Value Synthesis.

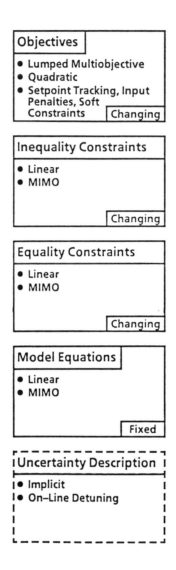

Figure 8: Quadratic Dynamic Matrix Control.

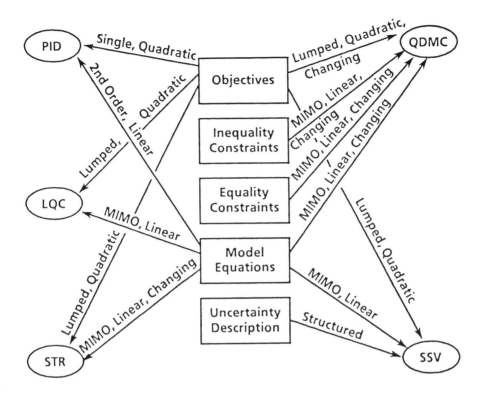

Figure 9: Summary of Elements of Unified Theory.

Control of Autoclave Processing of Polymeric Composites

Bradley R. Holt
Department of Chemical Engineering
University of Washington
Seattle, Washington 98195
(206) 543-0554

December 15-18, 1986

Abstract

Polymeric composite represent a relatively new and increasingly important material. A key step in one of the most important fabrication techniques involves curing the composite laminate under pressure and temperature cycles. The quality of the resulting component is highly dependent on this processing, a process which has traditionally been performed in an open loop manner. This paper provides a qualitative review of the production of polymeric composites components from prepreg. It examines the motivation and issues involved in applying feedback control to the processing of the components in an autoclave and qualitatively reviews the possible control strategies.

Introduction

Polymeric composites consist of a polymer resin binder with embedded fibers made from carbon, glass or other materials. The primary advantages of such materials include their high strength and stiffness (contributed by the fibers), their light weight, and the ability to directly fabricate complex shapes resulting in reduced assembly and machining. These factors make polymeric composites attractive in a wide range of industries, particularly those involved in transportation. The major disadvantage of such composites, particularly the high strength to weight advanced polymeric composites made from carbon or graphite fibers, is their high cost. This has tended to restrict their application to high value products or products where performance is the primary consideration, such as military aircraft.

Components can be made from advanced polymeric composites in a number of ways. Considering only continuous fibers, there are three primary fabrication techniques: Pultrusion, where resin embedded fibers are pulled through

27

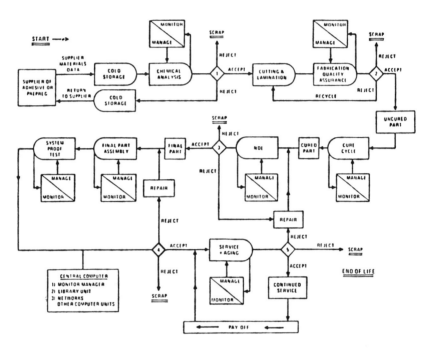

Figure 1: Acceptance Cycle of Composite Parts. (Kaelble, 1985)

a heated die in which curing takes place, Filament winding, where resin embedded fibers are wound around a mandrel in the shape of the component and then cured, and Lay-up/autoclaving where partially cured fiber resin strips, called prepreg, are layered over a tool (mold) in the shape of the component and subsequently cured in an autoclave with temperature and pressure cycles.

This paper will concentrate on the last of these fabrication techniques. It will review the procedures used to make components as currently practiced. It will illustrate the potential advantages and difficulties as well as possible approaches to applying feedback control to the curing cycle in an autoclave. And, it will suggest what the role of academic research should be in control and applied to polymeric composite processing. Although much of the discussion which follows applies to other systems, we will be primarily concerned with advanced polymeric composites made from thermoset resins using graphite or carbon fibers.

Component Manufacturing

The steps involved in manufacturing a polymeric composite component are illustrated in Figure 1.

The manufacturer receives his raw material in the form of prepreg tape.

Prepreg tapes consists of unidirectional fibers embedded in partially cured resin (typically 32-42% by weight resin). The tapes are typically 3 to 30 inches wide and have a thickness of 5 to 6 mills. They are produced by pressing hot, partially cured thermosetting resin into collimated fibers. The prepreg is then stored and shipped under conditions to prevent further curing of the resin.

The actual resins used depend on the application. One primary consideration is the temperature to which the finished component will be exposed. Examples of thermosetting resins suitable for temperatures up to 400 degrees F are epoxy resin systems such as Diglycidly ether of bisphenol A. Higher operating temperatures (<570 degrees F) are possible with polyimide resins. With the current emphasis on aerospace applications, the development of high temperature resins is an area of great interest.

After verifying that the prepreg is of acceptable state and quality for processing (step 1, Fig. 1), the prepreg tape is cut and layered to form the composite component. Since the tape fibers are unidirectional, the properties of the tape are anisotropic. Thus the tape is layered in a manner to provide the necessary physical properties. For example, the first layer of tape might be placed in a 0-degree orientation. The second layer might be laid in a 45-degree orientation and the third in a 90-degree orientation. As few as 10 or as many as 400 layers might be used, even within the same part. This procedure allows entire components (a wing for example) of widely varying thicknesses and complex shapes to be produced with minimal excessive material. The lay-up procedure is often done by hand in the aerospace industry (with relatively small volume production) though automated taped layers are becoming more common.

The third and final processing step to be considered is the curing in the autoclave. The component is placed in a vacuum bag enclosure such as that of Figure 2.

The major components of the bag include release materials and bleeder fabrics to separate the component from the bag and allow for the escape of volatile components and excessive resin during the curing process.

The cure process itself takes place in the autoclave with the laminate in the vacuum bag being subject to a series of temperature and pressure changes. An example of a curing cycle is one (of many) for PRM-15, a polyimide resin system (Darfler, S. C. and R. A. Buyny, 1986):

1. Apply a vacuum of 2 inches of Hg to the bag until a temperature of 300 degrees F is reached. Then apply a vacuum of 28" of Hg.

2. Heat the component at 2 degrees F/min. until 350 degrees F is reached.

3. Hold at 350 degrees F for 4 hours.

4. Heat a 6 degrees F/min. until 600 degrees F is reached.

5. When 460 degrees F is reached, pressurize the autoclave to 200 psig.

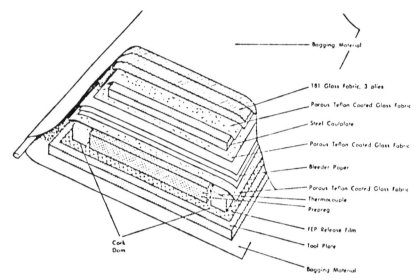

Figure 2: Vacuum bag set up. (Delmonte, 1981)

6. Hold at 600 degrees F for 3 hours.

7. Cool slowly.

In addition, a post cure cycle was also used in this study.

The goal of this processing is to obtain components free from defects with the best physical properties. Thus the initial temperature hold is to allow degassing while avoiding premature curing. The application of external pressure is to consolidate the plies, remove voids, and remove excessive resin (though pressure applied too soon will remove too much resin). The second ramp up in temperature is to start the curing reaction. In modern autoclaves, these cycles can be programmed in and the autoclave temperature and pressure maintained by feedback controllers. The actual properties of the component part are not measured, however, and the time-temperature-pressure profiles are usually based on trial and error experimentation.

Motivation for Control

There are a number of reasons to consider the application of feedback control to the autoclave process:

1. Because of variations between batches of prepreg (e.g. degree of cure) and in the environment in which the prepreg is exposed, for example, one component may have been laid out on a hot humid day, whereas another could have been laid out on a dry day. The former will have a higher water content and thus be more susceptible to void formation. The

current time-temperature-pressure profiles are designed by trial and error to be forgiving to such variations, though as they are experimentally derived there are no guarantees. In addition, by using the same cycle for all curing cycles, the physical properties desired may be compromised.

2. An examination of Figure 1 shows that the curing process in an autoclave is the third processing step. By this stage a considerable investment in the components will have been made, particularly in the lay-up of the composite. It is not unknown that the cost of the cured component exceeds one million dollars (Kaelble, 1985). This makes it important that the results from the curing cycle are satisfactory. Moreover, if good results can be guaranteed in the curing cycle, more complicated and larger components can be fabricated in one piece, thus saving assemble and machining costs.

3. Besides the need to insure satisfactory parts during the curing cycle, it is desirable to minimize the capital and operating costs. This means that the cure cycle should be as short as possible.

4. Since there is no flexibility in the processing, prepreg may have to be rejecting, which could have been used had another cycle been used.

5. Since the cure cycle is derived by trial and error experimentation, there is little hope of predicting how the cure cycle should be modified to account for different geometries or resins. If feedback control can be applied to the "appropriate" physical properties, it might be possible to cure a wider range of components without having to vary the control algorithm.

6. Last, and perhaps most important of all, the nature of the resins is restricted to "forgiving" resins. In the demand for higher performance materials those resins which may be able to meet the physical specifications, but are sensitive to the processing conditions, may not be considered; though, with an adequate feedback control system, reliable parts could be produced. Similarly, considerable effort on the design of the resign systems must be spent on designing "forgiving" resins rather than high "performance" resins.

Item 1 represents the primary cause of deviations whereas 2-6 represent the economic cost. Item 6 represents a particularly difficult problem. Since feedback control systems have not been used, only "forgiving" resins can be used. If "forgiving" resins are designed, only a minimal amount of control is necessary. The extent to which the circle will be broken, at least for aerospace applications, depends on whether "forgiving" resins can be designed which can meet the specifications called for in the next generation aircraft.

Recently the polymer composite community had begun to recognize the advantages offered by the application of feedback control, though as recently at 1983 control could mean replacing the timers used to cycle the autoclave with 8 bit processors in order to implement the autoclave cycles (Coulehan, R.E., 1983). In what follows we summarize the characteristics of the control problem, the strategies being used, and the results to date.

Process Control

The basic control characteristics for autoclave control of polymer composites are:

1. **Batch processes.**

2. **Constrained.** There are input constraint due to limitations on the rate at which the temperature can be raised (dependent on the thermal mass of the component being processed and the heater capacity). There are also constraints on the rate at which the pressure can be raised and the maximum pressure. In addition there are output constraints as the resin will degrade if too high of a temperature is obtained.

3. **Distributed.** The system may be considered distributed in two ways. First, multiple parts are often processed simultaneously in the autoclave. Due to obstructions in airflow, they may experience different environments. Second, one of the advantages of composite components is the ability to produce parts with complex geometrics and variable thicknesses resulting in complex transport behavior.

4. **Non-linear.**

5. **Limited input flexibility.** In the standard autoclave configuration the temperature and the pressure of the entire autoclave are the only adjustable parameters.

6. **Complex system with questionable – though improving – models.**

These represent a very different set of control characteristics than the standard continuous petrochemical process. There are three approaches that can be taken to the control problem. The first is to attempt to determine off-line the optimal pressure and temperature trajectories and then apply them open loop to the autoclave. The second approach is to apply feedback control from measurements on the composite to maintain those measurements on a specified trajectory. The third approach is to perform an on-line optimization.

The first approach is receiving most of the attention today. While it offers the least in terms of improving the quality of composites, merely replacing the experimentally determined trajectories for temperature and pressure with trajectories determined by fundamental analysis, it does provide a systematic way to obtain trajectories for any given composite step. The modeling is also the first step toward applying the other control techniques.

Recently a number of models to determine processing trajectories have been developed [for example: Mallow et al., 1986; Loos and Springer, 1983; Halpin et al., 1983; Lee et al., 1986]. Since the actual process includes 3 phases, mass transfer, heat transfer, multiple reactions and void formation, the models can be quite complex yet still represent simplifications of the actual process. For example, Halpin et al. uses a pseudo first order reaction rate though in reality as many as four competitive reactions can be taking place (Sichina and Gill, 1986). Note too that most of these models relate applied pressure and temperature to properties such as viscosity and internal temperature profiles. What is really required in models is relations to the ultimate strength, stiffness and other design parameters which must be inferred in these current models.

The second approach being taken in controlling autoclaves is to directly employ feedback control to insure that a profile is being followed. This will allow considerable improvement in adjusting for variations in the initial state of the laminate provided the right measurements are available. Since most investigators feel that temperature alone is not adequate for control, considerable research is being carried out on the development of new types of sensors. Examples include dielectric sensors for the measurement of viscosity and degree of cure [May, 1983; Day, 1986; Kranbuehl et al., 1986; Lane et al., 1986, and others] and the use of acoustic waveguides to monitor the degree of cure and viscosity by the attenuation of ultrasound [Harrold and Sanjana, 1986]. Note that it is possible to leave the measurement device in the final product provided it is small enough not to interfere with the polymer matrix (e.g., a thermocouple embedded in the middle of the matrix). Despite the effort in sensor development, little research has been done on what should be measured (in terms of providing the best indication of final product specifications) and very few examples of actual feedback control exist.

Finally, the most challenging yet potentially rewarding approach to autoclave control is to perform on-line optimizing control. Here measurements of the state of the component would be continually made and an optimal trajectory from the current state to the final product state could be determined based on a model of the process. This approach would solve many of the problems which motivated the use of feedback control, yet does not seem to be pursued by any research group.

This control strategy provides a number of interesting challenges as well. In particular, the extraction of a model suitable for on-line control from the fundamental models. Note that in a sense the requirements on such a model are less than the models used to determine the optimal trajectories because the feedback reduces the uncertainty - one no longer needs to know exactly how the applied temperature effects the glass transition temperature, but rather the general relationship. On a fundamental control level one is determining the control strategy for a batch, non-linear system with uncertain parameters and a high value product.

Another challenge arises when one adds that the control should improve with time as feedback from the results of each run is obtained. This represents an update of the model used in the control algorithm and would represent a non-linear adaptive controller.

A number of researchers have proposed a different approach to the control problem based on the use of expert systems [Servais et al., 1986, and Lee, 1986]. There they are essentially abandoning attempts to obtain a quantitative model of the process but plan to attempt to control the autoclave using quantitative information and rules of thumb. Such control strategies may work well in very uncertain environments, though this has not been demonstrated.

The final control strategy suggested here represents a quantum leap in potential performance. It is not, however, the last word as all of the discussions made here have avoided the distributed parameter nature of the problem.

The Role of Academia in Control Research

With reference to the discussion above on control in autoclave processing, we can point to how Universities can contribute to control research. First, University can develop and prove control concepts. Since industry is often unwilling to upset current production procedures without very good cause, the University can demonstrate and prove the concept. For example, in the composite processing, justification for the value of control and the suitability of the models can come from academia.

Second, the Universities can develop the basic theory behind control algorithms and procedures whether developed in industry or academia. Controllers, or rules of thumb which work, can lead to significant advances if properly understood.

Third, industry in general has a more immediate focus in their research. For example, in composites, while they are interested in controlling the autoclave process by employing feedback control to follow a specific temperature profile, they do not seem to be as willing to examine on-line optimization. This suggests that University research should be focused more toward the high risk but potential large pay-off problems.

A final area in which academic can contribute is in innovation and new ideas. Consider the use of autoclaves to cure composites. From a control standpoint this is a very clumsy tool with almost no "distributed" parameter control capacity. Since the University has no investment in large autoclaves, it is possible to consider altogether different approaches to the curing cycle, or at the very least, small variations such as individually controlled IR heat lamps within the autoclave.

While the academic investigators are pursuing these areas, industry can assist the universities in a number of ways besides direct assistance and support. One of the most important contributions which can be made is in the general area of communications. Too often industrial representatives suggest all of their control problem have been solved and that everything is proceeding in the "best" possible fashion. While perhaps painful to the company involved, it is very helpful for an academic to see what doesn't work in the real world and why. The more industry can "promote" their failures, the more likely they are to be solved and the entire control field enriched.

Summary

The control of autoclaves for curing of polymeric composites is an area where the application of feedback control offers a great deal of potential, both in improving the components made by the process and as a subject area for control research. It represents one of the more complicated batch processes and provides challenges in the treatment of constraints,uncertainty and nonlinearities. Moreover, due to the increased demands on the performance of such composites and the pressure to reduce costs, it represents an area where the only possible solution may be innovative processing employing active control.

References

[1] Coulehan, R. E., "Adapting a Microprocessor System to Autoclave Processing," *Proceedings of the European Meeting on Polymer Processing and Properties,* Plenum Press, New York.

[2] Darfler, S. C. and R. A. Buyny, "Cure Cycles for PRM-15 Effects of Imidization Times and Temperatures," 31st International SAMPE Symposium, 1986.

[3] Dave, R., J. L. Kardos and M. P. Dudukovic, "Process Modeling of Thermosetting Matrix Composites: A Guide for Autoclave Cure Cycle Selection," *Proceedings of the American Society for Composites,* Dayton, Ohio, 1986.

[4] Day, D. R., "Cure Control: Strategies for Use of Dielectric Sensors," 31st International SAMPE Symposium, 1986.

[5] Delmonte, John, *Technology of Carbon and Graphite Fiber Composites,* Van Nostrand Reinhold Company, New York, 1981.

[6] Halpin, J. C., J. L. Kardos and M. P. Dudukovic, "Processing Science: An Approach for Prepreg Systems," *Pure and Applied Chemistry,* 55, 1983, pages 893-906.

[7] Harrold, R. T. and Z. N. Sanjana, "Theoretical and Practical Aspects of Acoustic Waveguide Cure Monitoring of Composites and Materials," 31st International SAMPE Symposium, 1986.

[8] Kaelble, K. H., *Computer-Aided Design of Polymers and Composites,* Marcel Dekker, Inc., New York, 1985.

[9] Kranbuehl, D., et al, "Dynamic Dielectric Analysis: A Means for Process Control," 31st International SAMPE Symposium, 1986.

[10] Lane, J. W., J. C. Seferis and M. A. Bachmann, "Dielectric Modeling of the Curing Process," *Polymer Composites,* 7, 1986.

[11] Lee, J. W., J. C. Seferis and D. C. Bonner, "Prepreg Processing Science," *SAMPE Quarterly,* 17, (58), 1986.

[12] Lee, W. L., "Composite Cure Process Control by Expert Systems," *Proceedings of the American Society for Composites,* Dayton, Ohio, 1986.

[13] Loos, A. C. and G. S. Springer, "Curing of Epoxy Matrix Composites," *Journal of Composite Materials,* 17, 1983.

[14] Mallow, A. R., F. R. Muncaster and F. C. Campbell, "Science Based Cure Model for Composites," *Proceedings of the American Society for Composites,* Dayton, Ohio, 1986.

[15] May, C. A., "The Chemical Characterization and Processing Science of Composites," *Journal of Composite Materials,* 17, 1983.

[16] Servais, R. A., C. W. Lee and C. E. Browning, "Intelligent Processing of Composite Materials," 31st International SAMPE Symposium, 1986.

[17] Sichina, W. J. and P. S. Gill, "Characterization of Composites by Thermal Analysis," 31st International SAMPE Symposium, 1986.

Expert Systems in
Process Control and Optimization
A Perspective

D. B. Garrison, D. M. Prett, P. E. Steacy

Shell Development Company
Westhollow Research Center

December 15-18, 1986

Abstract

Recently the topic of expert systems in process applications has received a great deal of attention. The use of symbolic programming techniques to codify levels of human expertise has great potential in process control. Such systems can provide expert assistance, explanation, diagnosis, and a certain level of automated programming. There are, however, recognized limitations of this technology which should be a focus of research in artificial intelligence. This paper attempts to point out some of the strengths and weaknesses of expert system technologies with some emphasis on what we recognize as the potential and the problems associated with the use of these systems in real-time, process control applications.

Our experience in the design of knowledge-based systems has lead to two basic conclusions. The first is that knowledge-based systems can be a very powerful tool to accomplish cognitive tasks which currently require a human expert. The second conclusion is that knowledge acquisition and representation schemes are inadequate and that considerable research effort must be expended to create more expressive and natural representation methods. It is the refinement of knowledge representation techniques coupled with the use of learning which will allow the routine use of knowledge-based systems.

Introduction

In the past decade, Shell has embarked upon an aggressive mission to implement model-predictive control and on-line optimization to achieve the maximum profit potential from our manufacturing facilities. Optimization of complex chemical processes subject to economic criteria demands a control strategy which is viable over an extended range of operating conditions. The development of model-predictive control algorithms such as Dynamic Matrix

Control [1] has allowed implementation of control systems which are flexible enough to handle changing performance requirements and enforce operating constraints.

The optimization phase of the work has concentrated on the development of the required process representations and the design and implementation of improved mathematical programming algorithms. The modification of these algorithms has lead to significant improvements in speed and stability of numerical solutions for large-scale process optimization. On-line optimization is implemented with sufficiently high frequency to compensate for changes in the environment, product specifications, the volatility of the market, and the architecture and operability of the process.

To realize the profitability of on-line optimization, robust controllers must be available which can reject fast disturbances, enforce process constraints, and move the process smoothly to new operating points. This is achieved through the use of control strategies which have a strong basis in theory. Essentially, this theory is already in place. It is the difficulty the designer has in dealing with the complexity of the multivariable control problem that has encouraged the use by non-experts of ad hoc solutions of limited application. These solution techniques have redirected the focus of research in process control away from the consideration of the fundamental control problem [2]. Furthermore, there is a temptation to design control systems based on a presumed operating space. The solution of the optimization problem tends to extend this space [2-4].

The advent of complex control and optimization systems has amplified an old problem. How can we transfer technology to the field with the appropriate level of abstraction such that the technology is fully utilized?

Artificial Intelligence and Process Control

Chemical process control is becoming a more analytical activity. Fully instrumented plants require each operator to be responsible for a larger portion of the plant. Distributed computer systems provide data acquisition and control action but create an out-of-loop unfamiliarity for the both the operators and engineers.

Despite advances in theory and hardware, control experts still rely on intuitive process models developed through training and experience [3-5]. Engineers use this knowledge to develop explicit process models, identify appropriate control algorithms, specify constraints and performance objectives, and validate control designs. Operators apply this expertise to the supervision of control through signal validation, alarm diagnosis, and the proper response to extraordinary load changes and other unusual process events [6].

Recently, expert systems have been proposed as a means to model these activities. Research in artificial intelligence and cognitive psychology has been devoted to characterizing and codifying the skill of experts. In the field of process control this research has resulted in the design of expert systems to mimic the action of an expert operator or control engineer. Such systems strive to provide assistance for a full range of process operations including monitoring, control, interpretation, planning, and diagnosis [7-13].

We see the application of artificial intelligence as a step toward a more complete automation of the control room. As our world becomes more specialized it is critical that we provide diagnosis, advice, and explanation to the the non-specialist. There are, however, a number of research problems to be addressed before intelligent systems can be routinely constructed and implemented.

Heuristics in Control Applications

Control has sufficient leverage that the enthusiasm over the possibility of using expert systems to codify this expertise is understandable. Process control has often been cited as an attractive area for the application of knowledge based systems. Systems have been developed that claim to capture the expertise of a control engineer or master operator. There is a tendency, however, to build knowledge based systems which rely on purely empirical observation or that implement ad-hoc control techniques of limited applicability [14-19].

The phrase "capture the expertise of the operator" has appeared recently in the literature. An operator does have a level of expertise about the process operation but his knowledge is based primarily on empiricism. While heuristics ("rules-of-thumb") are important, they do not adequately represent all the information needed to build systems for process control applications.

Rule based systems can provide a methodology for technologically shallow organizations to improve upon poorly tuned PID control and implement constraint handling. The control strategy in such systems consists of an adaptive PID controller and a "safety jacket" of logical conditions. The rules for adapting the controller parameters and the safety jacket of constraints comprise the rule base for these systems [15-17]. We have coined the term *Automatic Manual Control* to describe such systems. The best known of these systems is EXACT, a rule-based, adaptive controller developed by Foxboro [17]. Other heuristic systems for rule-based adaptive control have been developed for reactors and other process systems [20,21], for diagnosing process alarms [22], and for designing distillation controls [23].

We are concerned that many of these systems ignore existing technologies because of the complexity of the multivariable control problem. The emphasis of research in the application of artificial intelligence to process control must be on the classification of the knowledge required to solve the control problem, the integration of hard (deterministic) and soft (heuristic) technologies, and the explication of the cognitive processes associated with design and diagnosis.

Technology Transfer

At Shell, we have developed explicit model-based control algorithms which are based on engineering principles and numerical mathematics. We have further developed on-line optimization systems capable of determining steady-state operating conditions by solving constrained non-linear optimization problems with economic objective functions [1-3].

These systems are certainly not devoid of heuristics. Consider the modeling tasks associated with various control activities. In a plant, several types of process models are used. Functional models that describe the major portions of a plant and the relations among them are used in logistics optimization. Detailed models that define the mathematical relations among process variables are used for on-line optimization. Expertise is required to select and validate these models and to choose appropriate physical property routines. Simulation and optimization based on these models requires the specification of objectives, constraints, and independent variables. The identification of a linear process model is required for control design. Finally procedural models must be determined to describe operation during start-up, shut-down, and load changes.

Optimization and control design rely on the expertise of the practitioner requiring assumptions and compromises, generalizations, and internal process descriptions. These cognitive processes should be modeled to create expert apprentices in order to aid transfer and acceptance of advanced technology in the field. The integration of expert systems and symbolic computing with advanced control and optimization techniques should be the focus of our research in expert systems.

Expert Apprentices

Advanced technologies in the on-line environment are implemented as software systems. The complexity of these systems has motivated the need for an integrated, high-level environment for control activities. Expert apprentices are based on the idea of providing support for the non-expert user. A user may not be aware of the latest developments in control theory or may not have the necessary experience to synthesize that theory into an effective analysis and design approach. Furthermore, he may not be a frequent practitioner of system design or user of the desired software package.

Such interfaces to control and optimization software have been described in the literature. An example system for optimization design has been developed by NASA. The system provides an interface to the Automated Design Synthesis (ADS) program for general purpose optimization [24].

The authors found that an engineer usually has sufficient knowledge in his discipline but lacks the necessary experience in optimization to make the proper choice among alternates in setting up the optimization problem. The system provides guidelines to identify the best choice of strategy, optimizer, and one-dimensional search. The choice between 98 alternate hypotheses is governed by a 200 rule knowledge base. The rule base is divided into three parts, (1) for constrained problems, (2) for unconstrained problems, and (3) for constrained problems to be treated as unconstrained problems.

There are several benefits of expert interfaces to software control design. It relieves the control engineer of the task of troubleshooting and permits concentration on the design of new control algorithms. Furthermore, the use of the system will train engineers at manufacturing locations in the theory and use of these new techniques. These engineers can in turn provide valuable insight into the performance of the algorithms and propose improvements in the algorithms and the capabilities of the expert system.

Perhaps the best example of an expert interface for control design is the GE CACE-III system [25]. The authors have identified several activities in control engineering.

- Modeling the plant.

- Determining the characteristics of the plant model.

- Making changes in the plant to facilitate control.

- Formulating the elements of the control design.

- Checking to see if the design problem is well-posed.

- Executing appropriate design procedures.

- Performing design tradeoffs.

- Validating the design.

- Providing complete documentation of the final design.

- Implementing the final design.

This system was built to interface with existing control software in an integrated environment. This system contains rule bases for

- providing support in model development,

- guiding the user in entering design specifications and checking for consistency, completeness, and workability,

- dealing with specifications, constraints, and plant characteristics,

- deciding which design approaches will best solve the problem,

- updating the solution frame to reflect the corresponding additions or corrections and supervising tradeoff analysis, and

- governing the final system validation and converting the idealized control design to a practical implementation.

In chemical process control, several such systems have been developed [26-27]. We are currently developing a prototype system to provide expert assistance to design SISO PI or QDMC controllers. The system consists of a query and menu based interface to a rule base which allows the user to develop simple control designs. It is being implemented on a Symbolics machine using KEE and ZetaLISP communicating with a VAX running the control routines written in Pascal.

Symbolic Programming

Artificial intelligence research has provided a number of new software engineering tools for the implementation of expert systems. Rule based systems use first order predicate logic to describe behavior. Object-oriented programming allows for classification with inheritance.

The development of the LISP machine has provided an integrated software development environment with specialized tools for symbolic programming. The open architecture of the machine allows the software designer to access all levels of the system and provides a more natural interface between system software and application programs. New user interface technologies provide windows, menus, and more powerful help facilities [28].

While these programming techniques provide significant additional capabilities to the application developer, they still provide only primitive forms of knowledge representation. The deficiencies of current knowledge representation schemes often result in the use of meta-rules invoked to control inferencing. Without this embedded control, an expert system to provide advice to a process operator could easily consist of tens of thousands of rules from which an inference could not be made in real time.

The use of meta-rules is cited as a means of preventing performance degradation as the size of a knowledge base increases. The use of such techniques to impose a higher level of control is in direct conflict with one of the most compelling reasons for building an expert system — the separation of domain-dependent knowledge and program control. The coercion required to attain this control can create a system which is unmaintainable and inextensible.

An additional area in need of significant development before real time AI becomes a reality is the interface between symbolic computing and numerical computing. Most symbolic programming environments do not provide an effective interface between symbolic programs and existing software based on traditional programming paradigms. Most distributed control systems provide data highways which can support the use of a symbolic computing engine but the necessary task-to-task communication tools are either not available or require significant additional programming by the user to affect the necessary protocols.

It is quite likely that new inventions in knowledge representation will parallel the progression of past research topics in software engineering. The development of language constructs in the early days of computing is a ready example. The concept of a pointer type now commonplace in modern structured languages provided a more natural means of implementing and manipulating certain data structures. Similarly, new knowledge representation methods can provide a more natural way of expressing coded expertise. We must conduct research into this area in order to create systems with the behavior which we desire.

The Knowledge Engineer

To date, our experience in this project and others has lead us to question the effectiveness of the knowledge engineer. The process of debriefing the expert is complex and requires that the knowledge engineer attain a high level of expertise. Furthermore, the expert is culturally inclined to simplify his expertise during a communication session with the knowledge engineer. As a result, these sessions become training in the traditional sense. The assumption of a teacher/student relationship rather than an expert/interviewer relationship imposes control over the passage of information which is contrary to that which is desired. When this happens, the information exchanged tends to be of the novice type which severely limits the capability of the resulting expert system.

The time required to debrief the expert creates a serious lag between the algorithm development and the development of the accompanying knowledge base. Refinements and new discoveries can significantly affect the use of these algorithms and can cause premature obsolescence of the knowledge base. Furthermore, systems in the on-line environment tend to be process dependent and are not easily generalized. These observations clearly indicate that knowledge engineering must be supplemented with other methods of knowledge acquisition. One obvious solution is the use of automated knowledge acquisition systems traditionally classified as machine learning.

Machine Learning

Presently, there are few systems which learn about or adapt to their working environment by asking appropriate questions and by observing the way in which they are used. Thus a technique to reduce the labor and time required to develop an expert apprentice may be to design a system which can augment its own knowledge base by observing the actions of an expert user. The rules devised by such a system would be of the form "state-action-state". Moreover, techniques exist which allow such systems to learn from examples and to make inferences based on analogical reasoning [29].

Systems which provide any type of machine learning are difficult to build. There are a number of questions which remain to be answered.

- What learning mechanisms are appropriate for an expert apprentice for interactive design software?

- Does the knowledge required for design in the process industries differ from that in other design applications?

- How will the learned expertise be represented?

- What is the appropriate level of generality?

- Will training examples be created in the laboratory or taken from naturally occurring instances in the field?

- Can the system, once taught, then be used as a means of instruction or simply as a means of providing advice to the user?

- How will the availability of a learning apprentice interface effect the design of new user interfaces?

There is considerable leverage in providing a means by which an expert can incorporate his knowledge throughout the algorithm development/ implementation cycle. Machine learning can provide a framework for delivery of new technologies and a more efficient vehicle for technology transfer. In addition, those applications which are appropriate for artificial intelligence can be developed without the use of a knowledge engineer.

Conclusions

The expert system research in process control within Shell Development Company began some time ago. During that time, we have devoted some attention to reviewing the work of others and analyzing the successes and failures of expert systems in engineering applications. The following conclusions have been reached:

- Expert systems are to be viewed as an additional programming tool with certain inherent advantages and recognized limitations.

- The role of expert systems in control applications should be a means of disseminating technology and not a means of facilitating the implementation of ad hoc solutions of limited applicability.

- AI research in the engineering disciplines should concentrate on the explication of the cognitive processes associated with design, diagnosis, planning, and interpretation.

- One of the most immediate problems which expert systems can address is that of technology transfer through intelligent interfaces to control software.

- Current knowledge representation schemes are inadequate and often promote the use of explicit control over the knowledge base. Research is needed to invent methods which are natural and expressive. This research must be stimulated by the increase in cognitive modeling listed above.

- It is absolutely essential that expert system tools be integrated with existing software and that the interface be natural and transparent.

- Knowledge engineering is a grossly inefficient process. It must be supplemented with other techniques for knowledge acquisition.

- Learning systems have the potential to eliminate the knowledge engineer and automate the knowledge acquisition task. The technology is available to build the first generation of such systems.

The lure of expert systems is the promise of a higher level of abstraction for software engineering. This abstraction cannot be achieved and the potential of knowledge-based systems realized until the research topics mentioned above are addressed.

References

1. Prett, D. M. and R. D. Gillette, *Optimization and Constrained Multivariable Control of a Catalytic Cracking Unit*, **AIChE National Meeting**, Houston, Texas, 1979.

2. Garcia, C. E. and D. M. Prett, *New Developments in Model-Predictive Control*, **11th Interamerican Conference of Chemical Engineering**, San Juan, Puerto Rico, 1985.

3. Garcia, C. E. and D. M. Prett, *Advances in Model-Predictive Control*, **Proceedings of the Third International Conference on Chemical Process Control**, Asilomar, California, January 12-17, 1986.

4. Garcia, C. E., and D. M. Prett, *Design Methodology Based on the Fundamental Control Problem Formulation*, Shell Process Control Workshop, Houston, Texas, December 15-18, 1986.

5. Prett, D. M., T. A. Skrovanek, and J. F. Pollard, *Process Identification – Past, Present, Future*, Shell Process Control Workshop, Houston, Texas, December 15-18, 1986.

6. Eberts, R. E., (1985) *Cognitive Skills and Process Control*, **Chemical Engineering Progress**, December, 1985, pp 30-34.

7. Bañares-Alcántara, R., et al., *Knowledge-Based Systems for CAD*, **Chemical Engineering Progress**, September, 1985, pp 25-30.

8. Schwarz, K., *Data Acquisition and Control: A Look Ahead*, **The Industrial and Process Control Magazine**, May, 1985, pp 63-65.

9. Brooks, K., *AI Tackles Real-Time Process Control*, **Chemical Week**, September 10, 1986, pp 38-39.

10. Goff, K. W., *Artificial Intelligence in Process Control*, **Mechanical Engineering**, October, 1985, pp 53-57.

11. Herrod, R. A. and Papas, B., *Artificial Intelligence Moves Into Industrial and Process Control*, **The Industrial and Process Control Magazine**, March, 1985, pp 45-49.

12. —, *Expert Systems Aiding Control Engineers*, **Control and Instrumentation**, May 28, 1986, pg 41.

13. Sripada, N. R., D. G. Fisher, and A. J. Morris, *Application of Expert Systems to Process Problems*, **Energy Processing Canada**, Nov./Dec., 1985, pp 26-31.

14. Moore, R. L. and M. A. Kramer, *Expert Systems in On-Line Process Control*, **Proceedings of the Third International Conference on Chemical Process Control**, Asilomar, California, January 12-17, 1986, pp 839-867.

15. Årzèn, K., *Use of Expert Systems in Closed Loop Feedback Control*, **Proceedings of the 1986 American Control Conference**, Seattle, Washington, June 18-20, 1986, pp 140-145.

16. Carmon, A., *Applying Self-Tuning Control to Plant*, **Control and Instrumentation**, January, 1986, pp 81-83.

17. Beaverstock, M., Bristol, E. H., Fortin, D., *Expert Systems as a Stimulus to Improved Process Control*, **Proceedings of the 1985 American Control Conference**, Boston, Massachusetts, June 19-21, 1985, pp 898-903.

18. Johnson, R. R., T. Canales, and D. Lager, *An Expert System to Control a Fusion Energy Experiment*, **Proceedings of the 1986 American Control Conference**, Seattle, Washington, June 18-20, 1986, pp 1170-1175.

19. Clapp, N. E. et al, (1985) *Application of Expert Systems to Heat Exchanger Control at the 100-Megawatt High-Flux Isotope Reactor*, **Proceedings of the Control West Conference**, Long Beach, California, September 16-18, 1985, pp 208-211.

20. Fisher, W. R., Doherty, M. F., and Douglas, J. H., *Operating Heuristics for the Control of Complete Chemical Plants*, **Proceedings of the 1985 American Control Conference**, Boston, Massachusetts, June 19-21, 1985, pp 293-298.

21. Nisenfield, A. E. and M. A. Turk, *Batch Reactor Control: Could an Expert Advisor Help?*, **Intech**, April, 1986, pp 57-58.

22. Palowitch, B. L. and M. A. Kramer, *The Application of a Knowledge-Based Expert System to Chemical Plant Fault Diagnosis*, **Proceedings of the 1985 American Control Conference**, Boston, Massachusetts, June 19-21, 1985, pp 646-651.

23. Shinskey, F. G., *An Expert System for the Design of Distillation Controls*, **Proceedings of the Third International Conference on Chemical Process Control**, Asilomar, California, January 12-17, 1986, pp 895-912.

24. Rogers, J. L. and J. M. Barthelemy, *An Expert System for Choosing the Best Combination of Options in a General Purpose Program for Automated Design Synthesis*, **Engineering with Computers**, Vol. 1, No. 4, pp 217-227.

25. Taylor, J. H., D. K. Frederick, and J. R. James, *An Expert System Scenerario for Computer-Aided Control Engineering*, **Proceedings of the 1984 American Control Conference**, San Diego, California, June 6-8, 1984, pp 120-128.

26. Trankle, T. L., Sheu, P., and Rabin, U. H., *Expert System Architecture for Control System Design*, **Proceedings of the 1986 American Control Conference**, Seattle, Washington, June 18-20, 1986, pp 1163-1169.

27. Niida, K. and T. Umeda, *Process Control System Synthesis by an Expert System*, **Proceedings of the Third International Conference on Chemical Process Control**, Asilomar, California, January 12-17, 1986, pp 869-894.

28. Bromley, H. *LISP Lore: A Guide to Programming the LISP Machine*, Kluwer Academic Publishers, Boston, 1986.

29. Michalski, R. S., Carbonell, J. G., and Mitchell, T. M., *Machine Learning*, Morgan Kaufmann Publishers, Inc., Los Altos, California, 1983.

An Artificial Intelligence Perspective in the Design of Control Systems for Complete Chemical Processes

George Stephanopoulos, James Johnston
and Rama Lakshmanan

Laboratory for Intelligent Systems in Process Engineering
Department of Chemical Engineering
Massachusetts Institute of Technology
Cambridge, MA. 02139

December 15-18, 1986

Abstract

This position paper, reflecting our current research efforts, describes the characteristics of; (a) a comprehensive model for the process of plant-wide control system design, and (b) the computing environment necessary to carry out the various engineering tasks, e.g. goal setting, planning, reasoning, analysis, evaluation, and learning. In particular, it discusses the principal ideas, currently under implementation, which transfer large segments of the control system design process from the designer to the computer, and the remaining open issues for future work. Such mechanization of the design methodology relies heavily on recent developments in knowledgebased systems and other areas of artificial intelligence, which provide the scope for the systematic use of existing powerful analytic tools from control theory. Finally, it argues that new computing environments and programming styles are needed to develop the proposed prototype of human-aided control system design, which depart from the established models of computeraided engineering.

I   Introduction

The configuration of control systems for a complete chemical process is required to satisfy a multitude of diversified control tasks, such as (Morari, 1982; Govind and Powers, 1982; Stephanopoulos, 1983); (a) regulate production and product quality at the desired levels (Buckley, 1964; Douglas, 1982), (b) satisfy environmental regulations, (c) provide safe operation (Rivas, et.al. 1974a,b), (d) achieve optimum economic operation (Arkun and Stephanopoulos, 1980, 1981; Prett and Gillette, 1980), (e) allow for smooth execution of turn-down, normal or emergency shut-down, change-over operations (Lau, 1982), etc. This diversity of high-level goals makes the process of designing control systems for complete chemical plants quite informal and an activity in the province of expert designers. Indeed no general theory is available for the systematic modeling of the process that leads to the design of such control systems. The existing analytical and to a limited extent synthetic tools from control theory can only tackle isolated and fragmented issues, e.g. analyze the interaction of control loops (Bristol, 1966, 1977; Tung and Edgar, 1981; Gagnepain and Seborg, 1982; McAvoy, 1983; Grosdidier, et.al. 1985), design the control system for a given set of inputs and outputs (Garcia and Morari, 1982; 1985; Cutler and Ramaker, 1980; Mehra, et.al. 1982; Stephanopoulos and Huang, 1986), analyze the effects of model uncertainty on the stability and performance specifications of a local control system (Doyle, 1978; Doyle and Stein, 1981; Doyle and Morari, 1986), propose decomposition of control systems with minimum interaction, etc. Their formal and explicit nature does not allow them to address the informal and implicit sequence of tasks, which constitutes the process of control system design itself. Thus, there is no theory which can answer the following questions during the design of control systems for complete chemical plants:

1. Where do I start the design from?

2. What do I do next?

3. How do I formulate the needed assumptions and conjectures for the design to proceed?

The problem is further complicated because the available knowledge on the chemical process and its required performance comes at various levels of detail, including qualitative, semi-quantitative, or fully quantifiable information. Furthermore, a number of control design objectives become explicit at various stages of the design process and to a large extent depend on the course that the design procedure has followed up to that point, e.g. the identification of secondary measurements, the addition of redundant manipulations, the stipulation of secondary controllers, etc. Finally, as the control system design proceeds, both factual and strategic knowledge is accumulated, which may suggest revisions in the design strategy up to that point or/and define the course for the remaining portion of the design.

In the absence of a general theory on how design is done, one can only rely on the experts for the formulation of a concise design procedure. Indeed, expert designers bring in to the design process the following skills:

1. Formulation of mental models as to what tasks to consider first, next, and last.

2. Organization of plans and strategies on how to achieve certain goals.

3. Modeling of the chemical process and of the associated control systems at various levels of detail.

4. Stipulation of beliefs that the designer derives from the current description of the control system through means of analysis, simulation, heuristic assumptions, etc.

5. Disciplined use of assumptions and conjectures with the accompanying ability to systematically retract them if proven infeasible or conflicting.

6. Ability to accumulate knowledge, both factual and strategic, as the design proceeds and learn from it.

But, the expert designer's approach is subject to certain biases and limitations. First, his/her previous experience has created inherent preferences and biases, which inhibit the development of creatively new control systems. Second, he/she does not make full use of the available theoretical results and tools, thus making the design procedure less rational and only empirically justifiable. To overcome these shortcomings, extensive efforts have been undertaken to develop a computer-aided environment that will support the designer at the various steps of the design procedure. Thus, convenient graphic interfaces make the interaction between designer and computer very transparent. Database management systems provide easy means for the deposit and retrieval of data, intermediate results, past designs, and other data. The availability of a very large number of special-purpose routines allow the use of a growing number of analytical and, to a limited extent synthetic tools.

Despite the continuous enrichment of computer-aided control system design environments, the character of the overall design procedure remains the same, namely, "the human does the design and the computer provides the support tools, without understanding the design process, its rationale, or the design decisions." The drawbacks of the "computer-aided" paradigm are several and at times critical. They all stem from the fact that the design procedure is implicit in the designer's mind. To the extent that we can untangle and make explicit the design procedure, thus emulating the expert's own methodology, the process can be mechanized. But, in this case we are moving towards a "human-aided machine design" paradigm (Balzer, 1981; Scherlis and Scott, 1983; Mostow, 1985), where the computer, guided by the human, can carry out significant portions of the design by understanding the design process itself, its rationale and the reasoning behind a number of design decisions. This is the paradigm that we are currently developing for the design of control systems for complete chemical plants. Preliminary prototypes of this model, but with rather limited automation, can be found in the works of Taylor (1986), Niida and Umeda (1986) and Shinskey (1986).The benefits from the availability of such mechanized models of control system design are many and diverse:

1. Improvements in cost and reliability.

2. Explicit documentation of the design process itself, e.g. why certain goals were set during the design and how they were achieved, how design decisions were made, what assumptions and simplifications were invoked, what models for the processing units were used at various stages of design, what alternative control schemes were examined and why certain ones were selected over others, etc.

3. Explicit documentation of the control system for a given chemical plant, e.g. what controllers are used for material balance or product quality control and for what kind of disturbances were designed, what control configurations safeguard the plant against hard constraints, what is the structure to implement turn-down operations, etc.

4. Easy verification and modification of the resulting control system design. Having an explicit documentation of the intermediate design tasks, generated alternatives, rationale behind various design decisions, assumptions, conjectures, simplifications, etc. one can replay the design scenario and easily verify the validity of the control system designed or modify its design premises for further improvements.

5. The mechanized model of a control system design methodology offers an excellent depository for the organization of new empirical knowledge or/and the systematic incorporation of new theoretical results and analytic quantitative tools. Such inclusion of new knowledge will progressively increase the automation of the design procedure itself.

The design and development of a software package that will implement the "human-aided" paradigm for the design of plant-wide control systems, requires a drastic departure from established prototypes of computing environments and programming styles. Given that the structure of the design methodology is not and cannot be a priori fully specified, the familiar concept of imperative FORTRAN flow charts is of little use. Prototyping such a system requires extensive programming flexibility, modular and incremental program development and testing without the time-consuming cycles of debugging–editing–linking–compiling–running. Such environment is offered by the SYMBOLICS 3600 series computers. Furthermore, in order to provide transparent interfaces, graphic description of process flowsheets and control structures, automatic modeling of processing units and control configurations, representation of declarative knowledge and a flexible representation of procedural knowledge, we have adopted programming styles such as "objectoriented programming" and "data-driven programming" using Symbolics ZETALISP and IntelliCorp's KEE (Knowledge Engineering Environment).

The rest of this paper is structured as follows. Section II describes the characteristics of a model for the plant-wide control system design, ideas currently under implementation as well as issues still open for future work. Section III outlines the value of various artificial intelligence techniques in implementing the model of systematic control system design, while Section IV presents more details on the software engineering of the CONTROL DESIGN-KIT, which implements the automatic and interactive facilities of the design procedure.

II   A Model for the Plant Control System Design Procedure and Its Characteristics

Designing the control structure for a chemical plant is largely a process of integrating constraints imposed by; the functional specifications of the control system, explicit and implicit requirements on its performance (static, dynamic), limitations on the available technology (sensors, actuators, digital architectures), and the restrictions on the design procedure itself (time spent on the design, available expertise, cost). Our research objective is to mechanize this process to the extent that this is possible, thus transferring the design procedure from the human expert to the machine. In recent years, significant research work has concentrated on the rationalization of the design process and the outline of comprehensive models for it. Since artificial intelligence addresses the mechanization of large and complex, knowledge intensive tasks, most of the contributions have come from workers in this area (Adelson and Soloway, 1984; Balzer, 1984; Barstow, 1984; Freeman and Wasserman, 1983; Katz, et.al., 1984; Mostow, 1985). Transfer of these ideas to particular domains of engineering design, such as the design of control systems for chemical plants, is not trivial and should not be superficially approached through imitation. Basically, the design model to be employed will emulate the design procedure invoked by experts. To this end, Mostow (1985) has identified the following list of essential aspects of the design model, whose better understanding is indispensable:

1. The state of the design; i.e., the series of control system descriptions at various levels of detail.

2. The goal structure of the control system design. If the design is to be mechanized, it should be a purposeful activity with clear goals which guide what is to be done at each level of design.

3. Design decisions. Once a goal has been selected, the designer outlines several plans for achieving it. Selecting the best plan involves design decisions, which should be clear and explicit.

4. Rationales for design decisions. They justify the goal selection and the choice of the best plan to accomplish it. The need for explicit rationalization forces the designer to evaluate his/her reasoning and allows the design procedure to evolve systematically as new knowledge becomes available.

5. Control of the design process. Guides the designer as to what should be done next, i.e. what is the next goal, what approaches (plans) are available for achieving it, and how to select the best approach.

6. Learning in design. Emulate the designer's own ability to learn from the accumulation of factual and strategic knowledge.

Let us now examine how a design model as the above applies to the problem of designing control systems for complete chemical plants.

II.A   The State of the Control System Design

As the control system for a chemical plant is being designed, it goes through a series of descriptions with variable detail. Thus, initially we may have a control system describing only the material flow controllers, while in subsequent steps the control system description includes controllers for product quality control, safeguards against hard constraint violations, etc. Each intermediate description of the control system should be consistent with its immediate predecessor and successor systems. Thus, we need to be able to represent (model) partial developments of the control system in an evolutionary manner with increasing amounts of added knowledge. This can be achieved through the "inheritance" mechanisms of object-oriented programming (see Section IV).

To organize the transfer of inherited information among partial descriptions of the control system, we need to identify the generic transformations leading from one partial control system to the next. Two such mechanisms can be used in the particular design problem we are examining:

- **Refinement model.** According to this model (Mitchell et.al., 1981; Rich et.al., 1979; Kant and Newell, 1983), a certain component of the current control system description cannot be implemented through existing technology and is further decomposed (refined) to implementable descriptions. Characteristic example is the decomposition of large MIMO control systems to smaller-size MIMO control systems for which reliable and robust designs can be implemented.

- **Transformational model.** According to this model (Balzer et.al., 1976; Lam and Mostow, 1983; Scherlis and Scott, 1983), we operate simultaneously on all components of an intermediate control system design. This approach is used when no natural decomposition among the various control sub-systems exists. Typical application of this mechanism is the design of safety interlock control configurations.

Past research work has heavily favored the refinement model, because of its simplicity and explicit character of its intended goal. Some examples; (a) the multi-level, multi-echelon decomposition of the overall control design problem, (b) the refinement of MIMO into a structure of SISO control loops, (c) the sequential design of start-up procedures, etc.

It is of paramount importance that the description of the intermediate design states (partial developments of the overall control system) be very explicit, i.e. what are the control objectives, the primary and secondary measurements and manipulated variables, the rationale of the intended control configurations (e.g. feedforward, feedback, inferential, etc.), the relevant constraints, the series of assumptions and simplifications reflected on the current design, the models of the processing units, information pertaining to the sensitivity analysis of the current control system with respect to model uncertainty and other parameters, etc. Such explicit description of declarative and procedural knowledge reflecting the state of a current partial control system is very well served by the object- and rule-representations central to the design package under development.

The benefits of an explicit description and documentation of intermediate control system designs are several: Easy maintenance; verification; analysis through various theoretical tools; modification through new knowledge; explanation of its objectives and design decisions, etc.

II.B The Goal Structure of the Design Process and Its Rationale

It is common to characterize the design of control systems for a chemical plant as a series of design activities leading from specifications to design implementations. The intermediate steps describe partial control system descriptions, as discussed above. But, the various design activities are not randomly implemented. They are guided by specific mental models, or goals, postulated by the designer. All of the computer-aided design packages do not possess an explicit goal structure, and leave it to the designer to implicitly invoke it. It should be emphasized that a goal is not a description of the state of the next control system design. A goal simply prescribes (not describes) how the current control system should be changed to accommodate the goal's objectives.

In the system we are developing, the goal structure must be explicitly defined, in order to allow its computer implementation. In the absence of a generalized theory this can be accomplished only through idealized design histories, which emulate the design scenarios played out by expert designers. For example, the following sequence of intermediate designs represents an incomplete scenario for the synthesis of control structures for complete chemical plants:

- Design the controller structure which achieves the desired range of material balance control.

- Design the controller structure which achieves the desired range of product quality control.

- Test the consistency of the above two control structures and complete the design with the necessary additional controllers.

- Design the secondary control structure that will safeguard the process against the violation of a set of hard constraints.

- Design the control structure for normal and emergency shut-down operations.

- Design the control structure required for automatic start-up.

Each of the intermediate goals stated above, represents a high-level goal with no explicit methodology for its direct satisfaction, i.e. there is no known procedure of the type: "implement material balance control <specifications>." Consequently, each of the high level goals must be replaced by structure of sub-goals until the final set of sub-goals can be implemented through existing procedures; qualitative, semi-quantitative, or rigorous fully analytic. For example, the top goal

> "design the controller structure to achieve the desired range of material balance control"

can be replaced by a series of sub-goals, such as:

- **Sub-goal 1.** Determine the set of external disturbances which define the scope of material balance control and their characteristics, e.g. origin, frequency, magnitude, measurability.

- **Sub-goal 2.** Determine the allowable control structures for the implementation of material balance control, etc.

Each of the sub-goals can be in turn replaced by other sub-goals, and so on, until the final set of sub-goals is directly implementable by feasible transformations. For example, sub-goal 2 gives rise to further sub-goals such as: "Determine the scope of the require material balance control structure for the main process streams;" "determine the scope of material balance control structure for the secondary processing streams (external recycles, internal recycled flows)," etc. It is clear that the higher level goals of the goal structure are less precise and are based on informal empirical heuristics. At lower levels of the goal structure the design objectives become clearer and can be handled by rigorous analytic methodologies. For example, the design of specific control systems using internal model control (IMC) generates a series of sub-goals which are clearly and explicitly defined, such as; find the desired operational band width, estimate the range of expected model uncertainties, decide on the tolerable deterioration of performance, etc.

Although the goals and ensuing sub-goals seem to be generating a tree-like goal structure (Wile, 1983; Balzer, 1984; Kant and Newell, 1983), in fact that is only partially true and one can easily identify interacting goals with the following departures from the tree structure; goal conflicts, goal sharing, down-stream goal prerequisites. For example:

- **Goal sharing.** The sub-goal, "define a static model for the process," arising from sub-goal 1 above, helps achieve the following sub-goal, "determine the scope of the required material balance control structure for the main process streams," which arose from sub-goal 2 above.

- **Goal conflict.** Sub-goals "determine the scope of the required material balance control structure for the main process streams" and "determine the scope of material balance control structure for secondary processing streams" may be in conflict and no way exists that both be achieved (due to lack of sufficient manipulations).

As we discussed earlier in paragraph II.A, the successive states of the control system design are related to each other either through the refinement model or the transformational model. The goal structure for the refinement model is very simple and quite explicit, namely; "decompose an unrefined component of the current control system into two or more implementable components" (e.g. the decomposition of MIMO into a series of SISO systems). On the other hand, the goal structure of the transformational model is more complex and requires meticulous care for its precise description. Furthermore, the goal structure for the transformational model is not unique, it depends on the selected plan that emulates the designer's own choices. Thus, in principle we define alternative goal structures and the selection of the most appropriate is an essential design decision.

It is very important to notice that not all goals carry the same significance and importance. For example, sub-goals intended to collect the relevant information on the characteristics of the disturbances and the simplified model of a chemical process are more critical than the sub-goal of generating the material balance controllers for the main process streams. Explicit knowledge characterizing the relative importance of the various goals in the goal structure is indispensable, since it allows the automated design approach to be explicitly rationalized.

In the control system design package under development we are (or will be) providing a series of alternative explicit goal representations. For example:

- "Implement <functional specification>" is the form of the general functionality goals generated by the refinement model, wherever it is appropriate.

- Production-rules are used to encode explicit empirical or model based knowledge, which generates the goal structure in a dynamic manner.

- "Aspects" and "demons," which are features of the declarative representation of objects (processing units, controllers, etc.), are very convenient facilities, within the scope of object-oriented programming, for triggering the generation of goal structures.

- Constraint propagation techniques can be used to encode performance goals and to select among alternative goal structures.

Ultimately, an automated design procedure should be reflective i.e. reason about its own behaviour. This is in principle possible, given the fact that the design route is determined by the goal structure, which is explicitly known. For example, the computer program can be "aware" of the path that the design procedure is following towards the generation of control structures for a chemical plant and its rationale. Attaching various measures (e.g. expertise required, cost, time, etc.) on the segments of the design on its behaviour and alter the course of design, e.g. follow a quick shortcut control system design, if such is required by the designer.

II.C   Design Decisions and Their Rationale

Two general classes of decisions are involved during the design of control systems for chemical plants; decisions related to the design path, i.e. to the goal structure, and decisions pertaining to the selection of the best method for achieving a certain goal or sub-goal. Let us discuss them in more detail.

- **Design path related decisions.** Different designers follow different paths during the execution of a design problem, i.e. they formulate in their minds different sequences of top-level goals and different goal structures to implement them. Selecting the sequence of design steps and the goal structure to implement it, constitutes a very important design decision task. For example, material balance control structures can be generated before or after a multi-level decomposition of the control tasks and a multi-echelon decomposition of the chemical process (Morari, et.al., 1978). The ensuing sub-goals structures is quite different. If multi-level, multi-echelon decomposition has preceded, the designer has framed the scope of material balance control in terms of operating considerations, but he/she needs to strive for consistency among the control structures corresponding to different process segments.

- **Design methods related decisions.** In addition to the design path related decisions, the designer ought to select the methods through which a goal is reduced to sub-goals, or a sub-goal is finally implemented. Several alternatives often present themselves. For example, if the goal is "decompose the process into several processing areas," one could use economic sensitivity analysis or measures of static interaction among the control loops of the various process segments (Morari, et.al., 1978; Arkun and Stephanopoulos, 1981), etc. Depending on the route that one selects, different goal structures are generated. Furthermore, if the decision is to select the most appropriate method for implementing (designing) an inferential control scheme (Weber and Brosilow, 1972; Shinskey, 1979; Morari and Fung, 1981) , one could use among a series of candidate methods such as, static, dynamic, based on modal analysis, least-squares estimate of model sensitivity, etc.

In both cases the design decisions can be modeled as choice sets. The goal structure should be explicitly annotated with competing methods for achieving each goal.

But, if the design is to be an explicit and purposeful activity, design decisions must be supported by the following three sets of knowledge:

1. Assumptions, commitments, beliefs and mental models associated with the set of alternative design methods should be explicitly presented and modeled. These are often expressed in terms of; production-rules within the framework of an expert system, additional constraints, or procedures. Mechanisms, which explicitly reason on whether we should make a particular assumption or committment, are not as yet available.

2. To select among alternative methods, one needs to provide an explicit goal with its associated goal structure, which involves; selection of criteria, evaluation, choosing the most appropriate alternative method.

3. Certain decisions, although they seem to be acceptable at a given stage of the design process, are proved to be unsatisfactory at a later stage of the design and they should be retracted. "Demons" in the frame-representation of objects are an excellent mechanism for triggering this form of dependent backtracking.

The credibility of a mechanized control design procedure depends to a large extent on its ability to provide explicit explanations to a series of design queries, such as;

- why did the designer follow the specific design path?

- why the particular goal structure (design plan) was generated to implement a given top goal?

- why did the designer use a specific set of criteria to compare alternatives?

- why a set of assumptions, commitments, etc. is reasonable for the given problem?

In attempting to provide explicitly the rationale for the various design decisions, one is forced to clarify and systematically analyze the implied goal structure and the methods for selecting among alternatives. Therefore, as Mostow (1985) has pointed out; what is good for explanation is good for design. Unfortunately, explanations using formal proofs are far from present capabilities and one can only rely on less general forms of reasoning.

II.D   Controlling the Path of the Design Process

Deciding the sequence of top-level goals is the most crucial aspect of the design methodology at large. We mentioned earlier that, selecting to structure the material balance controllers before or after the decomposition of control tasks and of the chemical process has very profound effects on the ensuing design path. To control the path of the design process means to make explicit the decisions which determine the sequence in which the various goals are to be achieved. For simple, tree-like, goal structures the problem is rather easy and straightforward. That is why many earlier works on structuring control systems for chemical plants have implied tree-like design goal-structures.

The problem becomes more complex when one allows interacting goals in the goal-structure; goals which are cooperative, competitive, or interfering with each other.

The most important questions that an automated control design procedure is called to answer are (Mostow, 1985): (i) How can the relationship between two goals be automatically inferred? (ii) What control strategies are appropriate for a particular relationship? There is no generalized methodology to answer the first question, which presently can be resolved by production-rules of limited scope. On the other hand, general strategies have been proposed to answer the second question and selection among them is problem dependent with specific guiding rules (Mostow, 1985).

As the discussion above indicates, controlling the design process itself is still at a rather immature state of development. Several important reasons have contributed to that, and among them the following two are of particular concern:

1. Expert designers are not fully aware of all the goals and the methods they use in their mental processing. Explicit cognitive models are still some way from computer implementation.

2. Like human designers, an automated design procedure could present a series of goals in a compiled form, i.e. not all goals need to be explicitly modelled. But, in such case the explanation facilities are not adequate and the system will suffer in credibility.

II.E   "Learning" in Plant Control System Design

During the execution of a plant control system design, designers accumulate, (i) factual knowledge related to the state of the control configuration, its performance, the active design constraints, etc. and (ii) strategic knowledge related to the effectiveness of the design path that has been followed up to that point, the validity of their assumptions and commitments in making some critical design decisions, etc. Such knowledge is continuously updated and invoked at subsequent stages of the design process. Capturing and emulating such new knowledge is an indispensable facility of a mechanized design procedure.

Presently, the following mechanisms are available and could be easily implemented for the acquisition of knowledge during the design of control systems for chemical plants:

1. **Learning from search.** At various stages of the design one is confronted with several alternatives (goals, or methods for the implementation of goals). Searching through the alternatives, the designer "learns" the conditions under which certain alternatives fail and will not try them again in future designs. Such on-line "learning" leads to rules of variable scope and depth, and is rather straightforward to encode them in the form of production-rules of an expert system. Thus, we can "learn" rules for selecting economic criteria over static interaction measures for decomposing a process into processing areas, or selecting methods for the design of multivariable controllers, or finally selecting the direction (raw material preparation → reaction → product recovery → product refining, versus , product recovery → reaction → raw material preparation → product refining) for the design of start-up control structures (Lau, 1982).

2. **"Learning" from the designer.** Although the focus of this paper is how to mechanize the design process of plant control systems, the design is guided by the designer in an interactive manner and every automatic decision can be vetoed by the designer. In this a dialectic process between the designer and the computer, the designer can always introduce new knowledge, on-line during the execution of the design. Such mechanism of external introduction of knowledge by the human is necessary, because the designer can "invoke" more knowledge from the current state and progress of the design than the "learning from search" mechanism can automatically generate. This is due to the fact that "machine learning" is still at a rather early stage of development.

III Knowledge-Based Expert System Methodologies

The design of control systems for chemical plants is a highly knowledge-intensive task. As such it depends heavily on efficient techniques and methodologies for the representation (modeling), handling and manipulation (reasoning), as well as introduction and generation (learning) of knowledge. In this section we will discuss various mechanisms that we are currently employing for the above three tasks and their interface to the automated system for the design of control structures.

A knowledge-based system is a computer software package (Barstow, 1979), which acquires, represents and manipulates knowledge to carry out a task or tasks. The field of knowledge-based expert systems (Jonhson and Keravnou,

1985; Michie, 1982; Hayes-Roth, et.al., 1983) is a sub-group of the general class of knowledge-based systems, which also includes natural language, interpretation systems, and others. A prerequisite for the application of expert systems is the availability of significant amounts of knowledge. Such knowledge in the domain of plant control system designs comes from the expert designers, past designs, and control theory. Loosely speaking, the designer's expertise is used to provide an overall grasp of the domain's informal but fundamental principles, and the specific theoretical knowledge to provide the rigor of the final engineering decisions.

III.A   Knowledge Representation

Knowledge can be represented (modeled) in various schemes that lend themselves to implementation on a computer. Selecting the most appropriate forms for the representation of knowledge is quite critical for the design and development of a knowledge-based expert system. For the mechanization of the plant control design procedure we have found that the following forms of knowledge representation are appropriate, and for reasons to be discussed in subsequent paragraphs. Knowledge, that is the things we know, could be of declarative or procedural form. The latter could be loosely distinguished, for practical purposes alone, into behavioral and reasoning knowledge.

1. **Descriptive or Declarative Knowledge.** It captures definitions, facts, and relations. It is a simple declarative statement without any procedural elements in it. For example, expressions of the form, $f(x, y, z) = 0$, where x, y, z could be variables with numerical or symbolic values, represent declarative statements of the relationship among x, y, and z, without any indication as to the imperative form that will be invoked for the solution of a particular problem, e.g. $x = g(y, z)$. Declarative knowledge is abundant within the scope of the plant control design problem, e.g.: Mathematical models describing the processing units' operations, before the manipulations and the controlled variables have been selected; qualitative definition of the processing units' characteristics, like isothermal continuous stirred tank reactor with external cooling jacket; control specifications, etc. The principal mode of declarative knowledge representation we are using is that of a "frame."

   A "frame" is the composite picture of an object (real world object, or abstract object), which may represent the declarative knowledge available for; a class of similar objects, a sub-class of the above class, or a member of a given sub-class. The knowledge in each "frame" is stored in "slots" and "facets" and needs not be only of declarative character but could be procedural knowledge too. Figure 1 shows a typical example of a "frame" describing a polymerization reactor, as it has been developed using IntelliCorp's KEE. Using the "inheritance" mechanism, knowledge from a "frame" describing a class of objects can be inherited by the "frames" describing the sub-classes emanating from the given class, and in turn be inherited by the "frames" describing the members of one of the upstream sub-classes. The associative network capturing the conceptual knowledge structure is very easy to construct, modify and expand; three features of significant value in our work. Figure 2 shows the structural interrelationship among different "frames" in KEE, indicating the transfer of inherited knowledge.

"Semantic networks" can be used to represent structured knowledge and thus explicitly describe the interrelationships among various "objects." The construction of such networks is rather straightforward, using the inheritance mechanism of the "frames." Within the scope of the KEE system, the generation of "semantic networks" can be achieved graphically.

A special representation scheme has been developed to formalize the relations among magnitudes of various quantities and it is used for the automation of the order-of-magnitude reasoning at various steps of the control system design. The MARS (Magnitude-order Analysis and Reasoning System) formalism (Mavrovouniotis, 1986) has established seven primitive and irreducible binary relations among magnitudes, which are invoked during reasoning.

2. **Behavioral Knowledge.** It describes algorithmic knowledge and actions taken. In our system such knowledge is implemented through LISP procedures which reside in the "slots" of the "frames" describing various objects. Behavioral knowledge allows the execution of rigorous analytical results from systems and control theory. They permit the implementation of various methods such as: Examine the over-, under-, or exact-specification of control problems (Johnston, et.al., 1985); carry out the design of a control system using specific design methodologies; explore the sensitivity of control designs to model uncertainty and its effects on stability and performance robustness; examine the interaction among control loops, etc. Furthermore, LISP procedures are being used to trigger actions on the control design objects such as; initiate the static or dynamic simulation of processing units, draw the icons of the resulting control loops for a given unit, restructure the control configurations, define the control system design path (e.g. sequential or conditional development of the goal structure), etc. Since the behavioral knowledge resides in the "slots" of the "frames," it can be passed on from a parent object to a child object, thus allowing tremendous flexibility in the structuring of such knowledge.

3. **Reasoning Knowledge.** It describes the heuristics and strategies used in various design decisions, either to control the path of the design procedure (see Paragraph II.D) or to select among alternative methods in generating or implementing the structure of design sub-goals (see Paragraph

II.C). Two major modes are employed to represent reasoning knowledge; "production rules" and LISP procedures. LISP procedures are residing in the "slot" of "frames" and have the same features as those discussed in the previous paragraph. Of particular importance to our work is the series of LISP procedures used to define the strict semantic interpretation of the order-of-magnitude relationships introduced by the MARS formalism.

Production rules that we are employing within the scope of the KEE system have the general format:

```
( IF ( < Premise >
                .                    list of premises
                .
                .            )  Returns true
   THEN ( < Assertion >
                .                    list of assertions
                .
                .            )  Change the knowledge base
       DO < Action >
                .                    list of actions; LISP expressions
                .
                .            )
```

They are used to encode qualitative reasoning knowledge and permit a certain centralization of the reasoning process.

III.B   Inference Mechanisms and Explanation Facilities

Inference mechanisms are used to handle the existing knowledge in various ways. For example, a series of production rules asserts new facts which are introduced in the knowledge base. The strict or heuristic interpretation of the order-of-magnitude relationships of the MARS formalism allows the derivation of new magnitude relationships, which also become part of the knowledge base.

Within the scope of the plant control design system we have been using the following inference strategies, which are available in the KEE system:

1. **Forward chaining.** By asserting a fact not present in the knowledge base, we trigger the execution of production rules, LISP procedures, interpretation of order-of-magnitude relationships, etc. whose premises are matched by the facts existing in the knowledge base. New assertions result from this process and are inserted in the knowledge base. The same procedure is repeated until no new assertions can be established.

2. **Backward chaining.** Through this mechanism we attempt to satisfy a goal. If the stated goal is the consequent result of a reasoning model, the premises of this model are examined to find out whether they are part of the knowledge base. If not, these premises become the new induced

sub-goals and the same procedure is repeated, until a final resolution of the initial goal is reached (in which case the stated goal is asserted in the knowledge base), or the process reaches an impasse (in which case the goal cannot be achieved through the existing knowledge).

3. **Multiple chaining.** Quite often, the existing knowledge allows the parallel development of different reasoning chains. In the absence of concrete knowledge to break the conflict, one could use heuristic strategies of conflict resolution. But, such an approach has as inherent weakness the uncertainty on the validity of the invoked heuristic for the particular domain of knowledge. Carrying a small number of asserted alternatives could be wise at times. Our system uses the KEE worlds facility of KEE to propagate chains of parallel reasoning, until new assertions permit the selection of the most desirable path of reasoning. At that point, the "truth" of the rejected reasoning steps is lifted, while a "truth-maintenance" mechanism readjusts the assertions in the knowledge base. This facility is very valuable for the design of plant control systems, since quite often we are confronted with alternative strategies and design decisions with no concrete knowledge to make a rational selection.

The availability of KEE allows us to establish a limited, but very convenient form of linguistic communication between the designer and the computer. The "TellAndAsk" facility permits;

- linguistic description of the knowledge base,

- querying the knowledge base,

- linguistic assertion of new facts,

- explanation facilities for the reasoning reached by production rules,

- introduction of linguistic assertions in production rules and LISP procedures.

Thus, the designer can query, retrieve, or retract knowledge from the knowledge base such as rules, algorithmic procedures, design strategies, models, etc. Furthermore, he/she can activate backtracking of the production rules that were fired and receive "explanations" as to the reasoning that went on in reaching various conclusions.

III.C   Strategies for Controlling the Design Procedure

In Paragraph II.D of the previous section, we discussed the complications in controlling the design procedure, arising from the interaction of the design goals. In this short paragraph we outline a few alternative strategies (Mostow, 1985) that should be present in an automated plant control design methodology, and which we plan to explore in the course of our future research work.

1. **Cooperative design goals.** Execution should proceed in such a way that it exploits best their cooperative relationship. For example:

   - If a goal satisfies a condition present in the other, it should be executed first.
   - If a goal is more general than the other, it should executed first.
   - The easier goal (simpler sub-goal structure, or simpler methods for the implementation of sub-goals) should be executed first.

2. **Competitive goals.** It may be impossible to achieve both goals, in which case we need to establish a rational approach for achieving both with optimum trade-off.

   - If a goal completely dominates an other one of lesser importance, then sacrifice the latter goal.
   - Relax one of the two goals to a weaker version and estimate the resulting trade-off.
   - Compute the trade-off for both objectives and select the most suitable compromise.

3. **Interfering goals.** The interaction can take various expressions, implying different strategies for the execution of the goals. For example:

   - If a goal imposes fewer restrictions, execute that goal first.
   - Execute first the goal which presumes the satisfaction of very tight constraints; this is the most critical decision.
   - Combine goals into a single goal.

Learning the type of interaction among different goals is still the most crucial issue, something to be resolved with improved representation (modeling) of the goals and goal structures.

IV Computing Environment and Programming Style: The "Control Design-Kit"

The design and development of a software package that will implement the "human-aided" paradigm for the design of plant control systems requires a drastic departure from established prototypes of computing environments and programming styles. Prototyping such a system requires extensive programming flexibility, modular and incremental program development and quick symbolic testing of the programs' consistency and completeness.

The continuing saga of FORTRAN programming has had a tremendous impact on how computer-aided engineering is performed in our days. Significant problems have been solved through the intelligent construction of large and complex computer programs, of which the large scale simulators for industrial usage are excellent examples. Nevertheless, it has also created a "culture," which imposes rigid barriers for the development of new and more flexible software tools, that will take the process engineers into the creative domain of "discovering" new solutions to the design-oriented problems. The most emphatic manifestation of the FORTRAN programming's shortcomings is the slow and painful process of developing integrated software packages, which will be able to capture all the forms of available knowledge and "synthesize," among other systems, control structures for complete chemical plants. Furthermore, we are still far from any working system which would allow the expert designer to interact with a computer program and be able to "ask" the program questions, test design hypotheses, and evaluate the significance of his/hers own design assumptions and conjectures. Several prototype efforts along these directions are very restrictive and they can only encapsulate the questions and assumptions that the builders of the program have included.

We are all too familiar with the first principle of FORTRAN programming; develop the logical flow diagram. But, as soon we have done that, we have locked ourselves into a specific problem-formulation and problem-solution, and we need to rewrite the whole program or parts of it in order to tackle modified versions of the original problem. Such philosophy is counterproductive when it comes to design-oriented problems because; the design of process control configurations, operating strategies, etc. involves a loosely defined interaction between the designer and the current context of the state of the design. Thus, a preliminary analysis of the effect that various operating variables have on the economics of process operation will determine the next stage during the design of the control configuration for a chemical plant. Of course, one could try to capture all possible scenaria for all possible plant control design strategies and encode them into a FORTRAN program. But, this alternative is almost by definition impossible. Even more, it would not attract the designer's respect because it lacks facilities to "explain" the rationale of the decision making process, or even worst does not allow the designer's own knowledge, heuristics, and past experience to be used effectively for the simple reason that have not been anticipated by the program and have not been included in it.

To summarize the above discussion one could observe that:

"The work in plant control system design is composed of two primary activities:

1. The first one is trying to form a theory by asking what is going on here. To do that large amounts of quantitative (numerical, mathematical models, graphs, tables, etc.), or/and qualitative (characteristics of the systems, previous experience with similar ones, rules, heuristics)- diversified data are been called and utilized in a very contextual manner.

2. The second one defines what one is going to do about it, and normally

involves grinding large amounts of numerical computations, or long sequences of logical inferences.

To achieve both with some acceptable efficiency one needs to modify the computing environment and adopt different practices in computing.

IV.A   Computing Environment

The recent developments in expert systems and the corrolary results have shaped a completely different computing environment. Let us examine what are the fundamental features of this computing environment, and how they lend themselves to efficient practices of the art of programming.

1. **A uniform environment of software development and testing:** During the development of a large and complex computer program, one would like to have the capability of coding, debugging, editing, compiling and running in a smooth and uniform environment without the nuisance of exiting the main program and calling all the service routines to accomplish the various tasks, described above. Furthermore, the bypassing of tasks such as linking the newly compiled versions is very desirable. All such conveniences allow the designer to concentrate on the creative part of his/her work, rather than dealing with the explicit mechanical actions of running the computer environment. But, such convenience also implies the existence of a computing environment with a unified coding of all service routines, such as that offered by the SYMBOLICS 3600 series and other LISP machines, which have everything written in LISP itself. Thus, within the LISP environment one can debug, edit, compile and run the evolving software design package, without ever leaving the development environment. Such conditions for software development are essential for design-oriented problems, because quite often the desired modifications are revealed during the running of the program, and they do not involve just minor editing changes, but fundamental reconstruction of the program's data-base or/and logical inference strategy.

2. **Modular and Incremental Program Development:** Large and complex design packages can only be developed in distinct modules. The extent of modularity depends on the purpose of the package itself and the expected breadth and depth of the application domain. LISP offers an extensive modularity, all the way to the statement of simple procedures. Its essential "unstructured" character and lack of any involved syntactic rules, make it an ideal vehicle for the efficient construction of modular segments. The syntactic simplicity of LISP translates this modularity into an efficient incremental process of program development. Finally, modularity and incremental program development lead to easy constructs of "new" languages, which can be taylor-made for specific

needs, such as providing an intelligent interface (probably with elements of natural language) between the designer and the computer.

IV.B   Programming Style

Object-Oriented Programming is the most important component of the new breed of software tools for design purposes. It has been a natural outgrowth of the research work in artificial intelligence, but its impact is now being felt by many more applications, than just expert systems. The SYMBOLICS 3640 and 3650 computers, which we use for system development, support an object-oriented programming style based on the concept of "flavors" in the ZETALISP dialect of LISP. In addition we employ the "frame" representation of objects, as supported by the KEE system. ZETALISP flavors and the KEE frames provide predefined flavors and frames, respectively, to cover graphics, windows, pop-up menus, etc.

Furthermore, the use of Production-rules and LISP procedures has allowed us to implement data-driven, or result-driven programming. Such facilities permit the dynamic development of the executable program, a clear necessity for design oriented programs.

IV.C   The "CONTROL DESIGN-KIT"

The CONTROL DESIGN-KIT is a general purpose intelligent interface and database, especially constructed for process control and operations related engineering applications. It allows easy and transparent construction and manipulation of graphic engineering objects, data-models describing these objects, solution procedures for large sets of modeling equations, editors for the generation of heuristic or/and model-based rules for an expert system, a generalized formalism for order-of-magnitude analyses, etc. It was intended to provide all necessary facilities to support the needs of the new prototype intelligent system, needed for the design of control systems for chemical plants and the planning of plant-wide operational procedures. It has been developed on SYMBOLICS 3640 and 3650 computers, using ZETALISP and InteliCorp's KEE system. Let us now outline its basic features:

1. **Database:** It has been constructed in a highly modular form using "flavors" of the SYMBOLICS ZETALISP and the "frames" of KEE. For example, flavors or frames describe

   - processing units as pieces of equipment, or as units where physical or/and chemical phenomena take place,
   - various simple control loop structures,
   - chemical reaction schemes,
   - information on catalysts or/and solvents,

- design and costing equations for various processing units, etc.

Such a database allows extreme modularization, even of the objects themselves, transfer of messages among the objects (an object "learns" immediately of any changes in its immediate environment). Furthermore, it allows;

- easy retrieval of information in graphical form through pop-up multiple windows, or command menus,
- simple updating of its components (simple numbers, model equations, rules,etc.)
- easy maintenance,
- straightforward checking of its consistency and reliability.
- easy construction of relational, network, or hierarchical structures of information, depending on the needs of the particular problem.

2. **Graphic interface:** The basic layout of the CONTROL DESIGN-KIT screen consists of: a large design pane; menus with equipment, control loop structures, and other graphical objects; an operations menu; and an interaction window. In addition, during the execution of particular problems, several specially designed ephemeral windows and command menus are available to the user.

Graphic operations are accessed directly from the graphic objects, or through menu items, with movement of the mouse, and they include; retrieving an object, creating new graphic objects or composites of existing objects, connecting or disconnecting objects, scrolling, moving individual objects, zooming, "abstracting" or "expanding" objects, displaying data-models, graphs, tables, through ephemeral windows, etc.

The common means for storing information (graphic objects are represented by flavors), allows the CONTROL DESIGN-KIT to possess an automatic "built- in" recognition of the graphs, reason about them, and offer explanations. Finally, the graphic interface is supported by rudimentary natural language interface, specifically designed for each particular application, which allows; query processing, automatic modelling and other simpler amenities. Figure 3 shows the layout of the CONTROL DESIGN-KIT's interface during the construction of a flowsheet with control configurations.

3. **Automatic model construction:** From the beginning it was required that the user should not have to learn the details about the mathematical models and methods that underlie their solution. But instead, the user should be capable of creating the desired models through english-like statements specifying the characteristics of the model. The CONTROL DESIGN-KIT presently possesses such capability and allows the designer to do the following:

- Automation of well specified modelling operations such as; formulation of material and energy balances, characterization of unit operations (e.g. isothermal or adiabatic reactors), mathematical description of composite unit operations (columns with side products, trickle-bed reactors, etc.)

- Explicit, english-like naming conventions are available for all operating variables, process equipment design variables, process streams characteristics, etc.

- Construct new data-models, or modify existing ones, without requiring an explicit knowledge of the system's internal conventions. This is accomplished through a specially designed facility, which achieves an elegant and precise correspondence between english-like naming conventions and the internal system variables description.

4. **Equation solver:** It provides a capability for solving large sets of algebraic equations and it is composed of the following facilities:

   - **Steward's algorithm,** which exploits the structural characteristics of the set of equations, and partitions the system into a series of smaller blocks, which can be solved sequentially.

   - **Design variable selection algorithm,** using structural analysis of the set of equations and heuristic rules, to inform the designer; what is the freedom of the engineering problem at a local or a global level, what variables should not be specified because they render structural singularities of the problem, and what variables are recommended as degrees of freedom for the particular problem being solved. Figure 4 shows a screen damp during the selection of the design variables.

   - **Symbolic differentiation program,** which uses a simple pattern matcher that looks at a particular analytic expression and determines the type of the operator. The symbolic differentiator is used to identify the "attractive" design variables, and provides the derivatives for the Newton-Raphson algorithm.

   - **Newton-Raphson algorithm for the solution of equations.**

5. Reasoning facilities: IntelliCorp's KEE (IntelliCorp's Knowledge Engineering Environment), provides the shell for the construction of extended lists of production rules, and a variety of reasoning strategies. Such features enhance the capabilities of the CONTROL DESIGN-KIT by providing rich and multifaceted reasoning abilities with explanation facilities included.

V   Summary

Automating the design of plant control configurations is a problem operating in multiple spaces with a series of knowledge intensive tasks. The basic scope of the methodology, established by expert designers, is complemented by techniques from knowledge-based expert systems and rigorous analytical tools offered by the control theory. The development of such a system requires new computing environments based on LISP machines, and programming styles whose central features are; object-oriented programming and data- or result-driven procedural programming. Many problems still have to be resolved, before a complete prototype is operational, the most important of which are; the efficient control of the design process and the automation of "machine learning" during design.

References

[1] Adelson, B. and E. Soloway, *Tech. Rep. 342,* Dept. Comp. Sci., Yale Univ. (1984)

[2] Arkun, Y. and G. Stephanopoulos, *AIChE J.* 26, p. 975, (1980)

[3] Arkun, Y. and G. Stephanopoulos, *AIChE J.* 26, p. 779, (1981)

[4] Balzer, R., N. Goldman and D. Wilde, *Proc. 2nd Intern. Conf. Soft. Engng.,* p. 373, (1976)

[5] Balzer, R., *IEEE Trans. Software Engng.,* SE-7, p. 3, (1981)

[6] Balzer, R., Rutgers Workshop on Knowledge-Based Design Aids: Models of the Design Process, (1984)

[7] Barstow, D.R., *Knowledge-Based Program Construction,* The Computer Science Library (1979)

[8] Barstow, D.R., *The AI Magazine,* 5, p. 5, (1984)

[9] Bristol, E.H., *IEEE Trans. Aut. Contr.,* AC-11, p. 133 (1966)

[10] Bristol, E.H., 16-th IEEE Conf. on Decision and Control, N. Orleans, LA (1977)

[11] Buckley, P.S., *Techniques of Process Control,* Wiley, (1964)

[12] Cutler, C.R. and B.L. Ramaker, *JACC Proc.,* San Fransisco, (1980)

[13] Douglas, J.M., in *Chemical Process Control II,* D.E.Seborg and T.F.Edgar (editors), Engng. Foundation (1982)

[14] Doyle, J.C., *IEEE Trans. Aut. Contr.,* AC-23, p. 756, (1978)

[15] Doyle, J.C. and G. Stein, *IEEE Trans. Aut. Contr.*, AC-24, p. 4, (1980)

[16] Doyle, J.C. and M.Morari, in *Chemical Process Control III*, M. Morari and T.J. McAvoy (editors), CACHE-Elsevier (1986)

[17] Freeman, P. and A.I. Wasserman, Tutorial on Software Design Techniques, 4-th edition, *IEEE Comp. Soc.* (1983)

[18] Gagnepain, J.P. and D.E. Seborg, *Ind. Eng. Chem. Proc. Des. Dev.*, 21, p. 5, (1982)

[19] Garcia, C.E. and M. Morari, *Ind. Eng. Chem. Proc. Des. Dev.*, 21, p. 308 (1982)

[20] Garcia, C.E. and M. Morari, *Ind. Eng. Chem. Proc. Des. Dev.*, 24, p.472 (1985)

[21] Govind, R. and G.J. Powers, *AIChE J.* 28, p.60 (1982)

[22] Grosdidier, P., M. Morari and B.R. Holt, *Ind. Eng. Chem. Fund.*, 24, p.221 (1985)

[23] Hayes-Roth, F., D.A. Waterman and D.B. Lenat, *Building Expert Systems*, Addison-Wesley (1983)

[24] Johnson, L. and E.T. Keravnou, *Expert Systems Technology: A Guide*, Abacus Press (1985)

[25] Johnston, R.D., G.W. Barton and M.L. Brisk, *Comp. Chem. Engrg.*, 9, p. 547 (1985)

[26] Kant, E. and A. Newell, *AAAI*, p. 177 (1983)

[27] Katz, R., P.A. Subramanian and W. Scacci, *J. Systems and Software*, 4, p. 1 (1984)

[28] Lam, M. and J. Mostow, *Proc. IFIP WG 10.2 8-th Intern. Symp. Comp. Hardware Description Languages and Their Applications*, Pittsburgh, PA (1983)

[29] Lan, H., Ph.D. Thesis, University of Minnesota (1982)

[30] Mavrovouniotis, M., Techn. Report, LISPE-MIT (1986)

[31] McAvoy, T.J., *Interaction Analysis*, ISA Monograph, Research Triangle Park, NC (1983)

[32] Mehra, R.K., R. Rouhani, J. Eterno, J. Richalet and A. Rauolt, Chemical *Process Control II*, D.E. Seborg and T.F. Edgar (editors), p. 287 (1982)

[33] Mitchell, T., L. Steinberg, G. Reid, P. Schooley, H. Jacobs and V. Kelly, *IJCAI* (1981)

[34] Michie, D., *Introductory Readings in Expert Systems,* Gordon and Breach (1982)

[35] Morari, M., Y. Arkun and G. Stephanopoulos, *AIChE J.* 26, 220 (1978)

[36] Morari, M. and A.K.W. Fung, *Chem. Eng. Sci.,* 36 (1981)

[37] Morari, M., in *Chemical Process Control II,* D.E. Seborg and T.F. Edgar (editors), Engng. Found., p. 467 (1982)

[38] Mostow, J., *AI Magazine,* p. 44 (1985)

[39] Niida, K. and T. Umeda, in *Chemical Proc. Control III,* M. Morari and T.J. McAvoy (editors), CACHE-Elsevier, p. 869 (1986)

[40] Prett, D.M. and Gillette, R.D., *JACC Proc.,* San Francisco, CA (1980)

[41] Rivas, J.R., D.F. Rudd, and L.R. Kelly, *AIChE J.,* 20, p. 311 (1974a)

[42] Rivas, J.R., and D.F. Rudd, *AIChE J.,* 20, p. 320, (1974b)

[43] Scherlis, W. and D. Scott, IFIP Congress (1983)

[44] Shinskey, F.G., *Process Control Systems,* McGraw-Hill, N.Y. (1979)

[45] Shinskey, F.G., in *Chemical Proc. Control III,* M. Morari and T.J. McAvoy (Editors), CACHE-Elsevier, p. 895 (1986)

[46] Stephanopoulos, G., *Comp. and Chem. Eng.,* 7, p. 331 (1983)

[47] Stephanopoulos, G. and H.-P. Huang, *Chem. Engng. Sci.,* 41, p. 1611 (1986)

[48] Taylor, J.H., in *Chemical Proc. Control III,* M. Morari and T.J. McAvoy (editors), CACHE-Elsevier, p. 895 (1986)

[49] Weber, R. and C. Brosilow, *AIChE J.,* 18, p. 614 (1972)

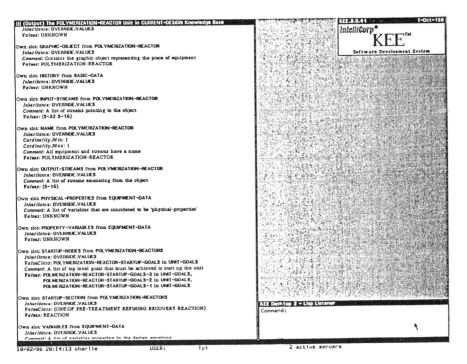

Figure 1: The "frame" Representation of a Polymerization Reactor Using KEE.

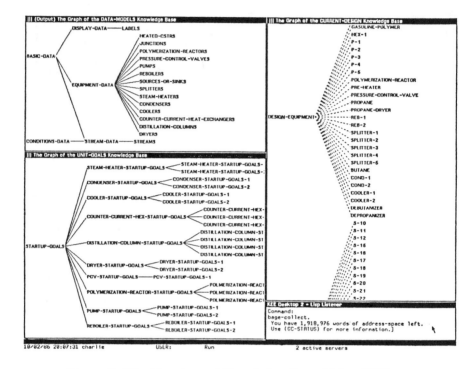

Figure 2: The Structure of Knowledge-Bases Using KEE.

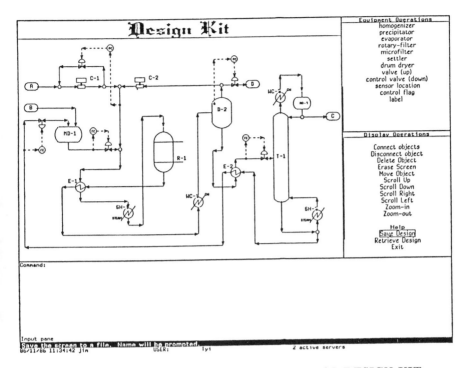

Figure 3: The Graphic Interface of the CONTROL DESIGN-KIT.

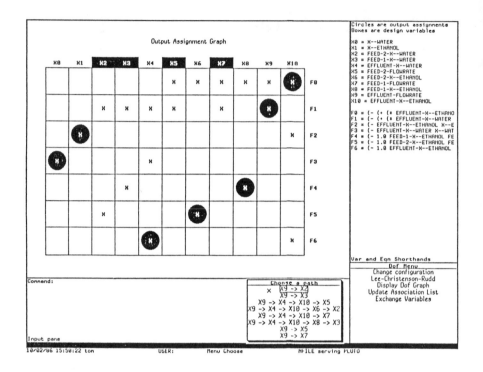

Figure 4: The Graphic Interface for Selecting Design Variables.

Process Identification - Past, Present, Future

D. M. Prett, T. A. Skrovanek, J. F. Pollard

December 15-18, 1986

Abstract

This paper reviews the field of process identification as practiced by control system designers. An attempt is made to extract general and important characteristics of various independent approaches in order to direct future research efforts. Emphasis is placed on the assumptions and compromises which must be made in practice in order to determine an adequate process model for control purposes. Finally, multivariable closed-loop identification of a simulated binary distillation column is presented in order to demonstrate the capabilities of current identification technology.

1 Summary

Implementation of modern advanced controllers such as DMC, QDMC, and IMC [1, 2] explicitly requires adequate process models. The most common approach to obtaining these models is to identify a process model using plant data. This report reviews the identification technology presently available. The common features of present identification techniques are described, including experimental design, data analysis, model structure, and parameter estimation. Several different identification techniques are compared from the perspective of the practicing control system designer. Finally, a rigorous simulation example is used to demonstrate that with a properly designed identification experiment the available identification techniques can be used to generate a process model from data collected under closed-loop conditions with several inputs manipulated simultaneously.

2 Introduction

With the emergence of modern advanced control techniques such as DMC, QDMC, and IMC [1,2], identification, the technology for obtaining the process model used by the controller, assumes a correspondingly important role. Broadly speaking, any technique which fits parameters in equations to experimental data can be considered an identification technique. From a practical viewpoint, for most processes model identification is probably the most time consuming step in designing and implementing advanced control technologies. One reason is that the controlled variables of typical process units are affected by unmeasured disturbances. A second reason is that the model equations used in the controller design only approximate the behavior of the process, so that these equations may fit the system only for a limited range of operating conditions.

The basic theory of identification is well developed, but utilization of process identification is often quite difficult in practice. Past experience suggests that perhaps the most critical step is proper design of the plant experiment. Identification experiments performed with no knowledge of the plant dynamics are unlikely to lead to accurate models. Most process data will be corrupted by unmeasured disturbances, although proper design of the plant experiment can minimize the effect these disturbances have on model accuracy. The objective of the subsequent model identification steps will then be to estimate a model of sufficient accuracy for control design in spite of the presence of unmeasured disturbances. For a well designed experiment with only low level disturbances, almost any identification technique will result in an accurate process model. The basis of comparison between different identification approaches should be how well they are able to work with process data from poorly designed identification experiments and data corrupted by higher levels of unmeasured disturbances. Most identification techniques used with virtually any set of process input-output data will yield a process model; the important issue is, of course, the accuracy of the identified model. As described by Prett and Garcia [3], the determination of the level of process model detail and accuracy required for a given control service is a very complicated task in itself. The steps of model detail determination must be integrated, at least philosophically, into the identification process. An overview of the framework for integrating the process model and the model uncertainty into a design procedure for process control is given by Garcia and Prett [4].

The objective of this report is to survey process identification technology from the perspective of advanced process control implementation in the Chemical Process Industries (CPI). From the evaluation of past and present technology and from an assessment of present needs, future research areas are proposed. Topics covered include the design of excitation signals, model types and how the assumed model types can impact the identification technique, the different techniques of parameter estimation which are available, and several different identification approaches. Rather than reviewing all available techniques from the literature and other sources, a limited number of identification techniques have been selected to display the range of choices, assumptions, advantages, and disadvantages typical in the area of process identification. Each technique description will include a discussion of its assumptions, advantages, disadvantages, and potential problems. The final section is a description of a multivariable closed-loop identification effort on a simulated binary distillation column which demonstrates available identification technology applied to a relatively complex system.

3   Model Identification Overview

This paper presents various steps in building a process model. The steps include

1. Experimental design,

2. System perturbation and data collection,

3. Data evaluation,

4. Model structure selection,

5. Model parameter estimation, and

6. Model verification.

It should be emphasized that these steps are part of an iterative procedure. Proper design of an experiment in step 1 requires some knowledge of the process model which is often not available until after the subsequent steps. At the model verification step, the user may decide to select a different model structure and re-estimate the parameters, or to re-design the experiment and re-run the experiment. Successful identification may require some or all of these steps to be repeated a number of times.

4   Process Data Collection

To obtain an adequate process model from input-output data, the process must usually be excited by perturbing the input(s) of interest. An adequate model must include in its input-output set all variables of interest in the control structure for which the model will be the basis. The type of excitation and other factors such as process disturbances and the tuning and modes of other loops will determine the type and quality of the data collected. The different types of input excitation are discussed in this section, as well as the proper design of an identification test which maximizes the probability of obtaining useful data.

4.1   Common Types of System Excitation

The most common forms of input excitation are step inputs and pulses of short duration relative to the system dynamics. The use of frequency response testing with sinusoidal inputs for excitation has been limited in the CPI by the large time constants inherent in many of the processes. Tests which use step inputs or pulses can usually be completed at least an order of magnitude faster than corresponding frequency response tests. A process model obtained from

step and/or pulse excitation can be converted into any other model format
of interest, so that step and pulse excitation are adequate for the majority of
identification purposes.

4.1.1   Step input

A step test is easy to implement and can be performed in a relatively short
period of time. The process is operated in an open loop, and allowed to reach
steady state. A step change is then made to the manipulated variable. For
stable processes the test should be allowed to continue until the output(s) of
interest reaches steady-state, which in the absence of disturbances will yield an
accurate representation of the steady-state gain of the process. A number of
step tests can be performed at different periods in time to facilitate verification
of the results.

There are several benefits to using a step test. The test easy to design and
perform and can be done in a relatively short period of time. In the ideal case
the test yields an accurate estimate of the steady-state gain of the process.
Finally, a model can often be obtained by simple inspection of the response
data.

The disadvantages of the step test are as follows. The data are susceptible
to disturbances, *i.e.* it is often hard to distinguish between the input and
any disturbances when fitting the model. Multiple tests may be needed to
verify the results. A good identification experiment often requires a step input
to the process of sufficient magnitude that the process output, the controlled
variable, is far from its desired setpoint for an extended period of time in
order to minimize the effect of noise and other unmeasured disturbances. In
summary, this type of test is short and simple, but often highly susceptible
to real world problems such as noise, unmeasured disturbances, operational
constraints, etc.

4.1.2   Pulsed input

A pulse test is a number of step tests in sequence such that, in general, the
process never reaches a steady-state. The pulsed input shifts the information
content of the test from the steady-state gain and low frequency dynamics
toward the high frequency dynamics. A special case of the pulsed input is
Pseudo-Random Binary Noise (PRBN). A PRBN signal has several properties which make it well suited for identification. A PRBN input is a good
approximation to white noise, a signal with the desirable property that all
frequencies are equally represented in its frequency spectrum. A PRBN signal
is not correlated with process disturbances, which reduces the effect of the
disturbances on the estimated model. Finally, PRBN test signals are easily
designed and generated so that the process can be excited over the frequency
range of interest.

Application of the PRBN test provides several benefits to the user. Unlike step testing, the test can be started before the process has achieved steady-state. The user can design the test to maximize model accuracy for a particular frequency band. The pulse test keeps the process near the nominal operating point, so testing can be done for a longer period of time without interfering with normal plant operations. The process can be operated under closed loop control, which of course is highly desirable from an operational viewpoint.

Several disadvantages can be listed for the pulse test. First, more information is required to design the test; a step test must often be performed to obtain a rough estimate of the dead time and settling time of the process. It is difficult to analyze the data from the test by inspection. It is also difficult to estimate the steady-state gain accurately if the switching time between pulses is too short. In summary, more accurate models can be obtained with pulse testing than with step testing, but a higher level of expertise is required to use appropriately.

4.2   Design of Test

Since the models resulting from an identification test are usually used to design controllers, there are some requirements that must be met by a properly designed test. All lower level and interacting controllers must be tuned properly before the start of the test. It is important to realize that the *basic* [1] process control system is considered part of the process during the identification step. Since different tuning of these *basic* controllers will result in a different process being identified, retuning after the test may invalidate the models. Any filters applied to outputs must be set at the values which will be used by the control system, because these will also become an integral part of the process model.

4.3   Multivariable Test Signal

The early approach taken to identify a multivariable process was to perform a separate response test for each input with all other inputs held constant. Operation of many process systems does not allow for keeping all inputs constant, so often a multivariable test is a necessity. The enormity of this task can be seen by considering the large number of inputs for a crude column or catalytic cracker, many of which for operation reasons can not be kept constant. Analysis of response data from a multivariable test will yield all models which make up the multivariable process. The major problem is that proper design of a multivariable identification experiment is considerably more difficult than design of a single variable identification experiment.

---

[1]The *basic* control system is that set of controllers that remain in closed loop mode independent of the control system for which the model is being identified. For example, a simple process may have a column temperature being controlled by steam flow to a reboiler. The model will relate temperature to steam flow. A basic loop will be the controller which adjusts a control valve based on the specified steam flow.

In general, simultaneous step tests on all inputs do not lead to acceptable results. The signals applied to the inputs must be statistically independent from each other. The reason for requiring that the inputs be independent from each other is quite intuitive. For example, consider a two-input, single-output system with step responses that are identical but opposite in the sign of the gain. Introducing a step change of the same magnitude on each input would result in the process output remaining constant. Obviously, the experiment would not yield useful information. A simple way to obtain independent test signals for multiple inputs is to generate a very long PRBN sequence and then divide it into equal portions for each input. This will guarantee that each input test signal will be mutually independent.

The same concerns that apply to a single variable test also apply to multivariable tests. The magnitude of the test signal must be large enough so that the effect of the test signal is higher than the noise level on the outputs. Also, the length and the switching period of the PRBN signal must be adjusted to the dynamics of the process, which requires some prior information about the process gain and settling time. A series of step tests may need to be performed on the process to get rough estimates of these parameters.

One further complication of the multivariable identification experiment arises when a multivariable process has widely different dynamics. In such a case a single PRBN signal may not be sufficient to identify all the dynamics. A PRBN signal that is appropriate for one subset of the process dynamics may be either too slow or to fast for the rest of the process dynamics. A separate test signal may have to be designed for some inputs to better match the known dynamics. However, it is not quite clear how to design the best signal for multivariable identification at this time. Multivariable signal design will be a topic of future research since experimental design plays such an important part in success of the overall identification process.

4.4   Correlation Analysis

Before attempting to identify a model from response data, it is a good idea to check if the data contains sufficient information to warrant proceeding with the analysis. One convenient method for doing this is correlation analysis. By calculating and plotting crosscorrelations between input and output signals, one can check whether the signals applied to the inputs are really mutually independent and whether there are significant relationships between the inputs and the outputs. Correlation analysis provides a quick check for proper test signal design and proper execution of the experiment and should be done before attempting the actual model identification. Input-output pairs that do not exhibit significant correlation should be left out of the fitting process, or the user may decide to re-test the process with a different experimental design. This step can help the user avoid the expenditure of a large effort in data analysis only to find that the data is inadequate for generating a process model which meets the end use requirements.

5 Process Representation

Most process systems are non-linear, but experience has shown that linear process models are good approximations of a large number of systems within the context of process control. Historically, the use of linear models has also been motivated by the fact that control system design has been unable to use non-linear models directly. Linear models result in control strategies that are easily implemented on standard control hardware and solvable with standard software (QP, LP). It should be noted that it is becoming clearer that working directly with the non-linear model often has control performance advantages for many process systems (*i.e.* batch control [5]) and future efforts in model identification research should not be hindered by past control system design limitations. Nevertheless, it seems quite clear that for the foreseeable future control system design will be dominated by linear system theory, and thus linear models will play a major role in model identification.

The dynamic relationship between the process input u and process output y for a linear system can be described by an nth order linear ordinary differential equation,

$$y^{(n)} + p_{n-1}y^{(n-1)} + \cdots + p_0 y = z_n u^{(n)} + \cdots + z_0 u + v \qquad (1)$$

where

$$
\begin{aligned}
y &= y(t) &= \text{process output,} \\
u &= u(t) &= \text{process input,} \\
v &= v(t) &= \text{process disturbance.}
\end{aligned}
$$

The variables y, u, and v are functions of time and the superscript n represents the nth derivative with respect to time. The objective is to identify the model structure and to estimate values for the parameters of the linear differential equation. The modeling process is complicated by several factors. Major problems include the model being an over-simplification of the underlying physical process and the measured process output being corrupted by unmeasured noise and other disturbances.

The model description is often modified to reflect the intended use. Control systems are often designed using Laplace transforms. A convenient form for (1) can be written as

$$[s^n + p_{n-1}s^{n-1} + \cdots + p_0]Y(s) = [z_n s^n + \cdots + z_0]U(s) + V(s) \qquad (2)$$

where $Y(s)$, $U(s)$, and $V(s)$ represent the Laplace transforms of the deviations of y, u, and v, respectively, from their steady-state values. The system is assumed to initially be at steady-state; processes not initially at steady state can be handled at the expense of more complex mathematics. Equation (2) should be familiar to most engineers with a background in classical feedback control.

A much more common description for models identified from process data is a discrete difference equation. Since process data are almost always collected by sampling the process system at regular intervals, a difference equation is a natural way of representing the process. A sampled linear model can be described by its impulse response as

$$y(i) = \sum_{j=1}^{\infty} g_j q^{-j} u(i) + v(i) \tag{3}$$

where

$$
\begin{aligned}
y(i) &= \text{process output at sample time } i, \\
u(i) &= \text{process input at sample time } i, \\
v(i) &= \text{process disturbance at sample time } i, \\
q^{-1} &= \text{backward shift operator } [q^{-1}u(i) = u(i-1)], \\
g_j &= \text{model parameters.}
\end{aligned}
$$

The time i is an integer value and is incremented by one each time the system is sampled.

The impulse response model (3) cannot be implemented, since an infinite number of parameters are required. Different model structures can be developed to reduce the number of parameters. Many process systems are open loop stable, so that the parameters g_j of the impulse response approach 0 as $j \to \infty$. Then a simple approach is to truncate the infinite series at some finite value N,

$$y(i) = \sum_{j=1}^{N} g_j q^{-j} u(i) + v(i), \tag{4}$$

for which $g_j \approx 0$ for $j > N$. For open-loop stable systems with a reasonable sampling time, N may range from 20 to 50. Even further reductions in the number of parameters can be obtained by using an Auto-Regressive Moving Average (ARMA) model of the form

$$A(q)y(i) = B(q)u(i) + v(i) \tag{5}$$

where

$$
\begin{aligned}
A(q) &= 1 - a_1 q^{-1} - \cdots - a_n q^{-n}, \\
B(q) &= b_1 q^{-1} + \cdots + b_m q^{-m}.
\end{aligned}
$$

Both (3) and (5) can be obtained directly from (1) for the many sampled systems which are implemented with a zero-order hold on the process input and a sampler on the process output.

Process systems often contain a dead time between the input and output. The dead time can be handled by increasing the model order, by explicit specification, or as another parameter to estimate. The dead time D could appear explicitly in the appropriate equations by replacing $U(s)$ with $e^{-sD}U(s)$ or $u(i)$ with $q^{-\tau}u(i)$ where τ is the integer quotient of the process dead time D divided by the sampling time.

No structure has been assigned to the unmeasurable process disturbance v up to this point. The disturbance is generally assumed to be generated by some stochastic process. Without any additional information, no estimate for v can be obtained. Usually, an attempt is made to estimate the present value of v from previous estimates for v. A general approach is to assume the disturbance is generated by a dynamic process such that $v(i)$ is generated in the same manner as $y(i)$, except that the input is now an unmeasured random variable. In this case,

$$v(i) = \sum_{j=1}^{\infty} r_j q^{-j} \epsilon(i) \tag{6}$$

where

$$\begin{aligned} v(i) &= \text{process disturbance at time } i, \\ \epsilon(i) &= \text{a random variable}, \\ r_j &= \text{disturbance model parameters.} \end{aligned}$$

Since the disturbance model parameters are unknown, they must be estimated along with the process model parameters. Like the previous discussion on the process model of (3), the noise model of equation (6) can be implemented as a truncated impulse response as in (4) or an ARMA model as in (5).

The process model is used to predict the process output at each time interval. For structures which include past values of the process output (*i.e.* (2) and (5)), the issue of how to obtain the past values of y arises, since the measured values of the process output include the disturbance component v. The *model output* $\hat{y}$ at time i can simply use *measured* values of y previous to time i to obtain an equation error representation as

$$\hat{y}(i/i-1) = f[y(i-1), y(i-2), \cdots, u(i-1), u(i-2), \cdots, \hat{v}(i-1), \hat{v}(i-2), \cdots] \tag{7}$$

where the function f depends on which process model structure is selected. The other common approach is to use a model error representation of the form

$$\hat{y}(i) = f[\hat{y}(i-1), \hat{y}(i-2), \cdots, u(i-1), u(i-2), \cdots, \hat{v}(i-1), \hat{v}(i-2), ...] \tag{8}$$

where the model output $\hat{y}$ at time i is based on previous value of the model up to time $i-1$.

6 Identification Accuracy Criteria

The objective of process identification is to obtain an *accurate* process model. The *accuracy* of a model can be defined in several ways, each resulting in a different problem formulation. Stated qualitatively, for a given process input, the model output $\hat{y}$ [2] should be a good approximation to the measured process output y. For the model output to approximate the measured process output, both an estimate of the process model parameters and an estimate of the unmeasured disturbance are required.

---

[2] $\hat{y}(i)$ is used to represent both the equation error representation, Equation (7), and the model error representation, Equation (8), of the previous section.

One basis for comparison is simply how well the model output matches the measured process output. The model output $\hat{y}$ is sampled at the same frequency that the original process measurements were obtained, and the two outputs are compared. The most widely used comparison criterion is the Least Squares (LS) formulation,

$$\min_{\hat{\theta}} \sum_{i=1}^{\infty} [y(i) - \hat{y}(i)]^2. \qquad (9)$$

Minimization of the squared error difference for a linear process model can be a linear or non-linear optimization problem, depending on the method used to estimate the disturbance v as well as the choice between an equation error and model error representation. Non-Linear LS (NLLS) algorithms can require 10 to 1000 times more computational effort than a comparable Linear LS (LLS) algorithm. That increased computational time may be a factor in on-line identification applications. A more serious problem is the question of convergence. NLLS algorithms are iterative algorithms that seek to improve the accuracy of the parameter estimates at each iteration. The problem with this approach is that the estimates may converge to some local minimum far from the desired global minimum. The major advantage of the NLLS algorithm is that conceptually any type of model structure can be handled in the same manner. For example, the process dead time can be explicitly estimated along with the other process parameters. Likewise, a non-linear model can conceptually be identified in the same manner as a linear model. As a general rule, most on-line identification algorithms use some modification of LLS. For off-line data analysis, NLLS algorithms are often used for their greater flexibility, often in conjunction with a LLS algorithm for generating initial parameter estimates.

A second method of identification is the Maximum Likelihood (ML) approach. A ML algorithm estimates model parameters that make the collected data most plausible. ML algorithms are attractive from a theoretical perspective since they provide the minimum variance unbiased estimate. The solution of a ML problem, however, can be quite formidable compared to the solution of a LS problem. Assumption of a linear model for both the process and the disturbances, as well as assuming that the input to the disturbance model, $\epsilon(i)$, is an independent normally distributed random variable, leads to a LS problem of the form

$$\min_{\hat{\theta}} \sum_{i=1}^{\infty} [\hat{\epsilon}(i)]^2 \qquad (10)$$

where $\hat{\epsilon}$ is the estimate of the unmeasured disturbance ϵ.

The limitations of the preceding criteria is that the requirements of the control system are not considered. In practice, closed loop performance determines the model adequacy, not the closeness of $\hat{y}$ to y. At the present time it is difficult (if not impossible) to translate control system requirements into a general quantitative identification criterion of a form as like in (9) or (10). The control system requirements can be integrated to some extent qualitatively, *i.e.* specifying the bandwidth of the PRBN signal based on the bandwidth requirements of the controller. Much research remains to be done on the integration of controller requirements into the identification process. The major focus of this effort involve an *a priori* acceptance of a degree of uncertainty in parameter estimates. Corresponding adjustments can often then be made in the control design stage for this uncertainty. In fact what is found is that certain levels of uncertainty are acceptable and others demand modification of the control system performance objectives in order to maintain the integrity and/or robustness of the completed design. In the above context, integrity and robustness denote a match or fit between the model uncertainty and the performance objectives of the control system.

7   Identification Approaches

Over the years different approaches for identifying process dynamics have been developed. The methods covered by these packages range from simple step testing and correlation analysis to multivariable techniques. The following sections discuss a subset of available approaches. Emphasis is placed on the advantages and disadvantages of different techniques. The ultimate identification package (yet to be developed) will be capable of identifying a multi-input multi-output non-linear process model under closed-loop operation based on some *a priori* specified control objectives.

7.1   Non-Linear Least Squares (NLLS)

7.1.1   Background

An early approach to model identification was simply to use one of the readily available optimization packages to obtain the *optimal* model parameters. The user could collect a set of process input-output data, hypothesize a model structure, then estimate the model parameters. Conceptually this approach is straightforward and easy to use, since a transfer function model can be directly estimated.

7.1.2   Assumptions

Conceptually, NLLS can deal with any model structure, linear or non-linear, of arbitrary order. However since this method was often developed and used by practicing control engineers, models used for fitting the data are often described by low order linear equations familiar to the control engineer. The Laplace transform of a second order process with deadtime (which is adequate for describing the dynamic response for the majority of process systems, with perhaps the major exception being inventory systems which are represented by ramp responses) is given by

$$\frac{Y(s)}{X(s)} = \frac{K\,e^{-Ds}}{as^2 + bs + 1} \tag{11}$$

where

$Y(s)$	= the Laplace transform of the output,
$X(s)$	= the Laplace transform of the input,
K	= the steady-state gain of the model,
D	= the model dead time,
a, b	= model constants.

The user is generally not required to develop a structure for the process disturbances. For the control system designer with classical training in process control, a common choice is simply to ignore the disturbance issue entirely.

7.1.3   Solution Technique

The model is obtained from input-output data by adjusting the parameters to minimize the sum of the squares of the residuals (output measurement − model output) of Equation (9). The model error representation output of Equation (11) is a non-linear function of the model parameters even when the model itself is linear, so a NLLS optimization is required. The optimization is generally performed in a batch mode using an iterative method. At each iteration, new parameter estimates are obtained which reduce the performance index of (9). When the parameters converge within some tolerance, the iterations are stopped.

7.1.4   Advantages

NLLS is a fairly easy approach to use in that there are few parameters for the user to specify. Model parameters of any structure, including non-linear models, can be identified. Process deadtime, which is fairly difficult to estimate directly in other identification techniques, is simply another parameter to estimate with NLLS.

7.1.5   Disadvantages

Reasonable initial parameter estimates are usually required. The optimization technique may have poor convergence properties for more than a few degrees of freedom. The optimization programs often yield only limited insight into the accuracy of the model. The computational load is very high relative to some of the techniques which use linear LS algorithms.

7.1.6   Summary

NLLS is a technique for using process data to obtain simple models using a non-linear optimization. The use of this approach may require more engineering judgment than the other identification approaches reviewed here, since model validation tools are generally not provided. Lack of model validation can be a serious drawback when identifying models from process data corrupted by process disturbances. The most serious problem is that use of this kind of technique appears to require little expertise, which is definitely not the case. Almost any set of process input-output data will yield model parameters, but much intuition is required to evaluate the resulting model.

7.2 Recursive Approximate Maximum Likelihood (RAML)

7.2.1 Background

Maximum Likelihood (ML) algorithms are quite attractive from a theoretical point of view. ML algorithms are efficient in that asymptotically they yield the minimum variance estimates for the class of unbiased estimators. The major drawback is that solution of a ML problem is generally quite difficult. As mentioned earlier, assumption of a linear process structure and a linear disturbance structure together with the assumption that the underlying distribution generating the disturbance is independent normally distributed leads to the LS formulation in (10). This is essentially the approach discussed in the classic work by Box and Jenkins [6]. A recursive ML algorithm can be formulated, RAML, which asymptotically yields the ML estimates [7].

7.2.2 Assumptions

The two primary assumptions associated with RAML are that the process can be described by a low order Auto-Regressive Moving Average (ARMA) model and the disturbance can be described by white noise passed through an ARMA model, resulting in a model of the form:

$$A(q)y(i) = q^{-\tau}B(q)u(i) + \frac{C(q)}{D(q)}e(i) + m(i). \tag{12}$$

If the noise which appears in the process output is Gaussian the parameters which are estimated will be the ML estimates. To guarantee stability of the estimation routine, the process input should be a persistently exciting test signal. The user must specify the model structure (order) and the dead time. In practice the user will find the model order and dead time using a trial and error procedure.

7.2.3 Solution Technique

The solution technique used is the RAML algorithm. Generally some method is required to obtain initial parameter estimates. Box and Jenkins describe a large number of analysis tools for identifying the model structure and evaluating model accuracy. Most of these tools are designed for off-line analysis.

Several analysis tools are available to the user to help in determining the minimal model structure. The first are plots of the autocorrelation of the residuals and the crosscorrelation between the input signal and the residuals. In general, the autocorrelation of the residuals is a test on the adequacy of the noise model. The crosscorrelation between the input and residuals is a test of the process model. If both correlations are insignificant then the model orders are high enough. Another tool for testing the order of the model is a pole-zero plot of the system transfer function. If there exists significant pole-zero cancellation in either the process or noise models, the order of the appropriate model may be reduced. A plot of the model parameters versus time as the identification progresses is available to determine if the model has converged. This, together with an estimate of the variance of each parameter, can aid in determining whether a given parameter is effectively zero. This information contributes to a decision to reduce the model order.

Several tools are available to test the goodness of fit. One is the objective function, a sum of squares of errors between actual and predicted response. Another is a simulation which plots the actual and predicted process responses together. The user can determine by visual inspection whether the model is adequate. Finally, the estimate of the variance of the estimated parameters gives some insight into the uncertainty of the estimated parameters for some specified confidence level.

7.2.4   Advantages

The RAML algorithm following the Box and Jenkins approach provides a number of checks for judging the adequacy of the model structure and the quality of the identified model. In addition, operation of the algorithm in a recursive manner allows the user to observe the trajectories of the parameters over time. Abrupt jumps in the parameter trajectories are often an indication of an abnormal disturbance to the process.

7.2.5   Disadvantages

The Box-Jenkins approach can be confusing to new users for two reasons. First, it requires a large number of parameters — transfer function model order, deadtime, noise model order — to be specified. A source of confusion is the overlap in function of several of the analysis tools, which often give conflicting indications to the model structure and accuracy.

Unlike the non-recursive ML algorithm, convergence of the parameter estimates from RAML can be very sensitive to the initial estimates. Poor initial estimates can result in the parameters failing to converge to any value, or simply very slow convergence. The RAML algorithm is particularly sensitive to the user specified deadtime.

Because of the model form, the technique does not lend itself to simple extension to multivariable systems. A 1 by 1 system requires specification of 5 parameter values by the user, the orders of the polynomials $A(q)$, $B(q)$, $C(q)$, and $D(q)$ as well as the deadtime. A 2 by 2 system requires specification of 20 parameters, a 3 by 3 system requires specification of 45 parameters, and so on.

7.2.6 Summary

The RAML algorithm has a number of attractive theoretical properties and has been found to give good results on a wide variety of actual process systems. Unfortunately application of RAML to multivariable systems is relatively difficult because of the large number of user specified terms. Even for single-variable systems, RAML is more difficult to use than the previously discussed technique, NLLS. However this expertise demand is actually required for both NLLS and RAML; NLLS only gives the misleading impression that little expertise is required. Hence although more demanding in execution, this technique can in fact reduce the potential for poor control system performance due to modeling inaccuracies.

7.3 Finite Impulse Response (FIR)

7.3.1 Background

Identification of an ARMA model as in Equation (5) using the measured values for the process output leads to computational difficulties. In particular, use of a LLS algorithm results in biased estimates of the parameters [8]. Using a Finite Impulse Response (FIR), Equation 4, results in a problem formulation for which the estimated parameters are unbiased for a large class of disturbances.

7.3.2 Assumptions

The process is described by an impulse response of the form described in Equation (3). No structure is assumed for the random disturbances. This method is essentially non-parametric in that no structure (*i.e.* model order) is assigned to the process.

7.3.3 Solution Technique

Parameters of a truncated impulse response are identified. The user must specify the order of the truncated series and the process dead time can be entered as an option. The user will generally specify the model order as an estimate of the 95% settling time for the process. The parameter estimates are obtained by using the LS criterion in Equation (9). The problem formulation leads to a LLS problem.

The tools described by Box and Jenkins for determining model order are not needed for a non-parametric model. Several tools exist to aid the user in evaluating the model accuracy. The objective function and the ratio of the output variance to estimated disturbance variance can be calculated. Plots of the model output versus the measured process output, the trajectories of the estimated parameters and steady state gain over time, and the step response of the output for each input can be checked.

Controllers based on non-parametric models (*i.e.* QDMC, DMC) can directly use the step response output of the identified model, although in most cases some smoothing of the step response will be required. Controllers based on parametric models (*i.e.* IMC) will require some model reduction scheme. One such model reduction scheme that has worked well in practice is a frequency-weighted Hankel-norm model reduction [9].

7.3.4   Advantages

The FIR technique has attractive theoretical properties. User-specified parameters are kept to a minimum. While deadtimes can be specified, models obtained without deadtime specification have been found to work well in practice for reasonable specifications of the process sampling time (*i.e.* the deadtime is smaller than 10 times the sampling time). The identified models are relatively insensitive to the specified order. The user does not need to specify a structure for either the process model or the disturbance model. The LLS algorithm used is robust and computationally efficient.

7.3.5   Disadvantages

Only stable processes can be identified with this package. Unstable systems, including those with ramp responses, cannot be identified directly. Only step response (non-parametric) models are obtained directly. To obtain a parametric model, the use of a model reduction technique is required. In general, a larger number of parameters must be estimated for a FIR model than for an ARMA model, which means the variance of the estimated parameters for the FIR model will be higher.

7.3.6   Summary

FIR is a simple, easy to use technique for identifying step response type models for both single-input and multi-input systems, combining the ease of use of NLLS with many of the model analysis tools of RAML. Unlike RAML, very few user supplied input parameters are required. Generation of transfer function (ARMA) models requires an additional step, model reduction.

7.4   High Order Arma Models (HOAM)

7.4.1   Background

The previous technique discussed, FIR, did not assume any kind of structure to the disturbances. Much of the work done in identification assumes a model for the process as in (12) where the disturbances are generated by white noise passed through a filter. A non-parametric identification approach can be developed similar in concept to FIR, but based on a high order ARMA model instead of a high order FIR model.

7.4.2 Assumptions

HOAM assumes that the process can be modeled adequately with a high order ARMA model,

$$A(q)y(i) = B(q)u(i) + e(i), \tag{13}$$

where $e(i)$ is generated by white noise passed through a filter. HOAM minimizes the equation error criterion of Equation (13) in order to estimate the model parameters using a LLS algorithm. It is well known that unless the product $A(q)e(i)$ is white and has zero mean, the model coefficients will be biased. By using a high order model, the bias problem is minimized in the sense that asymptotically, for model order n, the first n terms of the impulse response will be unbiased [10]. Dead time estimation can be done by increasing the order n to encompass the dead time.

Like FIR, the models generated by HOAM are non-parametric. As before, controllers that require a parametric model will generally require the user to perform some kind of model reduction.

7.4.3 Solution Technique

Solution of the equation error minimization problem is a LLS problem. A criterion for selecting the order of the ARMA model is the Final Prediction Error (FPE), defined as:

$$FPE = \left(\frac{N+p}{N-p}\right) \cdot det\left\{\left(\frac{1}{N-n}\right) \sum_{i=n+1}^{N} e(i)e^T(i)\right\} \tag{14}$$

where N is the number of data points, n is the model order, p is the number of parameters in the model, and $e(i)$ is the equation error from (14). The idea is to choose n such that FPE is minimized. As p increases the term $(N+p)/(N-p)$ increases, while the fit becomes better and the determinant decreases; as p decreases, the opposite occurs.

An optional step in the identification process is model order reduction [9]. The input to this is a high order model impulse response and the output is all of the lower order ARMA models. The user selects the model whose impulse response most closely matches the results of the high order model identification.

The final step is model validation. The same types of tools used by FIR can be used to evaluate model accuracy. In addition, the ARMA model can be checked for unstable poles.

7.4.4   Advantages

HOAM uses a LLS method for estimating model parameters, so it executes quickly and is well suited for multivariable systems. The main input parameter is the order of the model to be used. Even though model order is not very critical, a criterion can be developed to assist the user in selecting the order. The first n impulse response coefficients are asymptotically estimated without bias.

7.4.5   Disadvantages

HOAM assumes that the disturbances are generated from white noise passed through a linear filter, and may perform poorly with other classes of disturbances. Model reduction is required for control system design based on parametric models. Like FIR, the larger number of parameters to identify increases the variance of the estimated parameters.

7.4.6   Summary

HOAM is a multivariable identification technique using high order ARMA models. For systems where the unmeasured disturbances are Gaussian, the algorithm has attractive asymptotic properties. Use of non-parametric models simplifies the application of this technique.

8   Identification Example

Two different identification techniques are compared in this example, HOAM and FIR, to illustrate multivariable closed loop identification. A simulated deisobutanizer column, Figure 1, was used. The simulation is quite rigorous with each of the 21 trays in the column described by non-linear differential equations. It is a reasonable test of an identification algorithm since the process system is non-linear with a wide variety of disturbances.

The manipulated variables are the product draw stream from the accumulator (*FLOW1*) and the hot oil flow to the reboiler (*FLOW2*). The controlled variables are the top tray temperature (*TEMP1*) and bottoms temperature (*TEMP2*). The major disturbance to the column is the feed stream with variations in flow, temperature, and composition. *TEMP1* and *TEMP2* are under closed-loop control using PI controllers manipulating *FLOW1* and *FLOW2* respectively. For the identification experiment, independent PRBN signals were added simultaneously to the PI controller outputs to perturb the controlled variables around their nominal setpoints. The PI controllers were tuned for a sluggish response, since their primary objective is to compensate for long term disturbances that would result in the controlled variables drifting far away from their nominal setpoints. The objective of the identification experiment was to obtain the step responses of the 2 by 2 system.

Two identification packages were used to identify a step response for each input-output pair. Step response data were also generated experimentally by opening the accumulator draw and hot oil feed control loops and applying a step input. The results are compared in Figures 2, 3, 4, and 5, with the dashed lines representing the experimental step test results and the solid lines representing the identified model. It can be seen that the step responses of the models obtained from FIR match the experimental step responses more closely than the HOAM technique, but the final judgment as to which model is more accurate would depend on the control system requirements. The initial response — the first 20 samples — is quite good for most of the input-output pairs, while the steady-state gain is less accurately identified with both methods. This is not surprising, since the PRBN signal is bandwidth limited for both low and high frequencies. The poorest agreement between the step test and the identified process model is for the *FLOW2-TEMP2* pair. The problem here is that the final change in the steady-state value for *TEMP2* is below the level of the noise present. The obvious solution would be to increase the perturbation signal level, but the large magnitude of the initial response (approximately 25 times the magnitude of the steady-state response) limits that option. This input-output pair illustrates the conflicting problem of keeping the controlled variable between an acceptable range of values while perturbing the manipulated variable sufficiently to obtain an accurate model.

The accuracy of the model can be improved in several ways. A longer experiment will generally increase the model accuracy. The bandwidth of the PRBN signal can be shifted to increase the accuracy of the model at some specified frequency at the expense of decreasing the model accuracy over a different range of frequencies. The measured disturbances can be included as disturbance inputs to the model. The process input-output data can be filtered and/or detrended. It is easily seen that improvement in model accuracy requires a large number of design decisions to be made by the user. Hopefully, including the control system requirements into the identification process will reduce the number of decisions the user must make.

9 Conclusions

This report surveys the state-of-the-art of process identification. Identification is discussed from the perspective of the practicing control system designer.

Past experience indicates that the lack of success in identifying a model is seldom due to an inappropriate identification technique. It has been found from experience that the design of the identification experiment is often the most important step. The "goodness" of data used for identification is determined by several factors of which the signal to noise ratio is most important. The signal generated by the test must be significant compared to the noise level of the process so that a model can be extracted with minimal analysis effort. The multivariable, closed-loop identification example in this report indicates that a well designed test allows for the determination of an acceptable model with relative ease. Design of the identification experiment is an important issue towards which future research effort should be directed.

The identification techniques in use are based on a criterion of "goodness of fit" which is independent of the end use of the identified model (*i.e.* process control). The practice of identifying process models for the sake of fitting the data while the end use of the model is practically ignored should not be continued. The current techniques place much emphasis on getting a step response model which looks reasonable without really analyzing how the parameter variability indicated by the data can affect the control performance. For example, if the steady-state gains of a model obtained by applying two different identification techniques to the same set of process data differ by a certain amount, the assumption is made that one identification approach must be superior to the other. The key issue should be whether the accuracy of the identified models meets the requirements imposed by the intended use of the control system. The controller requirements might be such that the variability observed in the steady-state gains obtained using different identification techniques will have little effect on the closed-loop control performance. In such a case, the models may be required to be accurate only over some specified frequency range.

The identification technique of the future will demand that the user stipulate the control performance requirements for the process. These performance requirements will then be used to determine the model accuracy necessary for control. A test will be designed to extract the information necessary for determining the model parameters within the required accuracy. The model parameters will then be determined for use in the control algorithm.

An additional research objective will be the development of methods of parameter identification for non-linear first principle models, so that the variability of currently used linear models due to non-linearities can be removed. Again, this increased model detail must be motivated by the control requirements.

Finally, attempts to integrate the performance requirements of the control design and the process identification indicate the need to increase the sophistication of the technology in both areas. As with all new technology, this temporarily causes problems in technology transfer to the field. As described by Garrison, Prett, and Steacy [11], an area of much potential to facilitate this transfer from research and development to plant operations is the field of expert systems. The use of an expert system shell around advanced identification techniques will lead to a larger pool of technical staff capable of knowledgeably using state-of-the-art identification technology.

10 References

1. Prett, D. M., and R. D. Gillette (1979). "Optimization and constrained multivariable control of a catalytic cracking unit." AIChE National Mtg., Houston, TX.

2. Garcia, C. E., and M. Morari (1982). "Internal Model Control 1. a unifying review and some new results." Ind. Eng. Chem. Process Des. Dev., 24, 472-484.

3. Garcia, C. E., and D. M. Prett (1986). "Advances in industrial model predictive control." Chemical Process Control Conference — III, Asilomar, CA.

4. Garcia, C. E., and D. M. Prett (1986). "Design Methodology Based on the Fundamental Control Problem Formulation," Shell Process Control Workshop.

5. Garcia, C. E. (1984) "Quadratic Dynamic Matrix control of nonlinear processes: an application to a batch reaction process," AIChE Annual Mtg., San Francisco, CA.

6. Box, G. E. P., and G. M. Jenkins (1976). Time Series Analysis, Forecasting, and Control. Holden-Day, Oakland, CA.

7. Panuska, V. (1969). "An adaptive recursive least-squares identification algorithm." Proc. 8th IEEE Symp. Adaptive Processes.

8. Goodwin, G. C., and R. L. Payne (1977). Dynamic System Identification. Acedemic Press, New York.

9. Glover, K. (1984). "All optimal Hankel-norm approximations of linear multivariable systems and their L^{∞}-error bound." Int. J. of Control, Vol. 39, no. 6, 1115-1193.

10. Wahlberg, B. (1986). "On model reduction in system identification." ACC, Seattle.

11. Garrison, D. B., D M. Prett, and P. E. Steacy (1986), "Expert Systems in Process Control and Optimization: A Perspective,"Shell Process Control Workshop.

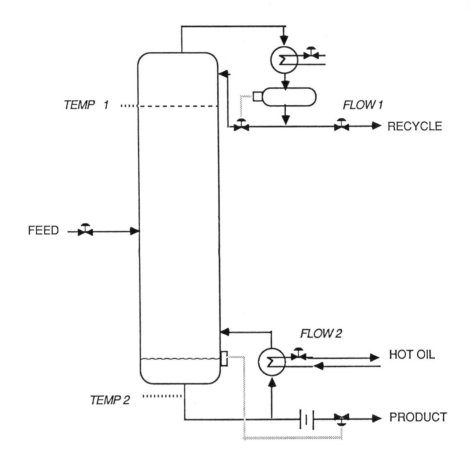

Figure 1: Identification Example Column.

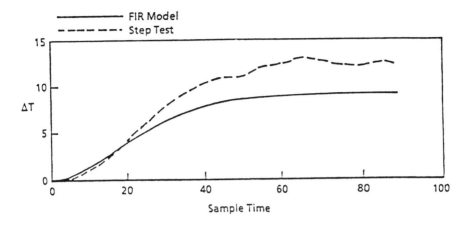

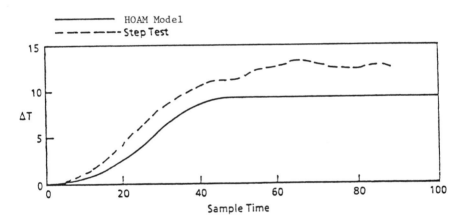

Figure 2: Step Response of *TEMP1* to *FLOW1*.

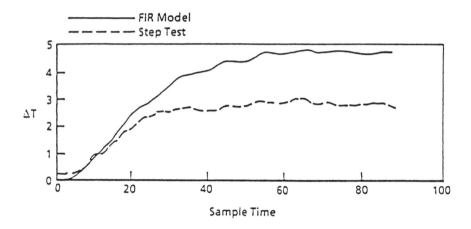

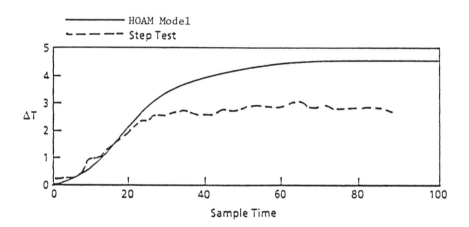

Figure 3: Step Response of *TEMP2* to *FLOW1*.

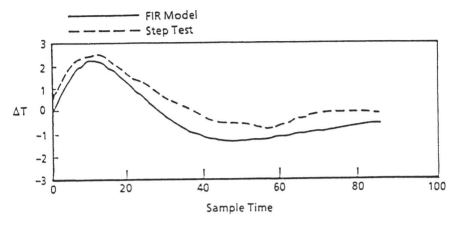

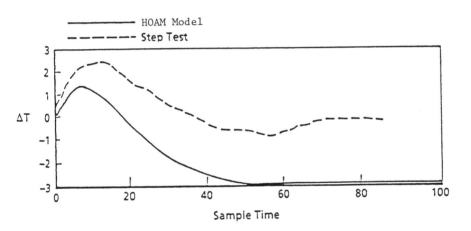

Figure 4: Step Response of *TEMP1* to *FLOW2*.

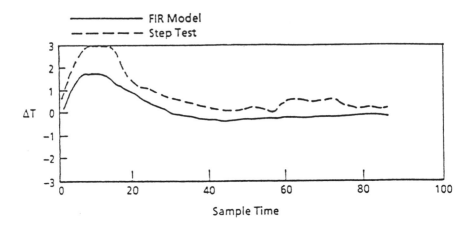

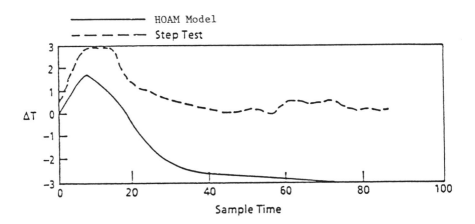

Figure 5: Step Response of *TEMP2* to *FLOW2*.

Adaptive Control Software for Processes with Significant Deadtime

C. Brosilow, W. Belias, C. Cheng
Case Western Reserve University
Cleveland, Ohio

Paper presented at 36th Canadian
Chemical Engineering Conference
Sarnia, Ontario

October 8, 1986

Abstract

A two-level adaptive controller uses an identified process model for deadtime compensation and an expert system to continuously adjust the closed loop response for a fast, stable response. The model identification is carried out a-periodically, as needed, without injection of a dither signal or other extraneous disturbances. Application of the control software to a laboratory head exchanger shows significantly improved performance over fixed parameter controllers.

I.   Introduction

As operating conditions change, process characteristics change. For many important processes, these changes are sufficient to make it very difficult to tune the process control system to operate satisfactorily throughout the operating range. When an experienced engineer is not available to tune such loops, it is common to have the operator assume manual control. This negates the benefits of automatic control, burdens the operator and often leads to poorer product quality.

Ideally, the function of an adaptive controller is to adjust the local behavior of the controller to match the local process characteristics. Tuning of an adaptive controller should, of course, not depend upon the changing characteristics of the process so that the adaptive controller can function over the entire operating region. In addition, the ideal adaptive controller adjusts its local behavior based on normal process responses to disturbances and process related setpoint changes. That is, the adaptive controller should not upset the process so as to carry out its tuning function.

There are currently two types of adaptive controllers on the market which do not rely on controller-induced disturbances to accomplish the desired tuning. These are (1) the self-tuning regulator controllers and (2) the expert, or rule-based, controllers. An example of a self-tuning regulator is Novatune[TM] marketed by ASEA of Sweden. Examples of the expert tuning approach are Foxboro's pioneering Exact[TM] controller and ControlSoft's Intune[TM]. A comparison of the two types of adaptive controller is beyond the scope of this paper, which is mainly concerned with the operation and performance of the Intune Controller.

The Intune controller[1] differs from the Exact Controller in that the Intune controller can be either a deadtime compensator, or a classical PI or PID controller. Intune produces a process model as a normal part of its operation. Also, it is highly likely that the expert rules used in the Exact and Intune Controllers differ significantly.

The operational philosophy behind the Intune controller is described in the next session. Section III gives an overview of the operation of the Intune software. Finally, the results of applying the P.C. version of Intune to a highly variable laboratory heat exchange system are given in Section IV.

II. Adaptive Inferential (Internal Model) Control

The structure of an inferential (alias Internal Model, Model Predictive, etc.) controller which accommodates control effort constraints is shown in Figure 1a. The rationale behind the structure in Figure 1a has been described elsewhere (Popiel '86, Brosilow '84) and so only a very brief review will be given here. Also, Morari '83 has elucidated the linear behavior of such control systems.

The model in Figure 1a will be restricted, for this discussion, to be a first order lag cascaded with a deadtime. That is

$$\text{Model} = \frac{K e^{(-Ts)}}{\tau s + 1} = \hat{G}(s). \tag{1}$$

The lag is then simply $K/(\tau s + 1)$ and the delay is T units long.

The role of the controller is to make the lag portion of the model follow the filter f, where the filter is also a first order lag. That is

$$f(s) = \frac{1}{\varepsilon s + 1}. \tag{2}$$

The control law which accomplishes the above is

$$m_k^* = \frac{e^{(-\nabla/\varepsilon)} f_k + \left(1 - e^{(-\nabla/\varepsilon)}\right)(v_k - d_k) e^{(-\nabla/\tau y_k)}}{K \left(1 - e^{(-\nabla/\tau)}\right)} \tag{3}$$

---

[1] Intune Controller Software can be purchased in a P.C. version or as a Source Code License from: ControlSoft, Inc., 7568 Briarcliff Parkway, Cleveland, Ohio 44130-6433. Telephone number (216) 234-5759

$$M_k = m_k^* \text{ if } m_L \leq m_k^* \geq m_U$$
$$= m_L \text{ or } m_U \text{ otherwise}$$

where

f_k = the current filter state

v_k = the current setpoint

d_k = the current distance estimate

y_k = the current model output

∇ = the sampling time

m_L, m_U = lower and upper bounds on the control effort

The first two terms in the numeration of (3) calculate the value of the filter output at time $k + 1$, the third term is the value of model output at time $k + 1$ if the control effort were zero over the time interval ∇, and the denominator term is the step response of the model in a time ∇.

Important properties of the control system of Figure 1a and equations (1), (2), and (3) are:

(A) The control system is stable if the process is open loop stable, the process gain does not change sign, and the filter time constant, ε, is large enough.

(B) There is no steady state error, if the control system is stable.

(C) When the model matches the process, the response from the setpoint to the process variable is the filter transfer function plus the deadtime.

The proofs of the above properties can be found in Morari '83. Methods for choosing the filter time constant for a fixed model, but an uncertain process (e.g. process gain, lag and deadtime lie between known limits) are given by Chen '84 and Morari '86. Qualitatively, the greater the uncertainty in the process model, the larger the filter time constant required to get acceptable control system performance. Acceptable performance is typically defined as having a damping ratio less than some preset limit for any process in the set of possible process.

The design methods of Morari '86 are conceptually the best that one can accomplish for a fixed model and a fixed filter time constant. However, for most local operating points (i.e. local process within the set of possible processes) the design will lead to a much more sluggish response to disturbances and setpoint changes, than if the filter time constant were chosen for that local operating point. Of course in this later case, it would be necessary to retune (i.e. change) the filter time constant for each operating point.

The first level (Autotune-1) of the Intune adaptive controller automatically adjusts the filter time constant so as to maintain a desired damping ratio. In this way, one achieves the best possible performance for the current model. The filter time constant is adjusted via a set of rules based on features of the response of the process variable and control effort to disturbances and setpoint changes. Figure 1b shows the trajectory tracking controller with automatic filter tuning.

The performance of the control system shown in Figure 1b is now limited by the mismatch between the model and the process. This mismatch can be reduced by identifying a model of the form of (1) for each local operating region. Techniques for identifying models are well established. For example, see Åstrom and Wittenmark '84. Since the only measurements available are those of the process variable and control effort, all of the identification methods rely on at least a qualitative model of the nature and frequency of the disturbances entering the process so as to recover a process model from the data. The disturbance model used by Intune is that the disturbances drift from one level to another relatively rapidly with respect to the time constant of the process, but that they enter the process relatively infrequently with respect to the process settling time.

Given the above disturbance model, a set of rules have been developed to determine when, and what data to send to the model identifier. This approach allows for rapid changes in the process model due to significant disturbances or changes in setpoint. Another set of expert rules analyze the model produced by the identifier and decide how to adjust the current model. User data such as limits on the possible model parameters can play an important role deciding on the validity of an identified model. The complete adaptive controller configuration is shown in Figure 1c.

III. Intune Software

A. How it Works

The Intune software implements the approach to adaptive control described in Section II. In the model based mode, the filter (alias controller lag) time constant is adjusted so as to achieve a desired damping ratio in the process variable response to disturbances or setpoint changes. The adjustment is made by an expert system acting on observed process response to setpoint changes and disturbances. The expert decides when to modify the filter time constant and when observations warrant updating the process model. No setpoint changes or dither signals are introduced to aid in the model or filter adjustments. In the PID mode, the filter time constant and model parameters are converted to the appropriated PID parameters.

B.   Tuning Parameters

As in any adaptive controller, the Intune system needs information relating to the process under control. The required information, or tuning parameters, are: (1) The noise band, (2) the "wait" time, and (3) the desired speed of response. The noise band is the magnitude of the expected error between the process variable and the setpoint due to measurement or process noise. The wait time is the maximum time that one would expect the process variable to remain outside of the noise band. The desired speed of response is either slow, medium or fast, which corresponds to damping ratios of 10%, 30% and 50% respectively.

Of the above, the noise band is the most important parameter. If the noise band is set too small, then Intune will retune the control system too often, and will attempt to adjust the filter and model so as to try to compensate for process noise, which often yields incorrect tuning parameters. Further, the retuning may result in cycling **of the adapter** between too tight and too sluggish tuning. The resulting control is still generally better than that achieved by a non-adaptive controller, but is not nearly as good as might otherwise be achieved. If the noise band is too large, the auto-tuning function will ignore data which could result in better tuning or a better model. The controller will not retune as long as the process variable remains in the noise band, and will ignore both sluggish and oscillatory behavior within the noise band. A small amount of time spent observing the process variable should allow one to choose an appropriate noise band. If in doubt as to magnitude of the noise, choose a value towards the high side of the range.

The purpose of the wait time is to prevent the control system from ever becoming overly sluggish. Choosing too short a wait period can cause the auto-tuning function to frequently reduce the filter time constant so as to speed up the system response, and return the process variable to the noise band within the wait period. This, in turn, can drive the control system into instability. Therefore, when in doubt about how long disturbances or setpoint changes are likely to keep the process variable outside the noise band, choose a conservatively long wait time. The configurator, described below, will suggest a wait time based on a user-supplied process model.

The response speed parameter should be set based only on how much damping one wants in the process variable. This parameter effects the amount of overshoot as well as the damping. A "slow" setting will give relatively little overshoot, but also results in a more sluggish response than the "medium" and "fast" settings.

C.   Configuring the Control System and Display

The P.C. version of Intune has a menu driven configuration disk to help the user set up the control system. The main section of the configurator menu are: (1) Hardware configuration, (2) Loop configuration, (3) Controller parameters,

(4) Operator display, and (5) Power failure recovery procedures.

The hardware configuration establishes the type of IBM computer and peripherals and the range for the input and output voltages. The loop configuration establishes the input/output engineering units, the input conversion rules (e.g. linear, square root), the type of process variable filtering, process variable alarms and setpoint limits.

The controller parameter menu first establishes if the user wishes to implement model based, PID or PI control, and whether the control action is forward or reverse acting. If model based control is selected, the user is prompted to supply any information available on initial model parameters, and limits on the allowable model parameters. The user must also supply a sampling period which cannot be less than 1/100 of the model deadtime or time constant. If the user selected PID or PI control, he is requested to supply initial values for the control gain and the integral and (if selected) derivative time constants. If the user has neither initial estimates of the model nor PI parameters, then the controller will start in manual mode. To enable the user to obtain initial model or PID parameters from the process, the software provides a "learn" mode.

The learn mode uses a relay controller similar to that suggested by Åstrom and Hagglund ('84). The user specifies control effort bounds and the control effort cycles between the bounds until the expert is satisfied that enough data has been collected to estimate model parameters. The controller automatically returns suggested PID and/or model parameters. The learn mode is entered from the engineer's interface described below.

The operator display consists of a bar graph display, a process variable trend, and a control effort trend (c.f. Figure 2). The range of the process variable trend, the time scale of the trend and the colors of the display are set in the configurator. There are three time scales which can be set independently in the configuration. These scales, referred to as short, medium and long, are accessed through the operator interface as described below.

The configurator also allows the user to determine mode of operation of the controller should the computer have to reboot itself due to a power failure. There are four options for the mode of recovery: (1) manual mode, (2) automatic mode (i.e. fixed parameter PID or model based), (3) Autotune-1 (i.e. automatic adaption of the filter time constant or controller gain) or (4) the last saved mode prior to the power failure. The software automatically saves the controller state about every fifteen minutes for use in power failure recovery.

D.   The Operator Interface

The menu for the operator interface as shown in Figure 2 is obtained by simultaneously pressing <alt> and <8>. In Figure 2, the mode key is highlighted. Pressing the <return> will display the various modes of controller operation: manual, auto, auto 1 (shorthand for Autotune-1) and Auto2 (Autotune 2).

The current mode is always highlighted. The highlight is changed to a different mode by pressing the $<\leftarrow>$ or $<\rightarrow>$ cursor keys. Pressing <return> changes the controller mode to that which is highlighted.

Returning to Figure 2, pressing the $<\rightarrow>$ cursor moves the highlight to the SP (setpoint) box. Pressing <return> brings up the display!

Current SP xx.xx

Desired SP

The desired setpoint is highlighted. The desired setpoint can be changed incrementally up or down by pressing the $<\uparrow>$ or $<\downarrow>$ cursors. Alternately, the setpoint can be entered from the keyboard by pressing <del> (which clears the desired setpoint display) and entering the setpoint. Pressing <return> causes the desired setpoint to replace the current setpoint.

The controller output, CO in Figure 2, can be changed when the controller mode is manual. The display and methods of changing the controller output are the same as for the setpoint.

The trend (trd in Figure 2) produces the following menu:

Short Medium Long

Again, the right and left cursor keys move the highlight from the current trend length to the desired trend length. Recall that the actual time spans for the various trends were preset using the configurator.

The report box (rep) brings up the menu

Summary   Summary + Lst

File A: File B: File C:

Summary causes a summary of the data on the current trend to be printed. The summary includes: The time initiated, the average value of the process variable (PV), the standard deviation error of the PV the maximum absolute difference between the setpoint and the process variable and the time at which it occurred, the average deviation and control effort (CO), the minimum and maximum control effort and when they occurred and the standard deviation of the control effort.

"Summary + Lst" prints, in addition to the above 218 samples of the trend curve giving the time, the value of the setpoint, value of the process variable and control effort. This latter data can be written to a floppy disk in the A or B drives or to the hard disk (C drive) by selecting File A:, File B: or File C:.

The operator acknowledges alarms by simultaneously pressing <alt> and <2>.

E.   Engineer's Interface

The main menu of the engineer's interface is shown under the simulated plasma display in Figure 3. The menu is entered by first pressing <alt> and <4> simultaneously and then entering a five character password followed by <return> in response to a prompt. The password is set in the configurator described previously.

The menu shown in Figure 3 is for a PID based controller. The only difference for a model based controller is that the label in the first box is replaced with contrlag (i.e. the filter time constant). Pressing return with the highlight on the PID box as shown in Figure 3 produces the menu

Prop Integral Deriv

Current Settings

Proportional xx.xx

Integral xx.xx

Derivative xx.xx

If Prop is highlighted, pressing <return> again allows one to insert a new proportional gain. Similarly, shifting the highlight to Integral or Deriv. and pressing <return allows one to change the integral and derivative time constant.

The model parameters (i.e. gain, time constant and deadtime) can be viewed and changed as above. The box labeled Process Parameters in Figure 3 is displayed only in the simulation mode. The user can change the simulated process parameters of gain, lag and deadtime in the same way as the model parameters. The Intune controller modules have no access to the simulated process parameters, and so treat the simulated process in the same manner as a real process.

The controller tuning parameters of noise band, wait time and response speed are accessed and changed through the Ctrl box (cf Figure 3).

The learn mode is activated through the LM menu. The user is requested to select the range of controller action as either = $\pm 10\%$ of scale (small), $\pm 20\%$ of scale (medium) or $\pm 50\%$ of scale (i.e. full range-large). Having selected the control bounds, the user initiates the controller cycling (or aborts). While collecting data as in Figure 4 (with medium control range) the user can continue, force an analysis using the data already collected or abort. The learn mode will terminate automatically when enough data has been collected to enable a good identification. If the learn mode was terminated by the user, the expert will judge whether the data is good enough to obtain model parameters. If not, a message will appear which so informs the user why the data was not good. Otherwise, the learn mode will return a model and a set of controller parameters. Upon termination of the learn mode for any reason, the controller returns to the mode it was in when the learn mode was started.

IV. Application of Intune to a Laboratory Heat Exchanger

A. Background

The heat exchange system, shown in Figure 5 and described briefly below, was selected to demonstrate the performance of the Intune controller because the operation of the exchanger and its local mathematical model are strongly dependent on its operating conditions. Also, a substantial amount of data on the performance of fixed model based controller is available (Parrish '82). A summary of this data follows the heat exchanger description.

The heat exchanger system (figure 5) consists of two steam condensers placed in series, with associated piping and instrumentation. Cooling water passes first through the horizontal and then through the vertical condenser before exiting the system. Steam is admitted independently to the horizontal and vertical condensers.

Each condenser consists of a single eight foot long, 3/4 inch O.D. stainless steel tube, enclosed in a three inch diameter pyrex tube. Steam condenses in the annular space and the condensate is drained through a steam trap. Water temperatures throughout the system may vary from 40 to 200 degrees Fahrenheit depending on the vapor space temperatures and water flow rate. A control valve is used to vary the water flow rate between 1.0 and 4.5 gpm. Steam is admitted to the condensers through individual control valves. Details of the heat-exchanger construction and materials are given by Shine ('80).

Parrish ('82) used the above exchanger system in the configuration shown in Figure 5 to control the outlet water temperature of exchanger B by manipulating the steam valve of exchanger A. Pulse tests responses of water outlet temperature to changes in steam valve supply pressure showed a variation in steady state gain of from 3 to 15 degrees F/psi when the nominal steam valve air pressure ranged from 4 to 11 psig and the nominal water outlet temperature ranged from 80 to 135 degrees F with water flow rates from 1.5 to 4.0 gpm. Similarly the estimated dead times between the valve signal and water outlet temperature measurement ranged between 7 and 25 seconds. Because of this wide variation in gain and dead time, a very sluggish control system was required in order to maintain stable operations everywhere within the possible operating range. Using an average model of

$$\hat{G}_p(s) = \frac{7.4e^{(-13s)}}{23s + 1} \frac{°\text{F}}{\text{psi}} \quad s \text{ in seconds} \tag{4}$$

Parrish ('82) found that a filter time constant of 50 seconds was required to give stable control everywhere. This tuning resulted in response times of between 180 and 280 seconds to more or less fully suppress various disturbances.

For small changes at set points near the ends of the operating range, a filter time constants of 10 seconds was sufficient to stabilize the control systems using the following models:

$$\hat{G}_p(s) = \frac{4.1e^{(-7s)}}{14s + 1}\frac{{}^{\circ}\text{F}}{\text{psi}} \quad \text{at } 80^{\circ}\text{F and a water flow of 4.5 gpm} \qquad (5)$$

$$\hat{G}_p(s) = \frac{10e^{(-25s)}}{20s + 1}\frac{{}^{\circ}\text{F}}{\text{psi}} \quad \text{at } 125^{\circ}\text{F and a water flow of 1.5 gpm} \qquad (6)$$

Using the above models and a filter time constant of 10 seconds gave response times of about 80 seconds for small disturbances to the steam pressure in exchanger B at the two operating points. If the models were accurate, the filter time constant of 10 seconds would lead to a response time of about 40 seconds. Inaccuracies in the fitted models probably are due to the nonlinear nature of the heat exchanger response, even for relatively small disturbances.

A plausible explanation of the widely differing heat exchanger models at high and low temperature operation is that air enters or is expelled from the shell side of the exchanger depending on the shell side steam pressures. At shell side pressures greater atmospheric, air is expelled, while air leaks into the shell when the pressure drops to sub-atmospheric. High (>120 degrees F) outlet water temperatures usually lead to greater than ambient steam pressures while low (<80 degrees F) water temperatures lead to sub-atmospheric steam pressures. Shifting concentrations of air in the shell could account for the observed gain variation and the changing effective dead time. The actual transportation delay of water through **both** exchangers is only six seconds at 1 gpm, which is substantially shorter than any of the observed delays.

B.   The Test Runs

The tests of the adaptive controller on the heat exchanger extended over a continuous period of 2 and 3/4 hours. Set point changes from low outlet temperature to high outlet temperatures were initiated to determine how well the controller would compensate for large changes in local behavior (i.e. model). Interspersed between setpoint changes is the use of the controller learn mode to get an independent estimate of the local models (i.e. independent of the controller model updating in the autotune-2 mode). Towards the end of the run the water flow rate was decreased from 2 gpm to 1 gpm to provide a controlled disturbance. Uncontrolled disturbances during the run were changes in the steam supply pressure, which influences the steam pressure in the shell side of both exchangers, and changes in the water supply pressure which changes the water flow rate. These disturbances occur naturally in the process, and were not induced.

The Intune controller was configured to operate in the model based mode. The noise band was set at 1.0 degree F, the wait time was 200 seconds and the response speed was set to "medium." The sample period of 1.5 seconds.

Figures 6 a&b show the first 1800 seconds (1200 samples) of the run. The controller is started in manual and at point A the learn mode is started using a control effort of ±20% of full span (i.e. ±2.4 psi). The learn mode terminated normally recommending the following model:

$$\hat{G}_p(s) = \frac{3.3e^{(-16s)}}{7s + 1} \frac{{}^\circ\text{F}}{\text{psi}} \quad s \text{ in seconds} \tag{7}$$

$$\text{Setpoint} = 65^\circ\text{F}$$

$$\text{Waterflow} = 2 \text{ gpm}$$

At point B, Figure 6a, the recommended model was accepted, a filter time constant of 6 seconds entered manually, and the mode changed to automatic (fixed parameter – not adaptive). The response of the outlet temperature is smooth until at C the set point is changed to 135 degrees F. The setpoint change causes a rapid rise in temperature, and the setpoint is reached in about 30 seconds. The outlet temperature overshoots the setpoint and breaks into oscillations. At D the setpoint is dropped to 68 degrees F and the system recovers (after a small undershoot) in about 80 seconds. A second setpoint change to 135 degrees F at E shows more clearly that the fixed parameter controller, which is well tuned at 65 degrees F, is unstable at 135 degrees F.

At point F (c.f. Figure 3) the controller is switched from auto to the autotune-2 mode. Figures 7 a&b are continuations of Figures 6 a&b. At G, the setpoint is again changed to 135 degrees F. After about two oscillations, the controller increased the filter time constant at H to 22 seconds. (The location of H is only approximate.) At point J the filter time constant was again increased, this time to 70 seconds and the controller changes the model to

$$\hat{G}_p(s) = \frac{7.8e^{(-15s)}}{9s + 1} \frac{{}^\circ\text{F}}{\text{psi}} \tag{8}$$

After several disturbances from unknown causes, the controller (near time K) changes filter time constant to 45 and the model to

$$\hat{G}_p(s) = \frac{31e^{(-15s)}}{25s + 1} \frac{{}^\circ\text{F}}{\text{psi}} \tag{9}$$

The large increase in gain over the previous model, at the same temperature setpoint, seems anomalous. However, later data indicates that this large gain increase may indeed have occurred. In any case, shortly after the model change shown above, the setpoint was changed (at point L) to 65 degrees F. The system responds quite sluggishly reducing the filter time constant as shown in Figures 8 a&b. After another unplanned disturbance at M, the model changed to

$$\hat{G}_p(s) = \frac{15.8e^{(-7s)}}{12s + 1} \frac{{}^\circ\text{F}}{\text{psi}} \tag{10}$$

and the filter time constant changed to 1 second.

The control system responds smartly to a 10-degree F. increase in setpoint, but is still somewhat sluggish when the setpoint is returned to 65 degrees F.

The model changes again somewhere between points N and P to

$$\hat{G}_p(s) = \frac{7.9e^{(-3.6s)}}{6s+1} \frac{^\circ F}{psi} \tag{11}$$

The filter time constant remained unchanged at 1.0. At point P, a 10-degree decrease in setpoint causes a rapid, but underdamped response and at R the controller increases the filter time constant to 4.7 seconds.

At point S (Figures 9 a&b) another increase in setpoint produces oscillations which are brought under control at T. At T the operator initiated the learn mode which suggested the following model:

$$\hat{G}_p(s) = \frac{10.8e^{(-14s)}}{11s+1} \frac{^\circ F}{psi} \tag{12}$$

At time point (U) the operator accepted the above model (which caused it to replace the previous model), entered a filter time constant of 4.7 and switched the controller to auto (i.e. fixed parameter) mode. Several minutes later, he reduced the setpoint to 65 degrees F. and at point W (Figures 10 a&b) reduced the water flow rate to 1 gpm. The outlet water temperature and controller output then begins to cycle indicating an unstable system. Changing the setpoint to 135 degrees F causes even more severe cycling.

At X the setpoint was changed to 65 degrees F and the controller placed in the autotune-2 mode. The water temperature returns to 65 degrees F somewhat sluggishly due to low limit saturation of the valve (Fig. 10b). However, it is clear from the control effort oscillation that at a filter time constant of 4.7 it is too small and at about point Y the controller increases the time constant to 64 and changes the model to

$$\hat{G}_p(s) = \frac{20.8e^{(-17s)}}{45s+1} \frac{^\circ F}{psi} \tag{13}$$

By point Z (Figures 11 a&b) the filter time constant has been reduced to 16 and by point AA, after the setpoint change, it has been reduced to 5.2. After appearing to settle out at the new setpoint, the system breaks into oscillations at BB. At CC the oscillations are halted by an increase in filter time constant to 38. At about point DD the model changes to

$$\hat{G}_p(s) = \frac{48e^{(-12s)}}{73s+1} \frac{^\circ F}{psi} \tag{14}$$

At point EE the filter time constant is reduced to 6.5 seconds.

Analysis of the Test Run

The response of the controller from time points Z to the end of the data has several interesting features:

1. At point BB, the control system, after having settled, apparently goes unstable for no obvious reason.

2. Only about two oscillations are required for the controller to respond to the apparent unstability and to suppress it by increasing the filter time constant to 38 at point CC.

3. At point DD, the controller produces a model which has an extremely high gain along with a large lag. The controller using this model and a filter lag of 6.5 (at point EE) is stable and brings the outlet temperature to its setpoint fairly rapidly.

4. The long tail to the response after EE is possibly due to a ramp disturbance or the valve sticking. Notice that the controller output is ramping up, but the outlet temperature is hardly changing.

From the above, it would appear that the process gain might indeed increase over time as indicated previously by the difference between the model produced at point K (c.f. – equation 9) and that produced by the learn mode at point U (c.f. – equation 12). Thus, the model given by (14) which seems to have too high a gain and lag may be more accurate than would first appear. The fact that a filter time constant of 38 is required to suppress the instability using the model given by (13) and that the controller functions smoothly with the model given by (14) and a filter time constant of 6.5, further supports the accuracy of (14).

Summary

In summary, the above tests demonstrate that the controller in autotune-2 mode effectively stabilizes the response at all operating conditions without unduly sacrificing performance. The models obtained from the autotune-2 mode are consistent with the previously observed behavior of the heat exchanger and consistently improve control system performance.

References

[1] Åstrom and Hagglund, "Automatic Tuning of Simple Regulators with Specifications on Phase and Amplitude Margins," *Automatica,* Vol. 20, page 645, 1984.

[2] Åstrom and Wittermark, *Computer Controlled Systems,* Prentice Hall, 1984.

[3] Brosilow, Zhao and Rao, "A Linear Programming Approach to Constrained Multivariable Control," *Proceedings of 1984 ACC,* San Diego, 1984.

[4] Chen and Brosilow, "Control System Design for Multivariable Uncertain Process," *Proceeding of 1984 ACC,* San Diego, 1984.

[5] Morari, "Internal Model Control – Theory and Applications," *Proceedings of 5th International IFAC/IMEKO Conference,* Antwerp, Belgium, October 1983.

[6] Morari and Doyle, "A Unifying Framework for Control System Design Under Uncertainty and Its Implications for Chemical Process Control," *Proceedings of 3rd International Conference on Chemical Process Control,* page 5, Elsevier 1986.

[7] Popiel, Brosilow and Matsko, "Coordinated Control," *Chemical Process Control – CPC III,* Proceedings of 3rd International Conference on Chemical Process Control, page 295, Elsevier 1986.

[8] Parrish, John, "The Use of Model Uncertainty in Control System Design with Application to a Laboratory Heat Exchanger Process," M.S. Thesis, Case Western Reserve University, 1982.

[9] Shine, S., "Development of Unit Operation Experiments: Heat Exchange and Gas Stripping," M.S. Thesis, Case Western Reserve University, 1980.

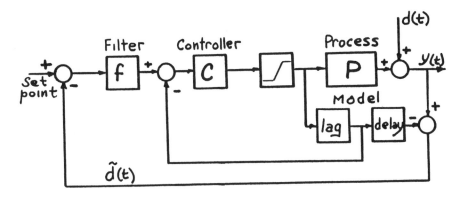

Figure 1a: Trajectory Tracking 'Inferential' Control.

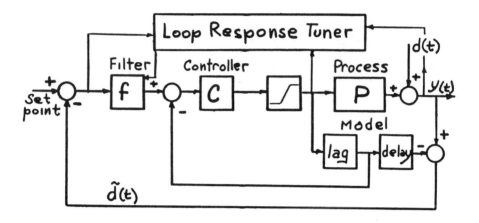

Figure 1b: Adaptive (*Autotune-1*) Trajectory Tracking 'Inferential' Control.

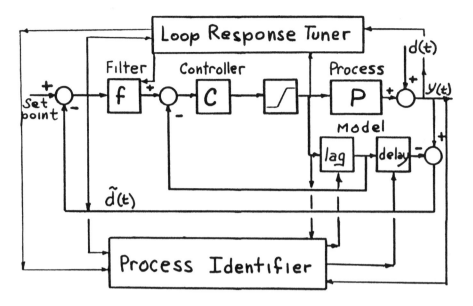

Figure 1c: Adaptive (*Autotune-2*) Trajectory Tracking 'Inferential' Control.

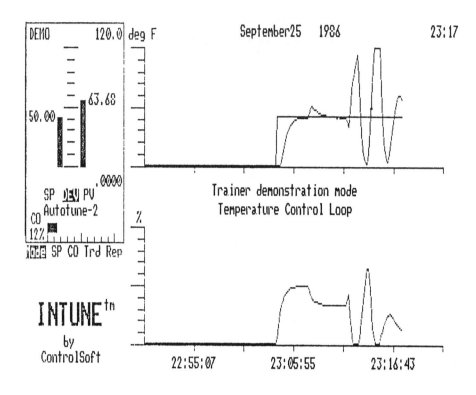

Figure 2: Operator's Interface.

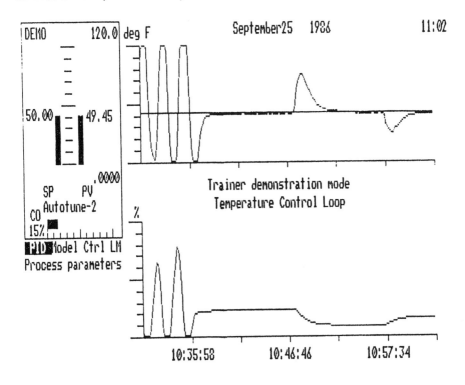

Figure 3: Engineer's Interface.

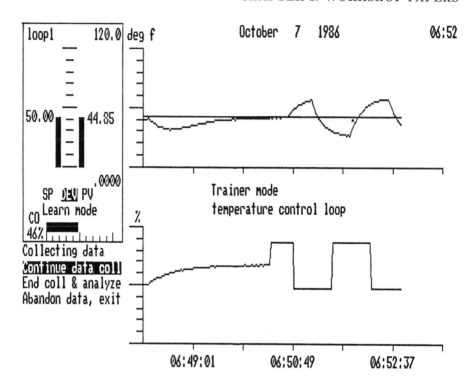

Figure 4: Learn Mode Operation.

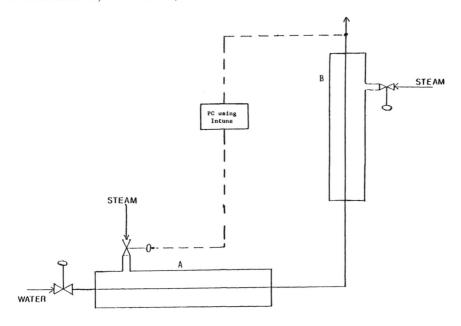

Figure 5: Dual Heat-Exchanger System.

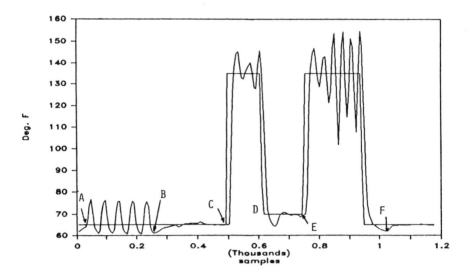

Figure 6a: Heat Exchanger Response - Setpoint and Measurement.

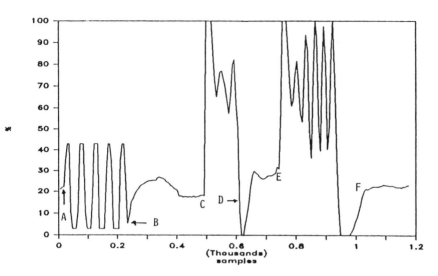

Figure 6b: Heat Exchanger Response - Controller Output.

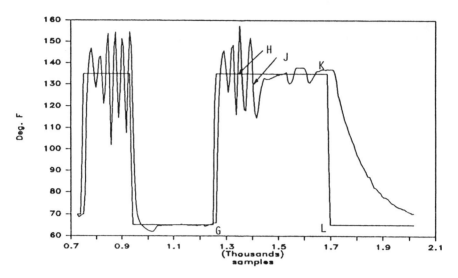

Figure 7a: Heat Exchanger Response - Setpoint and Measurement.

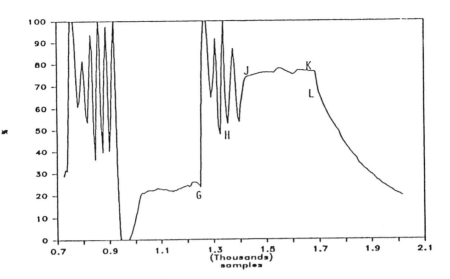

Figure 7b: Heat Exchanger Response - Controller Output.

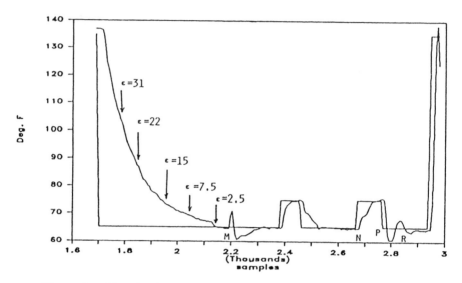

Figure 8a: Heat Exchanger Response - Setpoint and Measurement.

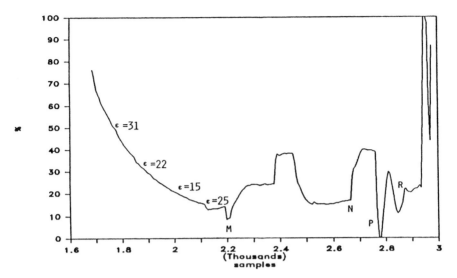

Figure 8b: Heat Exchanger Response - Controller Output.

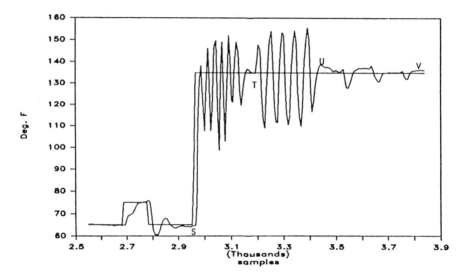

Figure 9a: Heat Exchanger Response - Setpoint and Measurement.

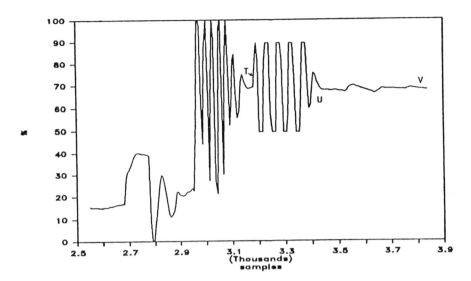

Figure 9b: Heat Exchanger Response - Controller Output.

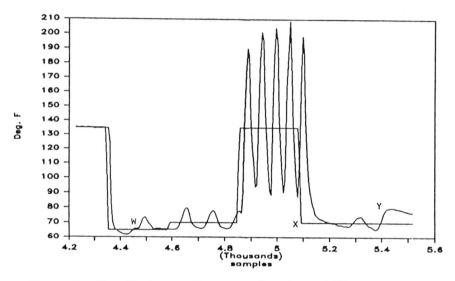

Figure 10a: Heat Exchanger Response - Setpoint and Measurement.

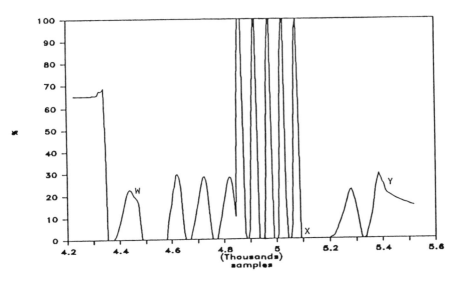

Figure 10b: Heat Exchanger Response - Controller Output.

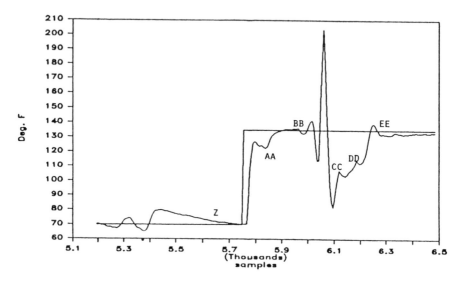

Figure 11a: Heat Exchanger Response - Setpoint and Measurement.

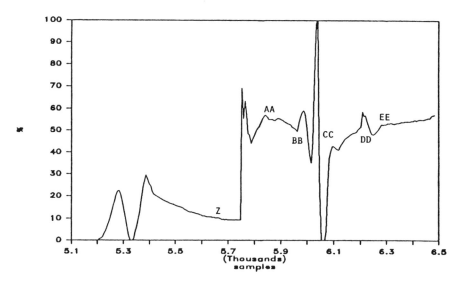

Figure 11b: Heat Exchanger Response - Controller Output.

A Systems Engineering Approach to Process Modeling

Odd A. Asbjørnsen
University of Maryland
Department of Nuclear and Chemical Engineering.

December 15-18, 1986

Abstract

This paper considers process modeling from a variety of view-points, reasonably characterized as a Systems Engineering approach. First, it is recognized that Expert Systems and practical rules of thumb have been the design tools for a large majority of process control loops, where the requirements for precision and sophistication of the process models were not pronounced at all. Those control systems have been developed as a practical learning process where experience and recommendations are accumulated in a set of Expert Advice, or "Recommended Practice." This recognition is attempted transferred to a similar approach to process modeling.

The starting point for a Knowledge Base for process modeling should be the physical, chemical and mathematical foundation of process technology, in the present case with particular reference to the process dynamics for control. Some of the fundamental principles of process operation, the conservation principle, are viewed from this angle. It is shown how this may lead to a somewhat different design of the control system, than if approached from a traditional state space model. The state space model is implicit in the conservation principle which is a balance of rates, but the conservation principle always leads to ratio control.

The complications of *a priori* developed deterministic models for the conservation dynamics in process operation, may be overcome by a merging of empirical models, obtained in plant operation, with the fundamental principles. It is shown that the residence time distribution is such a general empirical model for transport through production systems, and that the ARMA model is a logical discretization of the residence time distribution for continuous production systems. The ARMA model has the fundamental property that it describes time correlations and hence gives an indication of the time scale of the correlations, or in other words, the time frame of the process transients.

Finally, the paper considers the learning process from the basic principles, where general insight and knowledge in process behavior and special characteristics may be gained from a study of idealized process situations. Such a situation is the stirred tank or the gas tank, some of the most frequently encountered process equipments in chemical engineering. Here are demonstrated a few phenomena that are often overlooked in modeling, the difference between a gas tank and a liquid tank and the liquid tank with changing space velocity. Furthermore, it is also demonstrated that a dynamic approach to the phase separation near equilibrium in such a stirred tank may lead to results that are peculiar in relation to what is normally taken for granted.

The paper takes a general approach to most of the problems encountered in dynamic modeling of process technology, and points to the similarity between cases and even between entirely different systems. By this, the generality of the Systems Engineering approach is emphasized, and its wide applicability is demonstrated. Both in academic teaching and research as well as in industrial operation and experimentation, it is envisaged that a generalization and unified approach to the modeling of dynamics will lead to considerable savings in manpower, cost and time.

1   Introduction

One interesting and typical feature of practical process control on a simple, single loop level, is that Expert Systems have been in operation for the design of the control loops for a long time, in fact from the very childhood of process control. From very simple conceptual ideas as to how to control a particular unit operation are practical experience and modifications gained by a learning process, and the results (expert advice) are compiled in manuals referred to as "Recommended Practice of Instrumentation," either as handbooks [1] or as *in house* documentation [2], to be used in engineering or plant design work. Once the unit operation is specified and its instrumentation recommended are the design of the control system often reduced to a specification of control and instrumentation equipment, hardware as well as software. The tuning and checking out of the control loops are left to the commissioning and start-up teams. Truly, this approach has clear shortcomings in the fact that only limited subsets of the possible solutions to the control problems are recommended. However, the main reasons for adopting this procedure are related to budget constraints and time constraints in the engineering and construction [3]. Furthermore, it has been experienced that the majority of single loops do not require a frequent retuning, relative to process disturbances they are fairly robust. It is actually more relevant to get the human element out of the loop, because the majority of process disturbances and malfunctions are caused by human manipulations with control moves and parameters.

The somewhat conservative nature of the development of process control is partly explained by the economics, the slow dynamics of the process and plant developments and the feedback of information to the various stages of development, research, design and operation [3,4], as indicated in Figure 1.

A strong competitive edge for a corporation would be to master the form of information flow in Figure 1, in terms of quantity, quality and relevance of

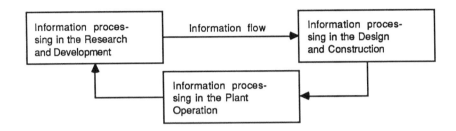

Figure 1: Information flow in the industrial learning process.

the information. This applies to the technical fields in general as well as to the particular case of process control. The Academia certainly has a role to play in this information flow, but it should be realized that the industrial point of view would concentrate on the efficiency of the learning process in the feedback loop within the corporation itself, and the steady improvement of corporate knowledge in the areas of research, design and operation with due recognition of the intellectual, psychological, technical and financial constraints [3] of the corporation.

For example, if a concept to control an operation is introduced without psychological acceptance and intellectual understanding on behalf of the plant management, one often finds that the development work has been wasted, at least temporarily. If a control algorithm requires extensive expertise in implementation, update and maintenance or sophisticated adaptation and tuning, they are usually not received with great enthusiasm by the operators and the management of the plant. The scope and comprehension of plant operation are sometimes different between the plant management and the control expert. Safety, reliability and human interface usually rank higher than error criteria and closed loop performance, except when overall plant economics, efficiency and quality control are at stake. Advanced control concepts are difficult to accept plant wide, unless they have proved or indicated improvements in some of those major issues. Steady state optimization has proved to be more rewarding than dynamic optimization in long terms.

On the other hand, process control would never make any progress if no one would be willing to take the lead and experiment in the industrial world. Most of the academic research in process control have addressed the insignificant problem of the single loop at the lowest level of the control hierarchy, while the chemical plant management is more interested in plant wide control and optimization and in safety measures like equipment condition monitoring and prediction. A fruitful interaction between academia and industry in this particular area is hampered somewhat by lack of industrial experience on behalf of the faculty, and by limited insight and theoretical experience in large systems modeling and simulation on behalf of the industry. Some of the lack of experience on both sides may be overcome by simulation, where insight into phenomenological behavior of industrial production systems may be gained, without the high cost of running industrial experiments.

A good academic example in this context, is the optimal control of an ammonia synthesis loop [5]. A complete nonlinear simulation of the process dynamics with a predictive optimal control scheme with an economic objective function, gives a good insight indeed into the major control and stability problems of such a loop, in particular the effect on the performance of constraints on states and manipulators as well as of different prediction and control horizons. Furthermore, a good overall picture is obtained of the gains of dynamic versus steady state optimization. Such studies and systems developments for simulation of real industrial processes are finding their way into the traditional academic courses and classrooms as projects parallel to the lectures [6]. In those projects are covered both the managerial aspects of project planning and control, as well as the technical aspects of operation, control, optimization and cashflow economics of the plant.

It has been pointed out by several authors of textbooks and papers [7, 8] that one of the major obstacles for a widespread application of advanced computer control in the chemical industry is a partial lack of understanding and appreciation of the complexity of the chemical process dynamics. It is certainly not the basic physical phenomena that are not understood, all the papers on theoretical models are clear indications of that. It is rather the enormous complexity caused by the large number of such basic processes and physical phenomena interacting in the production plant. This complexity and lack of basic data, as for example in reaction kinetics and in vapor-liquid equilibria, leads to an increasing degree of uncertainty in the structure as well as in the parameters of the dynamic models of the plant. The frustration that simple and basic models for the dynamics do not seem to apply, is sometimes a major factor why process designers and operators are frightened away from more sophisticated theories and the use of dynamic simulation as a general tool in design [9].

This situation can only be improved by the chemical engineers themselves, and in particular the University Academia. Process control should be considered an integrated part of process design and operation, utilizing tools that are common to several other fields of the profession, as well as to engineering in general. One such tool is the theory and technique of optimization, another is the general theory of dynamical systems, their operation and control, in particular under uncertainties and constraints. The main responsibility of the chemical engineers remains beyond any doubt in the area of process dynamics and the application of general systems tools to the modeling of the processes and their dynamics. An emphasis of chemical engineers on those general theories that are dealt with by other professions seems to be missing the main target, namely to bring to the forefront the understanding and modeling of the processes in design and operation. This involves a general systems engineering approach, from which both the general analysis and synthesis of chemical engineering unit operations and processes as well as their control will benefit.

There are several examples from process control where a starting point in the process itself brings a much better solution than a starting point in the general control engineering field, and this paper will show some of those examples. They are related to the perception of the nature of the process operation and the formulation of the purpose of the control. Examples are also shown, where a control engineering approach is the wrong one as compared to a process engineering approach. One example is the decoupling of the process interactions by the controller algorithms, while a simple transformation of the process equations makes them decoupled. Another example is the unsuccessful extension of the steady state relative gain concepts into similar concepts of dynamic relative gains.

The fact that control variables and rates of change are constrained when a process moves from one state to another is a fundamental property of process operation which has to do with safety margins and tolerances built into the process during the design. These concepts are crucial, but seldom recognized by the process designers. The control engineers mislead the designers by an emphasis on the little problems around a single loop, leading the designers attention away from the global movements and rates of changes of the plant state variables. Attempts are being made later to provide a model background for the plant wide control by plant experimentation, leading to a purely empirical model. However, the empirical approach is unsatisfactory because of reliability problems outside the region of experimentation, and it is absolutely necessary that the empirical approach merges with the basic approach to modeling within a systems engineering framework.

The fact that a systems engineering approach to the general teaching of chemical engineering has not yet reached a widespread application, is a clear indication of a conservative attitude in the chemical engineering profession. A classical example is found in the fact that process control is still treated as a special topic, and not as an integrated part of process design and operation. More and more fields are developing in the direction of integration and the use of quantitative methods and models in a systems engineering context, economics and operations research are good examples.

2   Systems Engineering: Some General Thoughts

Systems Engineering should be thought of in the first place as a discipline of a very general nature, comprising the various applications as subsystems. One such subsystem is control engineering, basically with real time computer control as the main theme, and another subsystem of this again is the process control, as applied for example to the control of chemical plants. But the process control is again a subsystem of another hierarchy of subsystems, namely the systems engineering tasks of design (specification), construction and operation of the chemical plant, where the interface with investment and cashflow economics is essential.

Another subsystem with similar interfaces as the control engineering is the information management system which comprises storage, retrieval, processing and communication of information. This system has interfaces also with the project management system for design and construction of plants, may be even more with plant operation and its management when the plant has been built. The information management system for plant-wide control and optimization has a strong interaction with the cashflow economics of the corporation, and has even time frames of dynamics in the change of information that is much shorter. The design and analysis in all these examples use Systems Engineering tools, sometimes even without a recognition of their multi-disciplinary character, and sometimes with different interpretation of the term systems engineering.

In this paper an attempt is made to identify some general aspects of systems engineering as a unifying discipline with such a multivariable outlook, with particular reference to mathematical modeling of chemical processes. First, some of the most general concepts in systems engineering are emphasized by a few conceptual headlines, and the various areas of specific applications are mentioned.

Cause And Effect Analysis

Cause and effect analysis is probably the most important ingredient of modern Systems Engineering, now merging into Expert Systems and Artificial Intelligence. This headline covers general concepts of cause and effect relations from qualitative statements to more rigorous and quantitative models of a numerical nature. The cause and effect analysis is the fundamental principle for control systems modeling, for statistical hypothesizing and for general consequence analysis often performed in political decisions. Into this general topic, one see the merging of several topics of operational mathematics for continuous and discrete systems, as well as the statistical tools for hypothesis testing, as applied to logical and mathematical modeling.

The Theory Of Learning And The Feedback Principle

The fundamental idea behind Knowledge Engineering, or more generally Artificial Intelligence, as they may apply to an industrial corporation, is to expand a Knowledge Base of the corporation for further use in modeling, reasoning, inference, decision and diagnosis. This is similar to the feedback mechanism of learning in the corporation, where knowledge and expertise is gained through practical training and learning in the life cycle of projects in management, research and production.

Optimal planning of experiments for hypothesis testing is another example of systems learning which takes a model for cause and effect as a hypothesis. The modification of this hypothesis from the experiments is a result from a learning process in feedback. But this is also very close to the model adaptation on modern adaptive control, where the model identification and parameter estimation is performed in a feedback fashion. The theory of learning is also the underlying principle for project planning and control, and is very close to human behavior in a social system where the individual adapts itself to the environment. Hence, there are several applications that might benefit from a general treatment of the theory of adaptation (learning) and feedback.

Prediction, Decision And Control

A fundamental feature of decision and control, is that the degrees of freedom exist only for the future. The past is irrecoverable. This inevitably requires some kind of prediction of the future for control, which is now seen applied to an increasing extent in the process industry. Modern control is often based on a model prediction of the future behavior of a plant using rigorous models for the plant dynamics and the past history of data for inputs and outputs. There is very little difference between this application of models for predictive control and scenarios and budgets for financial management, prediction and control. The theory behind prediction, decision and control is even a crucial concept for human behavior in a social system where the humans interact.

It is therefore felt that the theory of prediction, decision and control forms a platform for several applications. In the literature and in practical industrial applications, one may see covered here, general problems like accuracy and reliability of predictions (model uncertainty), general algorithms for decision and control as well as interpretations of their effects on the system stability and general behavior.

Information Flow And Communication

Information flow and communication become vital to most large projects where millions of dollars are at stake. A prerequisite for an efficiency improvement in this area is a good database design for information storage, retrieval and combinations. The theory and structure of relational databases are essential under this heading, but so is also the logical design of coding conventions, keyword thesaurus and icons for human interaction. Humans are basically visual in their information processing, their interpretation of pictures is several orders of magnitudes faster than the interpretation of alphanumeric information. The communication between human beings and information systems is shown to be enhanced tremendously by graphics like icons, windows and symbols.

However, communication between systems is equally important, as accentuated for example in distributed computer control of large chemical plants, or distributed information processing in large engineering projects. Response time and synchronization are essential topics both in a system communication with systems, in systems communication with a human and in humans communication with a system. The human communication has demonstrated an ambiguity and a lack of uniqueness, leading to misunderstanding and in many cases with disastrous results. Nonambiguity in coding and presentation of information is essential under this heading. Syntax, standards and conventions for communication are important specifications.

Multivariable Objectives, Objective Functions And Optimization

The basis for decisions is a set of targets proposed to be reached by efforts or control actions over a certain time horizon. In process control this is referred to as a control horizon, and in economic planning it is referred to as the budget horizon. The main problem of setting the targets is different in the various applications. Sometimes the targets are qualitative statements, but in other applications they may be quantified and expressed in the form of objective functions, which again are mathematical models of the performance of a system.

This area also has a lot of subsystems of applications, but has a main feature common to all. A general formulation of the objective functions is not realistic without simultaneous definition of the constraints on resources and achievements. This is indeed similar to the problems solved in optimal control of dynamical systems, and in optimal design of processes, plants, networks etc. The concepts of objective functions are even found in optimal design of experiments and in model discrimination and model parameter estimation.

The solution of the optimal decision and control problem as well as of the problems of model discrimination and adaptation is relying heavily on the theory of optimization, which now becomes one of the essential cores of the Systems Engineering discipline. Optimization, as used in engineering, relies on models for the objectives as well as for the the process equality and inequality constraints, but the same theory and philosophy of optimization are found even in systems with qualitative logical models (yes/no models) as applicable to the human behavior in social systems. Modeling is seen to be a fundamental issue even there.

Reliability, Uncertainty, Safety And System Failure

Most of the decisions and control actions taken in technical systems assume a reliable description of the cause and effect relationships. Failures or deterioration of components have a dramatic effect on such cause and effect relations, and represent an unpleasant uncertainty if they are not incorporated in the cause and effect analysis. But failures are generally stochastic in nature, and this headline represents a merging of stochastic modeling with deterministic modeling, by letting the model parameters have a certain stochastic variation with very large amplitude (0 or 1). Therefore, the application of modeling and statistics to this particular field, common to so many systems problems, is worth a separate heading. The topic applies equally well to technical systems, as for example to the construction and operation of plants, as to management systems where the interplay between information systems and humans is the main issue. Failures may be caused by equipment breakdown, by equipment malfunction and even by miscommunication to humans.

Human Behavior And Artificial Intelligence

The human element, say manager, operator, or general user in the information system for management decisions, plant operation and control, plant design and construction, is so crucial, that the topic of human behavior is singling out as a general topic of major importance. The interaction between a human being and a system functions in two ways, the human interpretation and re-action to information presented by the system and the system interpretation and reaction to human decisions and inputs.

This is an extremely complicated area of modeling of the learning process for system development, because the human behavior is dependent on culture (memory effects) and often disturbed by uncorrelated nonobservable exogenous variables. Experiments on human behavior may not be conclusive, in many instances in fact misleading. However, it is essential that the cause and effect relationships mentioned is given a general treatment and not necessarily as a subset of computer science in the form of user interface design, albeit very important in that area. In order to make the analysis of human behavior a valuable heading under Systems Engineering framework, it is essential that the general concepts like cause and effect relations, information interpretation, human inference, logic, deduction and decisions follow along the same lines as established in other research topics, as for example in Artificial Intelligence. This subject has indeed such a close relation to human behavior, that it seems logical to merge those under the same heading.

Project Work

One of the essential activities in practical Systems Engineering in industry is the project work. In this work, there are other aspects of human interaction that will distinguish one project team as a good one and another as a bad one. These aspects are for example the ability to lead and to follow, the total commitment to the project and the ability to communicate, verbally and literally, to illustrate, highlight and convince. In most university programs, a rational approach to the training of the students in future industrial teamwork of a reasonable size, project work and management, is virtually non-existing. One practical approach may be to introduce larger team projects in parallel with some of the courses or independently [6]. The purpose of such project tasks is to train and experience the students in management functions, in project planning and control, in resource allocation, task definition and time planning, as well as to get a deeper insight into the contents of a course or in systems engineering in general. Such project experiments have been run fairly regularly in connection with Modeling and Systems Simulation courses and Plant Design courses in Chemical Engineering, and the response from the students is very enthusiastic. Everyone gets a chance to oral and written presentations, to critique, to experience of psychological challenges from the team in a management function etc. [6].

3   A Qualitative Assessment Of Cause And Effect Relationship

The early stage of plant design and synthesis hardly gives room for advanced control system modeling or analysis. The most frequently used methodology is a traditional qualitative analysis of cause and effects [10,11], which has now gained an explosive popularity, first in "HAZOPS" [12,13] (HAZard analysis and OPerabitily Study) and later in the so-called Expert Systems. It may be typical for the academic world that their work hardly reflects the simple, heuristic and practical everyday rules applied by the industry. The large majority of academic papers are sometimes very remote from the practical application to industry, even to the interest of the industry. It must be a paradox for the authors to know, that there are millions of loops out there in industry that are designed according to unpublished "tools" for the practical life. Those are more or less based on industrial rules of thumb and practical experience (Expert Systems). By neglecting such rules in academic arrogance, that particular field of practical process control are about to be taken over by thrifty salesmen and computer experts on Artificial Intelligence and Expert Systems, in the same way as Statistical Process Control is taken over by statisticians and thrifty salesmen to the top management.

Here is one set of such traditional old heuristic tools and rules that have been in use [10,11,12] long before Bristol's Relative Gain Array [14]. Those rules are still in use to a very large extent:

- Based on practical experience or physical insight, set up a qualitative matrix (table) of process cause and effect relationships including all the measured process outputs, all the available manipulators and the known disturbances. Characterize the relationships by (+) (positive gain), by (−) (negative gain), by (0) (no effect) and by (?) (uncertain effects) and its dynamics by fast, medium and slow response.

- Pair the variables by the simple rules: Pair only the variables where a clear (+) or (−) appear. Never pair variables with (0) and avoid variables

Figure 2: A Countercurrent Heat-exchanger.

with (?). If that is desirable, seek more expert advice. Support your decision of pairing by well proven experience from similar processes.

- Consider each controller setpoint as a new input variable. Carry out the cause and effect analysis for the closed loop. If desirable to use set-points for control (cascade) ensure that the inner loop has faster dynamics than the outer loop.

- Make sure that the controller action, (+) or (−), is such that the controller action is opposite the process action, i.e. the product of the two actions is always negative in a negative feedback loop.

- Select the manipulators to control the material and energy balances and the quality of the product. Design the manipulators to cope with estimated disturbances. Balance control or conservation control favors ratio control.

- If more than one main flow manipulator is installed in a production line, only one can take care of the main flow. The others are for temporary material balance control which requires intermediate storage facilities. A distillation column is a good example on this.

With the heat exchanger in Figure 2 as an example, one may summarize the Expert Statements to the cause and effects and the process matrix as follows:

- If the flowrate q_1 increases, then all outlet temperatures will move towards the inlet temperature of the flow q_1.

- If the flowrate q_2 increases, then all outlet temperatures will move towards the inlet temperature of the flow q_2.

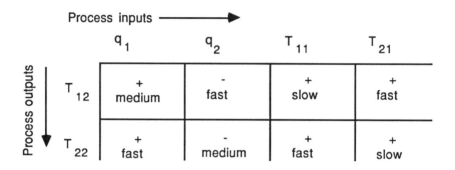

Figure 3: A qualitative cause and effect matrix for the heat exchanger

- If the inlet temperature T_{11} increases, then all outlet temperatures will increase.

- If the inlet temperature T_{21} increases, then all outlet temperatures will increase.

Even a superficial insight into the process dynamics will qualify for the following statements about the process dynamics, but they may be somewhat modified by the construction and operation of the exchanger. Provided the heat exchanger has countercurrent operation, one may state:

- The output temperature closest to the inlet flow or temperature under change, will change fastest (short response).

- The outlet temperature farthest away from the inlet flowrate or temperature under change, will show the slowest changes (long response).

The process matrix will be a two by four table, where the convention above may be used. One has to distinguish the two flows as hot and cold, because the response will depend on that. Let q_1 be the hot stream and q_2 the cold. Then the results will follow as shown in Figure 3.

Even if such simple procedures may not satisfy the academic requirements to a rigorous and quantitative treatment, they are still used to a very large extent, in particular at the early stage design of a plant design, because rigorous methods are not economically justifiable at that stage. If the project is continued and the economic frame of the project permits, nonlinear simulation of the suggested control structure may be performed [9].

4   A More Quantitative Assessment, The Relative Gain Matrix

Inspired by the popularity of the qualitative process matrix for the cause and effect analysis, attempts were made [14] to fill in numbers for the steady state cause and effect sensitivities in terms of relative gains. The process sensitivities are readily available through the steady state simulators now used for plant designs, and certain guidelines as to the pairing of loops are given by the Relative Gain Matrix [14]. It is important to remember that these guidelines do not ensure stability or loop performance for transients.

Let the output be y_j and the input u_k and the corresponding vectors (where the elements y_j and u_k are removed) be $\mathbf{y}_j$ and $\mathbf{u}_k$ respectively. Then the relative gain is the ratio between two sensitivities, one with all loops open, and one with perfect control in all other loops. The conceptual idea behind the Bristol interaction analysis is very simply stated as a relative comparison of sensitivities by the ratio of gains, the gain between one input and one output with no feedback control applied to any input, and the gain between the same pair of input and output when all other inputs are subject to perfect control. Perfect control in this connection means that the other outputs that are under control do not change:

$$\lambda_{j,k} = \frac{\frac{\partial y_j}{\partial u_k} \mid \mathbf{u}_k = \text{const}}{\frac{\partial y_j}{\partial u_k} \mid \mathbf{y}_j = \text{const}} \tag{1}$$

In mathematical terms the ratio of gains is a ratio between partial derivatives, one where all other inputs are constant, and another where all other outputs are constant. One requirement for this analysis is that the process matrix in question is square, and mathematically speaking invertible. In steady state, let the relationship between input and output be described by:

$$\mathbf{Ay} = \mathbf{Bu} \tag{2}$$

Provided $\mathbf{A}$ is nonsingular, the sensitivities are given by the matrix $\mathbf{Q} = \mathbf{A}^{-1}\mathbf{B}$. If $\mathbf{B}$ is nonsingular, it is possible to calculate the the vector u_k that would make $\mathbf{y}_j$ equal to zero once the output variable y_j is left uncontrolled and the input variable u_k is left unchanged. Then the relative gain is the ratio between two determinants:

$$\lambda_{j,k} = \frac{\det(Q)}{\det(Q_{j,k})} \tag{3}$$

where the matrix $\mathbf{Q}_{j,k}$ is the reduced matrix left of $\mathbf{Q}$ when row j and column k are removed.

In state space notation [15] one may attempt to extend the relative gain concept to dynamics by writing the model:

$$\mathbf{x} = (\mathbf{A} - s\mathbf{I})^{-1}\mathbf{Bu}; \; \mathbf{y} = \mathbf{M}(\mathbf{A} - s\mathbf{I})^{-1}\mathbf{Bu} = \mathbf{Q}(s)\mathbf{u} \tag{4}$$

In principle, it is possible to introduce dynamic relative gains, but that would require an inversion of the reduced plant dynamic matrix $\mathbf{Q}_{j,k}(s)$ the difficulty and limitations of which have been discussed extensively in the literature [16]. The physical and mathematical realizability of an inversion of the transfer function matrix $\mathbf{Q}(s)$ in general is highly questionable. That is why the crucial assumption of a stable and perfect feedback control is not realistic in dynamic sense, due to stability requirements and to physical constraints of the plant.

The reason for choosing perfect control as the reference, was to make the analysis independent of the controllers in the feedback loops when they were closed. However, perfect control can only be obtained in steady state, and it takes an integral action in the controller to achieve it. In dynamic sense, the relative gain depends upon the controller performance, or its transfer function, which again is designed according to the stability requirements of the other loops. Dynamic relative gains cannot be derived from the process transfer functions alone, because perfect control implies an inversion of the transfer function matrix of the process, which is practically unrealizable in most cases. The dynamic relative gain with perfect control is a fake, and that is why the method of dynamic interaction analysis based on the structured singular values of the process transfer function matrix [17] is a more rational approach.

5   Further Improvements In Modeling, The Conservation Principle Models

From a process engineering point of view, one may state that traditional control engineering treats that the role of dynamic conservation control very lightly, while it is probably the most useful principle in the process control and the principle most plant managers are familiar with and concerned about. The use of the conservation principle for control is as old as process control itself, but it may be argued that its potentials are not fully utilized. Furthermore, it is probably the most important physical principle that could be said to make the control of continuous production processes somewhat special. In any case, it is an example where the design of the control algorithm takes the process technology as a starting point rather than the more traditional control theory.

Continuous production processes are governed by a few very fundamental principles, but with a variety of specialities in their applications. One of those principles, and probably the most important one, is the conservation principle. This general principle may again be considered a balance between rates, expressed as a product of a specific rate vector $\mathbf{r}$ and a vector of weighting factors $\mathbf{a}$:

$$\mathbf{a}^{\mathrm{T}}\mathbf{r} = 0 \qquad (5)$$

This approach leads one to a systematic study of the relevant rates, as for example in a heat conservation system:

Rate of accumulation of heat:	$r_a = \dfrac{d(\int_V T c_p \rho dV)}{dt}$,
Rate of inflow of heat:	$r_i = \sum (T c_p \rho q)_{i,j}$,
	j = stream number,
Rate of outflow of heat:	$r_e = \sum (T c_p \rho q)_{e,j}$,
	j = stream number,
Rate of production of heat by reaction:	$r_p = \int_V \mathbf{r}^T (-\Delta \mathbf{H}) dV$,
Rate of heat conduction:	$r_c = \int_A \left(k \frac{\partial T}{\partial \mathbf{x}} \right) 1 dA,\ \mathbf{x} \perp A$,
Heat balance ($\mathbf{a}^T = [1, -1, 1, -1, 1]$):	$r_a - r_i + r_e - r_p + r_c = 0$.

Steady state conditions are special cases of conservation where the rate of accumulation is zero. But so are also all phase equilibrium processes, a special assumption where the rate of interface accumulation goes to zero. Chemical equilibrium processes may be considered similarly as a special assumption where the reaction rate weighting coefficients in **a** becomes very large as compared with the weighting coefficient for the rate of accumulation (equivalent to very fast reactions).

The rate processes and conservation principles are coupled via the intensive thermodynamic properties, leading to process interactions in multivariable control loops. However, it is shown [18,19] that some of these couplings may be removed by natural invariance phenomena found in stoichiometry and in dimensional analysis [20]. This phenomenon is characteristic for chemical production processes, but is very often overlooked in process models found in the literature. The fact that some rate processes are invariant to others offers interesting features both for control and for model identification and parameter estimation. This phenomenon has been experienced as extremely useful in industrial applications [21].

The conservation principle very often leads one to define and use extensive variables rather than intensive ones in multivariable process models and control. The conservation principles are usually decoupled and linear in those variables, and that may offer some advantages for model simulation and for control [20]. The conservation principle may be used very effectively for load disturbance compensation (Disturbance Rejection [7] or Disturbance Accommodation Control [22]) as used for example in the traditional ratio control. This requires simple arithmetic calculations of the extensive properties as for example products of a measured flow variable, some physical properties, measured or estimated, and some measured or estimated intensive properties. It is also important to notice as a general feature of the conservation principle and the rate processes that there is practically always a natural feedback between the level of accumulation and some of the rate processes, e.g. outlet flowrate. Sometimes this self-regulation is sufficient for the production control, and more sophisticated man-made control may be discarded, but some other times this self-regulation may lead to instability that needs to be stabilized and compensated for by automatic control [23].

As an example of dynamic material and energy balance control or conservation principle control, consider the operation of a stirred tank, as shown later in Figure 4, for simplicity without reactions. This is a classical textbook example [24] which has a lot of interactions between intensive and extensive properties, and which is usually approached by decoupling techniques that is really not required. If one approaches this problem from a systems engineering point of view of the conservation principle, one would use the balance of rates of accumulation, feed and effluent for control:

$$\mathbf{r}_a = \mathbf{r}_i - \mathbf{r}_e \tag{6}$$

In this form, the process model is linear. If the purpose of the control is to keep the level of accumulation constant, then one would control the rates such that the rate of accumulation is zero:

$$\mathbf{r}_i = \mathbf{r}_e \tag{7}$$

Obviously, this requires as many manipulators as there are balance equations, and one would have to calculate the required rates from a product of intensive and extensive properties. If this is done, the control problem is very much simpler than usually considered by chemical engineers. If one assumes that the effluent rate is measured or computed from the outlet measurements and that there are as many feed rate manipulators as there are balance equations, the calculation of those feed rates follows from a linear set of equations [20]:

$$\mathbf{C}\mathbf{q}_i = \mathbf{r}_e; \text{ or: } \mathbf{q}_i = \mathbf{C}^{-1}\mathbf{r}_e \tag{8}$$

where the matrix $\mathbf{C}$ contains the intensive properties of the feed streams. Those are normally not measured, and there are errors in that matrix. However, if one measures the intensive properties of the effluent stream, say $\mathbf{c}$, one may write a simple material balance for the control:

$$\mathbf{q}_i = \mathbf{C}^{-1}\mathbf{c}q_e, \text{ or: } \mathbf{q}_i = \mathbf{v}q_e \tag{9}$$

where $\mathbf{v}$ is a set of ratio coefficients, as in the traditional ratio control. If these ratios are wrong, this will be noticed in the dynamic balance because the rate of accumulation will no longer be zero. This rate may be measured as the product of the extensive volume of the stirred tank and its intensive properties which are the same as the effluents, say by differences over a convenient time interval:

$$\mathbf{r}_a = \frac{[V(k)\mathbf{c}(k) - V(k-1)\mathbf{c}(k-1)]}{\Delta t} \tag{10}$$

from which the ratios are adjusted according to a slow feedback mechanism, for example a simple P-I control algorithm. The quality of the output product may be included as a setpoint of the intensive properties:

$$\mathbf{e} = \alpha(\mathbf{c}_{sp} - \mathbf{c}) - \frac{(1-\alpha)\mathbf{r}_a}{q_e}, \ \mathbf{v} = \mathbf{v}_o + \left[k_p\mathbf{e} + k_i \int \mathbf{e}dt\right] \tag{11}$$

Here is $0 < \alpha < 1$ a relative weighting factor signifying the product quality control (a close to 1) or the load disturbance rejection via the rate of accumulation (α close to 0). One should keep in mind that the inclusion of the first term in the error signal takes care of the control of the level of accumulation in the tank, as this gets lost if $\alpha = 0$. The vector $\mathbf{v}_o$ comprises the initially estimated ratios. Any offset in these will be compensated for by the integral action in the feedback.

Even if this is an artificial and oversimplified example, it demonstrates four important advantages over traditional control engineering, which may qualify as practical Expert Statements or rules of thumb:

- Conservation principle control is basically linear and decoupled in the extensive variables, and the control law should be based on the conservation model, which is generally applicable to any system. The use of intensive properties in the model, makes the control nonlinear in those variables.

- Unknown disturbances, uncertainties or errors in the parameters in the conservation model are automatically compensated for by a slow feedback control utilizing the integral nature of the conservation principle. This is the simplest possible parameter update.

- For a stirred tank, conservation principle control is identical to the traditional ratio control and is ideal for disturbance rejection. It eliminates process interactions and the needs for decoupling of the control loops.

- For more complicated models for the rate of accumulation, the evaluation of the level of accumulation, the conservation principle and the stability of the control system becomes more complicated. However, the basic principles shown here still apply.

The conservation principle for control always leads to a combined feed forward and feedback control. The feed forward is a load compensation and the feedback is essentially an adaptive parameter adjustment. It seems that this concept would be worth while a more systematic research.

6 Merging Empirical Models With Fundamental Principles, The Empirical Residence Time Distribution For Accumulation Dynamics.

Even if models for the conservation dynamics in chemical plant are recognized as essential for control and for a proper understanding of plant operation, it has been realized that quantitative models are complicated and difficult to establish from first principle. In order to circumvent some of these complications are empirical residence time distributions for a production line introduced. This concept is used widely in reaction engineering [25], but it was shown early [26] to have a much wider application to modeling and to transfer functions. The concept is statistical in nature, and assumes the transportation through a system to be subject to a delay, which is the residence time. The outlet flow is then a simple integration over the distribution function $f(\tau)$ of the delayed inlet flows, provided there are no losses or transformations in the process:

$$n_e(t) = \int_0^\infty f(\tau) n_i(t - \tau) d\tau \tag{12}$$

The Laplace transform of Equation (10) identifies the residence time distribution as the transfer function between the rate of inflow and the rate of outflow [26], as long as the residence time distribution does not change with time:

$$N_e(s) = \left[\int_0^\infty f(\tau) \exp(-\tau s) d\tau\right] N_i(s) = F(s) N_i(s) \tag{13}$$

The residence time distribution is usually determined experimentally, but in certain simplified cases where the flow-pattern may be described, as for example stirred tanks, tubes, laminar flow etc., one is able to establish models from first principles.

The concept of residence time distributions may be extended to multiple components and multiple flows in and out of the production system, by a simple definition of a matrix of residence time distributions, or equivalently, a matrix of transfer functions:

$$\mathbf{N}_e(s) = \mathbf{F}(s)\mathbf{N}_i(s) \tag{14}$$

At first look, the residence time distribution may seem to have interesting properties for modeling, but it has indeed its limitations. The concept is not very easy to apply to systems where the residence time distribution changes with time. This rules out most of the practical industrial situations where the total flow-rate is under control and the residence time distribution therefore changes with time. The concept is insufficient for processes where the rates depend on position or other properties of the fluid, as for example with higher order reactions. It has it greatest power as a modeling tool for all those processes that are functions of residence time alone, as for example all decaying processes (radioactive decay and first order reactions). In those cases, a certain fraction of the entrance flow is consumed, dissipated or disappearing as a first order decaying process:

$$N_e(s) = \left[\int_0^\infty f(\tau) \exp(-\tau s) \exp(-k\tau) d\tau\right] N_i(s) = F(s + k) N_i(s) \tag{15}$$

However, as soon as the total flow rate is changing with time, as in many of the practical control problems, the concept loses its applicability even in the case of decaying reactions. The general conclusion and Expert Statement one may draw from these observations, may be summarized as follows:

- The conservation dynamics for a production process may conveniently be described by a residence time distribution, under certain limitations.

- The residence time distribution is inconvenient as an empirical model for all those cases where it changes significantly over time, e.g. flow-forced control systems.

- For processes that are functions of other position and time dependent variables also, and not just the residence time, the residence time distribution is inadequate as a model.

- The residence time distribution is ideally suited as an empirical model for processes where the distribution does not change with time and where the processes taking place in the production system is a function of residence time only, e.g. an exponentially decaying process.

It should be noted, that these expert statements are results of a simple application of basic principles of conservation, and not restricted to any particular process or reactor.

7   The ARMA Model

One of the most popular models for discrete linear systems, is the Auto-Regression Moving Average model [7], or the ARMA model. This model may be described as a weighted sum of discrete time values of input and output:

$$y(k) = \sum a_j y(k-j) + \sum b_j u(k-m-j) \tag{16}$$

At first sight, the model may seem totally empirical, but there are distinct dynamic relations in the model which may be regarded as a result of a general principle. One such principle is that most systems exhibit a dynamic nature which makes one observation correlated to a previous one. How far back in history such a correlation extends is a measure of the time scale of the dynamics. Another such principle is that the response of a system is most frequently observed to be related to a weighted time average of the sequence of inputs. Both these principles describe the interesting general fact that all dynamic systems have memory which fades away as time goes by. The ARMA model is easily extended to multivariable systems, and is seen to have a general applicability and be a theoretical foundation for dynamic systems.

The step from a residence time distribution to an ARMA model is very logical, as one may regard the function $f(t)$ as a weighting function for the various inputs, or as a Moving Average in the ARMA model. If one discretizes the integral in Equation (10), one may write:

$$n_e(k) = \sum_{j=1}^{\infty} b_j n_i(k-j); \; k = \frac{t_k}{\Delta t} \tag{17}$$

where the summation is a weighting of all previous feed-rates similar to the residence time distribution in histogram form:

$$b_j = f(t_j)\Delta\tau = f(j\Delta t)\Delta\tau \tag{18}$$

In order to generalize, one may regard this infinite set of weighting factors as a result of a ratio of two finite polynomials in the forward shift operator q for the time series in the input and output:

$$
\begin{aligned}
n_e(k) &= \sum a_j n_e(k-j) + \sum b_j n_i(k-j) \\
\Rightarrow \quad n_e(k) &= \left[\frac{\sum b_j q^{-j}}{1+\sum a_j q^{-j}}\right] n_i(k) = \left[\frac{P(q)}{Q(q)}\right] n_i(k)
\end{aligned}
\tag{19}
$$

which indeed is the discrete analogy to the transfer function. The ARMA model applied to the transportation through a production system may be regarded as a discrete version of the residence time distribution. Naturally, the same restrictions as mentioned above, are indeed valid for the discrete form and the same Expert Statements made above apply. In addition one may note:

- An ARMA model applied to the conservation dynamics is the discrete form of a residence time distribution.

- The discrete form of the residence time distribution may be expressed as an infinite series or recursively as a ratio of two closed polynomials in the shift operator. This form has the least number of parameters or terms.

This examination of the residence time distribution, developed originally for chemical reactors, shows that the concept is general, and really is just another form of expressing a dynamic relationship between an input and an output. The fact that the residence time distribution may express a transfer function in continuous or discrete form is certainly an eclatant demonstration of that. This example also demonstrates the power of the systems engineering approach, without any particular regard to the specifics of the process in question.

8   Learning From The Basic Principles, First Order Processes

All first order single input single output (SISO) processes are described by the first order differential equation in one variable:

$$\frac{dx}{dt} = bu - ax; \text{ or: } \frac{dx}{dt} = a\Delta x \tag{20}$$

Here u is the input, x the output and Δx represents a special case where the driving force is a potential difference between input and output as in Figure 4.

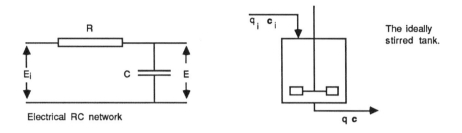

Figure 4: Physical similarities between a stirred tank and an RC-network.

In that case the parameter a is a phenomenological conductivity with the dimension time^{-1}.

In Figure 4 is shown a simple RC-circuit that complies with this model, where $a = (RC)^{-1}$ and Δx is the potential drop across the resistor R. The voltage signal out of the circuit is x. The analogy to a stirred tank is well known, in fact the foundation for parallel analog computations used extensively for process dynamics simulation in the sixties.

Multivariable first order processes of a similar nature may be described by the state equation:

$$\frac{d\mathbf{x}}{dt} = \mathbf{A}\Delta\mathbf{x} \tag{21}$$

Or, for more generally applicable input excitations $\mathbf{u}$:

$$\frac{d\mathbf{x}}{dt} = \mathbf{B}\mathbf{u} - \mathbf{A}\mathbf{x} \tag{22}$$

A very large number of traditional physical models for chemical engineering processes may be transformed to or expressed in some of these general forms. A few of the most common examples are given below.

8.1   The Single Phase Stirred Tank

The stirred tank is by far the most frequently encountered physical model for chemical engineering processes, and used for single phases, for several phases, liquid systems and gaseous systems. One would think that the analysis and modeling of such a common process would be completely exhausted by now, but there are still phenomena and similarities that are often overlooked in the textbooks and in the literature. Above all, a unified and systematic treatment of that common process is still not seen. This chapter represents an attempt to extract some of the common features of the stirred tank as a process model under different applications and different processes, and hence provide a starting point for the Knowledge Base in an Artificial Intelligence system for Process Modeling. Depending upon the process going on in it, the stirred tank may exhibit almost any peculiar dynamic characteristics of linear and nonlinear systems.

For example, consider even the simplest possible process where the fluid in the tank is a homogeneous, single phase. There is a great difference in the application to liquid phase processes and to gaseous processes. In the liquid phase, the holding time or space velocity usually changes due to changing volume, while the density is constant. In gaseous systems, the opposite is the case. The volume is constant, while the density changes, and the mass space velocity varies with time in gaseous systems too.

Single Phase Non-Reacting Liquid Systems

The dynamic conservation principle is generally applicable as stated in Equation (6), and the effluent term and the accumulation term are expressed by the total liquid flow and intensive properties. For non-reacting systems and reaction invariants, the component material balances are linear and completely decoupled in terms of the molar rates of the components:

$$\mathbf{n}_a = \mathbf{n}_i - \mathbf{n}_e \tag{23}$$

The molar rates may be expressed as products of intensive properties, say volume concentrations, and an extensive property, say volume:

$$\frac{d(V\mathbf{c})}{dt} = q_i\mathbf{c}_i - q\mathbf{c} \tag{24}$$

This balance equation is always associated with the total liquid flow balance:

$$\frac{dV}{dt} = q_i - q \tag{25}$$

Combining the conservation in Equation (22) and (23) is surprisingly rare in the elementary textbooks [27,28], but doing so, brings the stirred tank into the form which is identical to practically all first order processes with one common eigenvalue for all states of the components, the holding time or inversely, the space velocity:

$$\frac{d\mathbf{c}}{dt} = s_i(\mathbf{c}_i - \mathbf{c}) = \frac{(\mathbf{c}_i - \mathbf{c})}{\tau_i} = s_i\mathbf{I}\Delta\mathbf{c} \tag{26}$$

$$s_i = \tfrac{q_i}{V} = \text{Inlet space velocity;}$$
$$\tau_i = s_i^{-1} = \text{Holding time or mean residence time.}$$

The matrix $\mathbf{A}$ in Equation (19) is the identity matrix $\mathbf{I}$. It is important to notice the established facts in a Knowledge Base for the modeling of a single non-reacting liquid mixing process in an ideally stirred tank using the material conservation principle. Under such conditions:

- A stirred tank and an electrical RC-circuit has the same dynamic model, where the potential driving force is the difference between the inlet potential and the outlet potential of the intensive property in question, and the phenomenological conductivity is the inlet space velocity, analogous to the inverse of the RC time constant.

- For tanks with varying volume or liquid level, the inlet space velocity is inversely proportional to the volume, and hence also varies with time.

- For multicomponent systems without any reactions or phase change, the dynamic models are completely decoupled and each intensive property individually have the form identical to an RC-circuit with the same circuit parameter.

- The stirred tank has basically as many degrees of freedom for control as there are individual and different feed streams, plus the effluent stream. More effluent streams does not add to the controllability of the plant, they are just added up and treated as one effluent stream, because the intensive properties of all effluent streams are the same.

The Problem Associated With Heat Conservation

As soon as energy, for liquid systems essentially heat, enters the analysis, there will always be heat leakages or heat transfers as well as heats of mixing. Therefore, the heat balance for a stirred tank is bound to be different, and there may be a one-sided coupling between the conservation of heat and the conservation of materials. For a non-reacting system, the conservation of heat may be written:

$$E_a = E_i - E_e - E_t + E_p \qquad (27)$$

Most of these rates may be written in terms of the intensive property temperature. For example, the accumulation rate is:

$$E_a = \frac{d(V\rho u)}{dt} = \rho u \frac{dV}{dt} + V \left(\frac{\partial(\rho u)}{\partial T} \right) \left(\frac{dT}{dt} \right) \qquad (28)$$

where u is the internal energy per unit of mass, ρ the density and T the temperature.

In liquid systems, where the difference between internal energy and enthalpy is practically zero and where the physical properties ρ and c_p may be taken as constant, the terms in the conservation Equation (27) will be simplified to:

$$E_a = \rho c_p (T - T_{ref}) \frac{dV}{dt} + V \rho c_p \frac{dT}{dt} \qquad (29)$$

$$E_i = q_i \rho c_p (T_i - T_{ref}) \qquad (30)$$

$$E_e = q\rho c_p(T - T_{ref}) \tag{31}$$

$$E_t = AU(T - T_c) \tag{32}$$

$$E_p = q_i \Delta \mathbf{c}^{\mathrm{T}}(-\Delta \mathbf{H}) \tag{33}$$

where $\Delta \mathbf{c}$ is the vector of concentration differences across the tank, i.e. between inlet and outlet of the tank, and $\Delta \mathbf{H}$ is the vector of specific heats of mixing of the components. The contact area for heat transfer is A and the heat transmission coefficient to the cooling or heating system is U.

The general equation for the conservation of heat in a liquid system with varying volume is therefore described by:

$$\frac{dT}{dt} = s_i(T_i - T) - \left(\frac{AU}{V\rho c_p}\right)(T - T_c) + s_i \Delta \mathbf{c}^{\mathrm{T}}\left(\frac{-\Delta \mathbf{H}}{V\rho c_p}\right) \tag{34}$$

The relative significance of the terms in this equation may vary, and it is often up to the opinion of an expert to decide when some of them may be neglected. In the absence of an expert, reasonable decisions may be reached by the Artificial Intelligence System itself by an inspection of the range of values of the terms. By an order of magnitude calculation of the various terms under reasonable range of operating conditions for the plant in question, those orders of magnitudes may then be compared with the overall accuracy of the actual process representation by the whole model used for simulation or control. Based on the result of that comparison, terms may be neglected.

As an example of such simplifications, take the ideally adiabatic condition (i.e. the heat transfer coefficient $AU = 0$) for completely ideal solutions with no heats of mixing (i.e. $(-\Delta \mathbf{H}) = 0$). Then the conservation of heat is the same as the conservation of charge in the RC-circuit, and the differential equation model is simply:

$$\frac{dT}{dt} = s_i \Delta T \tag{35}$$

which is the same form as the model for the RC-circuit and the model for the component material balances.

When heat enters the picture, the Expert Statements may be expanded as follows:

- For ideal liquid solutions under adiabatic conditions of a stirred tank, the dynamic model for the heat balance is identical to the composition balance.

- Heat transfer and heat of mixing change the differential equation for the temperature. The heat transfer couples the stirred tank dynamics to the cooling or heating system or to the environment, while the heat of mixing couples the heat balance to the composition balance. The composition balances are still invariant to the heat effects.

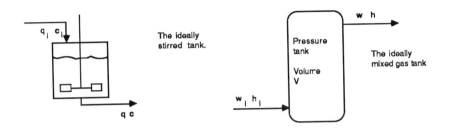

Figure 5: Holding tank for liquids and gases.

Non-Reacting Single Component Gaseous Systems

When the stirred tank approximation is used for gaseous systems, the simple picture drawn for liquid systems changes dramatically, even for the conservation of total mass. The simple total conservation of mass was given by Equation (25) for liquids, but for gases, there will be an interaction between pressure and temperature. How important this interaction is, is again a question for an expert.

In gaseous systems a holding tank will be characterized by its volume, its pressure and its temperature. There are basically two quantities one has to be concerned about in the application of the conservation principle, the total energy and the total mass, while the volume is constant. The physical similarity and differences between gaseous systems and liquid systems are indicated in Figure 5.

Since the density for gaseous systems changes, the logical basis for the intensive thermodynamic variables is a mass basis. The conservation of total mass is simply:

$$w_a = w_i - w \tag{36}$$

and the conservation of energy is practically identical, when a non-reacting adiabatic tank for a single gas component is considered:

$$E_a = E_i - E_e \tag{37}$$

The various terms in those equations are (note that the volume is constant):

$$w_a = V\frac{d\rho}{dt} = V\left[\frac{\partial \rho}{\partial T}\frac{dT}{dt} + \frac{\partial \rho}{\partial P}\frac{dP}{dt}\right] \tag{38}$$

$$E_a = V\frac{d(\rho u)}{dt} = V\left[u\frac{d\rho}{dt} + \rho\frac{\partial u}{\partial T}\frac{dT}{dt}\right] \tag{39}$$

$$E_i = w_i h_i = w_i c_p (T_i - T_{ref}) \tag{40}$$

$$E_e = wh = w c_p (T - T_{ref}) \tag{41}$$

where it should be emphasized that the accumulation term must be expressed by the rate of change of the internal energy in the tank, and not the enthalpy. Since the volume is constant, no work is performed by the tank, and work can not be accumulated in the tank [29].

The simple form of the conservation principle as seen for liquids no longer applies to gaseous systems. The gas storage tank is basically described by two nonlinear first order differential equations, when pressure and temperature are used as states. The choice of these states makes the equations coupled, even if the energy balance and the total mass balance are not. One gets the following set of equations, where the last expressions are valid for an ideal gas:

$$\frac{dT}{dt} = \frac{s_i \left[h_i - h + (h-u)\left(1 - \frac{w}{w_i}\right) \right]}{c_v} = s_i T \left[\kappa \left(\frac{T_i}{T} - 1 \right) + (\kappa - 1)(1 - \frac{w}{w_i}) \right] \tag{42}$$

$$\begin{aligned} \frac{dP}{dt} &= s_i (\frac{\partial P}{\partial \rho}) \left[\left(\rho - T(\frac{\partial \rho}{\partial T}(\kappa - 1) \right) (1 - \frac{w}{w_i}) - (\frac{\partial \rho}{\partial T})\kappa(T_i - T) \right] \\ &= s_i P \kappa \left[\frac{T_i}{T} - \frac{w}{w_i} \right] \end{aligned} \tag{43}$$

Here κ is the adiabatic exponent, the ratio between the specific heats at constant pressure and at constant volume, c_p / c_v. From the equations above, one should note the general expert advice to the Knowledge Base of gas tank modeling:

- A general model of an adiabatic gas tank has a nonlinear form where both temperature and pressure should enter as state variables. The model is derived from total mass and total energy balances, even for a single gas component.

- The equations expressing total mass and energy conservation are decoupled if extensive state variables are used.

- All physical and thermodynamic state properties should be expressed on a mass basis.

- The gas tank has basically the same degrees of freedom for control as the liquid stirred tank, all possible inlet streams for composition control in addition to the tank outlet stream. More outlet streams essentially only widen the range of control of the tank outlet.

8.1.1 The Solid Sphere With Perfect Heat Conduction

A frequently encountered dynamical system is the solid sphere. The spherical particle with approximately perfect internal heat conduction, is recognized as a system in the field of catalyst production, fertilizer prilling, thermocouple

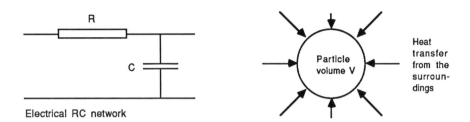

Figure 6: Physical analogy between heat transfer and an RC-circuit

joints and many others. This system may also be treated as a stirred tank, or equivalently as an RC-circuit, in the modeling of the dynamic heat balance, provided the perfect internal heat conduction holds as a reasonable approximation. This is characterized by a dimensionless number which is the ratio between the solid heat conduction and the heat transfer to the particle. The system considered is shown schematically in Figure 6.

The dynamic model of this simple physical approximation is again given by the differential Equation (19):

$$\frac{dT}{dt} = s\Delta T \tag{44}$$

where s is now the "space velocity" of temperature (it may be regarded as analogous to a traveling velocity of temperature into the center). For spheres, cylinders and slabs, this quantity is:

$$\text{Spheres: } s = \frac{6h}{d\rho c_p}; \text{ Cylinders: } s = \frac{4h}{d\rho c_p}; \text{ Slabs: } s = \frac{2h}{d\rho c_p}; \tag{45}$$

where h is the heat transfer coefficient, d the diameter or thickness, ρ the density and c_p the specific heat of the solid.

The conditions under which the stirred tank approach is valid, may best be analyzed by the transfer function approach [30]. The heat conduction in the sphere may be described by the parabolic differential equation:

$$\xi^2 a \left(\frac{\partial T}{\partial t}\right) = \frac{\partial \left(\xi^2 \frac{\partial T}{\partial \xi}\right)}{\partial \xi}; \xi = \frac{r}{r_o}; a = \frac{r_o^2 \rho c_p}{k} \tag{46}$$

where k is the heat conductivity and r_o the particle radius.

The transfer function from the environment temperature to the surface temperature of the sphere is simply [30]:

$$\frac{T(s)}{T_o(s)} = \left[1 + b(\sqrt{as}\coth\sqrt{as} - 1)\right]^{-1} ; b = \frac{k}{hr_o} \qquad (47)$$

and one may deduce from this expression, that the crucial parameter is the ratio between the two transport properties:

$$\beta = \frac{k}{hd} = \frac{b}{2} \qquad (48)$$

Even the results of this simple exercise is worth including in the Knowledge Base, together with an indication as to when the model is valid:

- A solid particle may be modeled as a stirred tank for heat, provided practically perfect heat conduction applies to the solid. If an effect of the heat conduction on the "stirred tank" time constant of less than 2.5 percent is required, this condition may be expressed as: $\frac{k}{hd} > 10$.

8.2   The Stirred Tank As A Gas/Liquid Separator

Gas/liquid separation is probably the most common unit operation in process technology, and the most frequently used model for the dynamics is again a stirred tank, one for the liquid phase and one for the gas phase. Such a gas/liquid separator is shown schematically in Figure 7.

The conditions at the interface between the phases is a gray area of process knowledge, but there are two major, but different, assumptions that most experts agree on. The first is the assumption of equilibria between the bulk of the phases and the second is the assumption of limited interfacial transport rate:

- Thermodynamic equilibrium is established throughout the bulk of the phases, i.e. the interface transport is infinitely fast.

- There is a certain limitation in the transport near the interface, but the actual interface is still in thermodynamic equilibrium.

A systems engineering approach to the two phase mixing stage identifies the conservation principle as the major rule (use superscript g for gas and l for liquid), at first sight to be a linear and decoupled set of rates:

$$\mathbf{n}_a^g = \mathbf{n}_i^g - \mathbf{n}_e^g - \mathbf{n}_t \qquad (49)$$

$$\mathbf{n}_a^l = \mathbf{n}_i^l - \mathbf{n}_e^l + \mathbf{n}_t \qquad (50)$$

One obvious conclusion that is drawn from this formulation, is that the total conservation equation is invariant to transport phenomena and equilibrium conditions:

$$\mathbf{n}_a^g + \mathbf{n}_a^l = (\mathbf{n}_i^g + \mathbf{n}_i^l) - (\mathbf{n}_e^g + \mathbf{n}_e^l) \qquad (51)$$

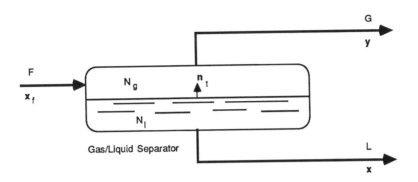

Figure 7: A Gas/liquid Separator

However, as soon as the rates are identified in terms of extensive and intensive properties, a number of crucial phenomena associated with the separator dynamics will appear.

The Introduction Of The Intensive Properties

Interface phenomena, such as transport and equilibrium conditions, are basically modeled as functions of intensive properties, mole fractions or concentrations. If mole fractions are being used and $\mathbf{n}$ is a vector of molar rates, then the mole fractions become nonlinear functions of the extensive properties. The total molar flowrates and hold-ups are:

$$
\begin{aligned}
F &= \mathbf{1}^\mathrm{T}(\mathbf{n}_i^g + \mathbf{n}_i^l); \\
L &= \mathbf{1}^\mathrm{T}\mathbf{n}_e^l; \\
G &= \mathbf{1}^\mathrm{T}\mathbf{n}_e^g; \\
\tfrac{dN_g}{dt} &= \mathbf{1}^\mathrm{T}\mathbf{n}_a^g; \\
\tfrac{dN_l}{dt} &= \mathbf{1}^\mathrm{T}\mathbf{n}_a^l; \\
&\quad \mathbf{1} \text{ is the unitary vector.}
\end{aligned}
\tag{52}
$$

Then the overall molar balance is expressed as a first order differential equation, similar to the total liquid balance in Equation (25):

$$
\frac{d(N_g + N_l)}{dt} = F - L - G
\tag{53}
$$

Inserting the intensive properties into the dynamic conservation principle and utilizing the general over all balance, give two coupled differential equation systems for both phases:

$$
N_l \frac{d\mathbf{x}}{dt} = F\mathbf{x}_f - L\mathbf{x} - \mathbf{n}_t - \mathbf{x}\left(\frac{dN_l}{dt}\right)
\tag{54}
$$

$$N_g \frac{d\mathbf{y}}{dt} = -(F - L)\mathbf{y} + \mathbf{n}_t + \mathbf{y}\left(\frac{dN_l}{dt}\right) \tag{55}$$

The crucial input variables are the feed and liquid draw-off flowrates. It is also seen that the effect of the dynamics of the liquid molar hold-up is not trivial. Usually that effect is ignored, but that is not necessarily obvious.

In the case of true equilibrium between the phases, one may use an equilibrium relationship as an algebraic equality constraint to eliminate $\mathbf{x}$ or $\mathbf{y}$. A very frequent relationship is the proportionality model, which in matrix form may be written:

$$\mathbf{y} = \mathbf{K}\mathbf{x} \tag{56}$$

where $\mathbf{K}$ is a matrix of equilibrium constants or distribution coefficients. That matrix is a diagonal matrix for ideal systems where the interactions between the component activities are low.

In the case of bulk phase equilibrium, it is possible to eliminate the interface transport by a simple addition of Equations (54) and (55) and still obtain a unique solution by a substitution of Equation (56) in the resulting equation:

$$\frac{d\mathbf{x}}{dt} = (\mathbf{K}N_g + \mathbf{I}N_l)^{-1} \left[F(\mathbf{x}_f - \mathbf{K}\mathbf{x}) - \left(L + \frac{dN_l}{dt}\right)(\mathbf{I} - \mathbf{K})\mathbf{x} \right] \tag{57}$$

This equation contains a lot of general information about the nature of phase separation dynamics, which may qualify for some Expert Statements:

- In separation stages, there may be interactions between the composition dynamics, but these interactions are caused by thermodynamic interactions between the components, as expressed by the matrix of equilibrium constants or distribution coefficients. Usually, these interactions are small.

- Whether or not the gas phase accumulation dynamics is significant for the overall dynamics of the equilibrium stage, is entirely up to the equilibrium coefficients. If the diagonal elements in the equilibrium matrix $\mathbf{K}$ are much larger than 1, then the gas hold-up may be more important than the liquid hold-up.

- The dynamics of the total molar conservation in both phases is coupled to the dynamics of the conservation of components in a much more complicated way than just a time varying space velocity, as was the case with a single phase conservation dynamics.

The Effects Of Interface Transport Limitations

As soon as the transport between the bulk of the two phases becomes limited, the simple dynamic model developed above breaks down, because the equilibrium conditions are not met and Equation (56) can not be used without modifications. One of those modifications, is to introduce an equilibrium "efficiency" by means of a vapor efficiency [31]. From a systems engineering point of view, the way this is done in the chemical engineering literature is very primitive, and without any generality. In this paper and a previous one [31], the concept of vapor "efficiency" is approached from the general point of view of dynamic systems, namely pseudo-stationarity of a very fast dynamic mode.

The effect of transport limitations may be included as a phenomenological relationship between a transport flux and a driving force, which is the general Ohm's law applied to interface transportation. The driving force is the deviation from true equilibrium of the bulk of the phases, and the transport properties are included in a square matrix $\mathbf{T}$ of transport numbers:

$$\mathbf{n}_t = \mathbf{T}(\mathbf{y} - \mathbf{Kx}) \tag{58}$$

From this simple model, it is easily seen how true equilibrium is approached, the transport numbers go to infinity and the driving force goes to zero, while their product remains finite and unique. The limit of equilibrium is expressed by the driving force equal to zero. But in the equilibrium limit, the interface transport can not be determined by Equation (58), having a non-essential singularity at the equilibrium point. Clearly, the interface transport is a dynamic phenomenon affecting the basic principle of conservation, but now both phases have to be treated separately, and the set of system equations is expanded to the double:

$$N_l \frac{d\mathbf{x}}{dt} = F\mathbf{x}_f - L\mathbf{x} + \mathbf{T}(\mathbf{y} - \mathbf{Kx}) - \mathbf{x}\frac{dN_l}{dt} \tag{59}$$

$$N_g \frac{d\mathbf{y}}{dt} = -(F - L)\mathbf{y} - \mathbf{T}(\mathbf{y} - \mathbf{Kx}) + \mathbf{y}\frac{dN_l}{dt} \tag{60}$$

For simplicity, let the molar hold-up in the liquid phase be constant and let the expanded mole fraction vector comprise both phases' mole fractions. Then the system dynamics may be described by a partitioned set of equations:

$$\frac{d\begin{pmatrix}\mathbf{x}\\\mathbf{y}\end{pmatrix}}{dt} = \begin{bmatrix}-(LI+\mathbf{TK})/N_l\\\mathbf{TK}/N_g\end{bmatrix}\begin{bmatrix}\mathbf{T}/N_l\\-(GI+\mathbf{T})/N_g\end{bmatrix}\begin{bmatrix}\mathbf{x}\\\mathbf{y}\end{bmatrix} - F\begin{bmatrix}\mathbf{x}_f/N_l\\0\end{bmatrix} \tag{61}$$

$$G = F - L$$

where it is clearly seen, as $\mathbf{T}$ approaches infinity, that the limit corresponds to a degeneration of the dynamic equations to:

$$\begin{aligned}(\mathbf{y} - \mathbf{Kx}) &= 0\\-(\mathbf{y} - \mathbf{Kx}) &= 0\end{aligned} \tag{62}$$

In the range close to equilibrium, the picture is best viewed when the system is reduced to a binary system [31]. The transport property is then conveniently described by an interfacial resistance to transport, which would be the inverse of the transport conductivity T, say $\varepsilon = 1/T$. Those conditions are generally described by a 2x2 systems matrix $\mathbf{A}$:

$$\frac{d\mathbf{z}}{dt} = \mathbf{A}\mathbf{z} \quad ; \quad \mathbf{z}^T = [x, y]$$
$$a_{11} = \frac{-(L+k/\varepsilon)}{N_l} \quad ; \quad a_{12} = \frac{1}{\varepsilon N_l} \tag{63}$$
$$a_{21} = \frac{k}{\varepsilon N_g} \quad ; \quad a_{22} = \frac{-(F-L+1)/\varepsilon}{N_g}$$

the eigenvalues of which are always real and negative:

$$\lambda_{1,2} = (a_{11} + a_{22}) \frac{\left[1 \pm \sqrt{1 - \frac{(a_{11}a_{22} - a_{12}a_{21})}{(a_{11}+a_{22})^2}} \right]}{2} \tag{64}$$

As the transport resistance ε goes to zero, a first order approximation in e of the eigenvalues may be written as:

$$\lambda_1 \approx -\frac{[kN_g + N_l + \varepsilon(LN_g + GN_l)]}{\varepsilon N_g N_l} + \frac{[L + kG + \varepsilon LG]}{[kN_g + N_l + \varepsilon(LN_g + GN_l)]} \tag{65}$$

$$\lambda_2 \approx -\frac{[L + kG + \varepsilon LG]}{[kN_g + N_l + \varepsilon(LN_g + GN_l)]} \tag{66}$$

The first of those eigenvalues is associated with the speed by which equilibrium is established, and it is clearly seen that this eigenvalue goes to negative infinity as the transport resistance goes to zero, due to the first term. Physically speaking, this means that the dynamics of phase equilibrium becomes infinitely fast as the resistance goes to zero, which is logical.

The second of the eigenvalues derived above goes to a negative finite value as the transport resistance goes to zero. This is also logical, because this eigenvalue is associated with the accumulation dynamics of the combined two phase separation stage. In the case of true equilibrium, these dynamics degenerate to what is described by Equation (57) for a binary system:

$$\lambda_2 \approx \frac{[L + kG]}{[kN_g + N_l]} \tag{67}$$

The two dynamic modes related to the two eigenvalues may be decoupled by an orthogonal transformation through the eigenvector matrix. By this transformation, the dynamic mode of equilibrium dynamics may be studied separately and decoupled from the conservation dynamics. In terms of the original mole fraction variables, the decomposed modes may be written:

$$\xi_1 = x + u_1 y \text{ and } \xi_2 = u_2 x + y \tag{68}$$

"Pseudo-equilibrium" may now be defined as an approximation to real equilibrium, by an assumption that the equilibrium dynamics are extremely fast relative to the other mode of conservation. This approximation is equivalent to assuming $\xi_1 \approx 0$ in Equation (68). In systems engineering, the assumption

of "pseudo-equilibrium" is equivalent to letting the stationarity approximation (as applied to very fast radical kinetics in reaction engineering [32]) apply
to a very fast mode of the dynamics. By this, the number of differential
equations in time is reduced by a replacement of an algebraic equation resulting from letting the decomposed state variable associated with the very fast
mode (corresponding to a measure of the departure from equilibrium) be zero.
When this general principle is applied to the specific system considered here,
a "pseudo-equilibrium" constant k' is the result:

$$\xi_1 = 0 = x + u_1 y; \quad y = k' x; \quad k' = -\frac{1}{u_1} \tag{69}$$

Inserting the actual system parameters, gives the "pseudo-equilibrium" constant in terms of the transport resistance:

$$k' = -\frac{a_{21}}{(\lambda_1 - a_{11})} = \frac{k}{\left[1 + \frac{\varepsilon(G - \frac{LN_g}{N_l})(1+\alpha+\frac{\alpha^2}{2})}{(1+\alpha)^2}\right]}; \quad \alpha = \frac{kN_g}{N_l} \tag{70}$$

or, in terms of a vapor efficiency as used in traditional chemical engineering
textbooks [33]:

$$k' = nk = \left[1 - \frac{\varepsilon(G - \frac{LN_g}{N_l})(1 + \alpha + \frac{\alpha^2}{2})}{(1+\alpha)^2}\right] k \tag{71}$$

but it should be emphasized that this vapor efficiency is totally different from
the concepts used in those textbooks. The traditional chemical engineering
definition of a vapor efficiency is entirely empirical and has so little relationship
to fundamentals, that it is sometimes required to define different efficiencies
for the two phases. This is of course not required here, the efficiency presented
is a true measure of the deviation of the "pseudo-equilibrium" from the true
equilibrium, in a general and proper systems engineering context.

As it is seen, this vapor efficiency goes to 1 as the transport resistance
goes to zero, which is logical. What may be a surprise to some however, is
its relationship to the parameters of the conservation dynamics. This result
would never be obtained by the traditional chemical engineering approach,
which shows how powerful a more general systems engineering approach would
be. The result is logical however, and the physical interpretation is fairly
transparent. As it is seen from Equation (71), the vapor efficiency depends
heavily on the gas and liquid flowrates and the ratio of molar hold-ups in the
gas phase and in the liquid phase. Writing:

$$y = nkx \tag{72}$$

where n is a vapor efficiency, it is seen that the "pseudo-equilibrium" is pushed
towards a richer composition in the gas phase (to the left), than what the

true equilibrium would predict, when the vapor efficiency is greater than 1. Similarly, the "pseudo-equilibrium" is pushed towards a leaner composition in the gas phase (to the right), when the efficiency is less than 1. This is all controlled by the space velocities in the two phases, i.e. the term:

$$\beta = \frac{G}{N_g} - \frac{L}{N_l} \tag{73}$$

which is the difference between the outlet space velocities in the two phases.

If the space velocity in the gas phase is larger (the residence time shorter) than the space velocity in the liquid phase, then the "pseudo-equilibrium" is pushed to the right (β negative). On the contrary, if the space velocity in the gas phase is smaller (the residence time longer) than the space velocity in the liquid phase, then the "pseudo-equilibrium" is pushed to the left (β positive). This fairly fundamental result seems quite obvious and has been known for a long time [31], but apparently ignored in the chemical engineering practice. The result is in direct analogy to chemical reaction kinetics (forwards and backwards reaction rates) and their equilibria or "pseudo-equilibria" (equal rates forwards and backwards) which may be pushed to the right or to the left in the stoichiometric formulae.

The two phase separation stage certainly differs from a single stage in much greater complexity, and even when the conservation of energy and its coupling to the composition variables have not been considered, as in this case. The learning from the present set of examples is therefore much more difficult to generalize. However, a few statements may be made, which might qualify as Expert Statements:

- The general conservation principle applied to the total inflow and outflow of a multicomponent phase separation stage is invariant to the equilibrium conditions and to the interface transport kinetics.

- Provided the bulk phases behave as completely stirred tanks (ideally mixed in each single phase) then the deviation from equilibrium may be described by a phenomenological interface transport resistance. However, if interfacial transport resistance is significant, then the ideal mixing assumption is usually questionable.

- "Pseudo-equilibrium" of a separation stage described by ideal mixing and interfacial transport resistance, may be described by a vapor efficiency. The "pseudo-equilibrium" is pushed to the right or to the left of the true equilibrium, depending upon the space velocities of the two phases that are separated. If the space velocity of the gas phase is higher than that of the liquid, then the "pseudo-equilibrium" is pushed towards the liquid phase. If the opposite is true, then the "pseudo-equilibrium" is pushed towards the gas phase.

Phase separation is probably a good example where both thermodynamic relations, conservation dynamics and transport kinetics should be considered, and where a general systems engineering approach, not a specialized unit operation approach (as for example in distillation), should be taken in order to expand the learning and solicit experience from the fundamental principles. Even the brief examination presented in this paper discloses weak and controversial points in the traditional treatment of phase separation. Phase separation is an area where the complications are severe enough, where the insight still sometimes is superficial and where fundamental principles and industrial experience would merge in a constructive dialog beneficial to expand the knowledge base of both industry and academia. The vapor efficiency controversy presented here is a good example.

9 Learning From The Basic Principles, The Stoichiometric Law

One of the most fundamental principles for the modeling of chemical processes is the stoichiometric law. Any model, empirical or otherwise, that does not take full advantage of this principle is incomplete, and one still sees examples of modeling in the literature where the stoichiometric law is not fully utilized. From a systems engineering point of view, the stoichiometric law represents a special structure of the linear algebraic equations used to establish the extensive conservation model, making some conservation equations invariant to chemical reactions expressed as the rate of production term.

The rate of production term in the balance of rates in the general conservation principle expressed in Equation (5), appears in the sum:

$$\mathbf{r}_f - \mathbf{r}_e - \mathbf{r}_a - \mathbf{r}_l + \mathbf{r}_p = \mathbf{0} \tag{74}$$

where the index f stands for feed, e for effluent, a for accumulation, l for loss, say to the environment, and p stands for production. The vector $\mathbf{r}$ stands for the rate of change or flow of any extensive property, say a component, heat or general energy. The stoichiometric law applies to $\mathbf{r}_p$ as expressed by the matrix $\mathbf{N}$ of stoichiometric numbers and a vector of basic production rates $\mathbf{r}$ [34]:

$$\mathbf{r}_p = \mathbf{N}\rho' = \mathbf{B}\rho; \text{ where } \rho = \mathbf{C}\rho' \tag{75}$$

where $\text{rank}(\mathbf{B}) = \text{rank}(\mathbf{C}) = \dim(\rho)$, $\mathbf{B}$ is the full rank part of $\mathbf{N}$, and:

$$\mathbf{C} = (\mathbf{B}\mathbf{T}\mathbf{B})^{-1}\mathbf{B}^T\mathbf{N} = \mathbf{B}^+\mathbf{N} \tag{76}$$

$\mathbf{B}^+$ is the pseudo-inverse or Penrose matrix of $\mathbf{B}$. The matrix $\mathbf{N}$ is a rectangular matrix of stoichiometric numbers from the basic reaction mechanisms proposed:

$$\mathbf{N}^T\mathbf{m} = \mathbf{0} \tag{77}$$

where $\mathbf{m}$ is a vector of chemical formulae for the components expressed as a combination of elementary atoms a [35] and a matrix $\mathbf{A}$ of atomic numbers given by the molecular structure of the component in question:

$$\mathbf{m} = \mathbf{A}\mathbf{a}; \ \mathbf{N}^{\mathrm{T}}\mathbf{A} = \mathbf{0} \tag{78}$$

A fundamental property of the matrices of stoichiometric numbers $\mathbf{N}$ and atomic numbers $\mathbf{A}$ is that they are orthogonal, as expressed above.

The matrix $\mathbf{N}$ may have full rank ($\rho' = \rho$; or $\mathbf{C} = \mathbf{I}$ and $\mathbf{B} = \mathbf{N}$), or reduced rank ($\dim(\rho) < \dim(\rho')$; or $\mathrm{rank}(\mathbf{B}) = \mathrm{rank}(\mathbf{C}) = \dim(\rho)$) relative to the dimension of the basic chemical reaction or production rate ρ. The following Expert Statements are valid for all chemical reactions and their stoichiometry [34,35]:

- The rank of the stoichiometric matrix is always equal to the dimension of the basic reaction rate vector ρ, which again is less than the number of conservation equations, which is the dimension of the total production vector $\mathbf{r}_p$.

- If the rank of the stoichiometric matrix is reduced in relation to its dimensions, then the proposed reaction mechanisms are not basic, but have mutual linear relationships in their stoichiometry.

The general application of the very fundamental property of stoichiometry may lead to a decomposition of the conservation principle into two subsets, a chemical reaction variant where the production by the reactions is visible, and a chemical reaction invariant where the production by reactions is invisible. Let the full rank stoichiometric matrix $\mathbf{B}$ be decomposed into its square part $\mathbf{B}_1$ and its remaining part $\mathbf{B}_2$ and the rate vector in Equation (74) be partitioned correspondingly into $\mathbf{r}_1$ and $\mathbf{r}_2$, and let the reaction variant subset of rates be $\mathbf{x}$ and the reaction invariant subset of rates be $\mathbf{y}$. Then these two subsets obey the relations [34,36]:

$$\mathbf{x}_f - \mathbf{x}_e - \mathbf{x}_a - \mathbf{x}_l + \mathbf{B}_1\rho = \mathbf{0}; \ \mathbf{x} = \mathbf{r}_1 \tag{79}$$

and:

$$\mathbf{y}_f - \mathbf{y}_e - \mathbf{y}_a - \mathbf{y}_l = \mathbf{0}; \ \mathbf{y} = \mathbf{r}_2 - \mathbf{B}_2\mathbf{B}_1^{-1}\mathbf{r}_1 \tag{80}$$

This decomposition may not apply to the conservation principle in general, but, when applicable, it has a tremendous impact on the modeling, identification and control of plants and unit operations where chemical reactions are supposed to occur. For example, it may be possible to decompose the parameter identification such that one subset of observations may be used for the estimation of the transport parameters without the influence of the chemical reactions, but with the reactions still *in situ*.

Since the extensive reaction rate ρ is a function of the total volume or the total amount of catalyst as well as the intensive properties such as activities, pressure and temperature, a change in the variables in Equation (80) affects the reactions, but the reactions do not affect Equation (80). What has been obtained by the decomposition from a systems engineering point of view, is a reduction of the feedback of the reactions to the conservation equations.

The results of the general analysis above may be summarized as a set of Expert Statements for the conservation principle models if reactions are present:

- If chemical reactions take place within a process boundary, it may be possible to decompose the conservation equations for the process into a reaction variant subset and another reaction invariant subset.

- Transport parameters and flow patterns may be estimated in the reaction invariant subspace without any interference of the reactions.

- Control actions performed in the reaction invariant subspace still has an effect on the reactions via the relationship between the chemical reaction rates and the intensive properties.

9.1 An Application To The Residence Time Distribution Model

The residence time distribution introduced in Equation (12) as a dynamic model for transport systems may be used to model a reactor as well. The total rate of production must now be taken as the integral over the system volume of the intensive reaction rate, and Equation (12) is modified to read:

$$\mathbf{n}_{e,1}(t) = \int_0^\infty \mathbf{F}_1(\tau)\mathbf{n}_{i,1}(t-\tau)d\tau + \int_V \int_0^\infty \mathbf{\Phi}_1(\alpha, V)\mathbf{B}_1\mathbf{r}(t-\alpha, V)d\alpha dV \quad (81)$$

where the diagonal matrix $\mathbf{\Phi}_1$ is a matrix of life expectancies in the system for the components produced by the reactions. Provided the life expectancies as well as the residence times are the same for all components, the matrix $\mathbf{B}_1$ may be taken outside the integration, and the following pair of equations will result:

$$\mathbf{n}_{e,1}(t) = \int_0^\infty f(\tau)\mathbf{n}_{i,1}(t-\tau)d\tau + \mathbf{B}_1 \int_V \int_0^\infty \phi(\alpha, V)\mathbf{r}(t-\alpha, V)d\alpha dV \quad (82)$$

$$\mathbf{n}_{e,2}(t) = \int_0^\infty f(\tau)\mathbf{n}_{i,2}(t-\tau)d\tau + \mathbf{B}_2\mathbf{B}_1^{-1}\left[\mathbf{n}_{e,1}(t) - \int_0^\infty f(\tau)\mathbf{n}_{i,1}(t-\tau)d\tau\right] \quad (83)$$

Now, introducing the reaction invariant $\mathbf{y} = \mathbf{n}_2(t) - \mathbf{B}_2\mathbf{B}_1^{-1}\mathbf{n}_1(t)$, it is clearly seen that this property is a true reaction invariant satisfying the equation for the original residence time distribution as if no reaction took place:

$$\mathbf{y}_e(t) = \int_0^\infty f(\tau)\mathbf{y}_i(t-\tau)d\tau \quad (84)$$

The condition for this to be valid is that the life expectancies and the residence times are independent of what species is considered. This may be summarized again in an Expert Statement:

- The concept of reaction invariance applies to a general residence time and life expectation model for a reactor, provided the expected residence time and life time in the reactor are essentially independent of what component is being considered.

9.2 An Application To The Stirred Tank And The Tubular Reactor

The consequence of the stoichiometric law of reaction invariance has been studied extensively for the stirred tank and tubular reactors [36], and below is a summary of the modeling principles as they may apply to those types of reactors under isothermal and adiabatic conditions, and under conditions of heat exchange.

Isothermal Conditions

Under isothermal conditions, Equation (26) applies directly to the reaction invariants:

$$\frac{d\mathbf{y}}{dt} = \frac{(\mathbf{y}_i - \mathbf{y})}{\tau_i}; \, \mathbf{y} = \mathbf{c}_2 - \mathbf{B}_2 \mathbf{B}_1^{-1} \mathbf{c}_1 \tag{85}$$

while the reaction variants are modified by the intensive reaction rate of chemical production per unit of volume:

$$\frac{d\mathbf{x}}{dt} = \frac{(\mathbf{x}_i - \mathbf{x})}{\tau_i} + \mathbf{B}_1 \mathbf{r}; \, \mathbf{x} = \mathbf{c}_1 \tag{86}$$

The reaction invariants are always stable (in systems engineering terms, an asymptotic invariant) while the small scale stability of the reaction variant is examined by the system matrix $\mathbf{A}$ for the linearized reaction variant set:

$$\begin{aligned}
\frac{d\Delta \mathbf{x}}{dt} &= \Delta \mathbf{x}_i s_i + \mathbf{A} \Delta \mathbf{x} + \mathbf{J}_2 \Delta \mathbf{y}; \\
\mathbf{A} &= (\mathbf{R}_i - \mathbf{I} s_i); \\
\mathbf{R}_i &= \mathbf{B}_1 (\mathbf{J}_1 + \mathbf{J}_2 \mathbf{B}_2 \mathbf{B}_1^{-1}); \\
\mathbf{J}_2 &= \frac{\partial \mathbf{r}}{\partial \mathbf{c}_2} \\
\mathbf{J}_1 &= \frac{\partial \mathbf{r}}{\partial \mathbf{c}_1}
\end{aligned} \tag{87}$$

Equation (87) shows how the changes from the reaction invariant set is propagating sequentially into the reaction variant set but not the other way around. There is no feedback from the reaction variant set to the invariant set, and the stability of the isothermal reactor is uniquely determined by the eigenvalues of the reaction matrix $\mathbf{R}_i$. Let the eigenvalues of this matrix be $\lambda_{r,i}$. Then it is seen that these eigenvalues have to have real parts that are less than the space velocity of the reactor in order for the uncontrolled reactor to be stable.

It is also seen that the stability of the reactor is affected by the space velocity s_i [23,34,36].

The analysis above is easily extended to a tubular reactor with axial dispersion, as basically, only the differential operator is being changed. The two sets $\mathbf{x}$ and $\mathbf{y}$ are now obeying the relations:

$$(\partial/dt + s\partial/\partial\xi - \alpha\partial^2/\partial\xi^2)\mathbf{x} + \mathbf{B}_1\rho = \mathbf{0} \tag{88}$$

$$(\partial/dt + s\partial/\partial\xi - \alpha\partial^2/\partial\xi^2)\mathbf{y} = \mathbf{0} \tag{89}$$

where α is the axial dispersion coefficient and s the space velocity. The manipulations are entering the model equations, partly via the boundary conditions and partly via the coefficients like the axial dispersion coefficient.

Adiabatic Conditions

As soon as the reactor deviates from isothermal operation and the temperature changes, the conservation of energy needs to be considered along with the conservation of components, because the reaction rates are strong functions of the temperature. Due to the nature of the heat release or absorption by the reactions, the production of heat, or total change of enthalpy under constant pressure, follows the stoichiometric law:

$$\Delta H_r = -\Delta \mathbf{H}^T \mathbf{r} \tag{90}$$

Under adiabatic conditions only a minor modification of the system matrix is required, as shown [23,34,36] in the literature, provided the adiabatic exponent is very close to unity, as in liquids. If this is not true, then both pressure and temperature need to be included as state variables, as for gases shown in Equations (42) and (43), and the heat balance may no longer be transformed to a reaction invariant.

For simplicity, let the fluid be a liquid where the adiabatic exponent is unity. Then the conservation of energy in adiabatic mode is reaction invariant, and is just an additional equation that may be added to the set of reaction invariants. This is done by the introduction of the variable y_T in the set of reaction invariants. By a simple analogy to the stoichiometry applied to heat [23,34,36], this variable is defined as:

$$y_T = T - \nu^T \mathbf{B}_1^{-1}\mathbf{c}_1; \; \nu = -\frac{\Delta \mathbf{H}}{\rho c_p} \tag{91}$$

where ν is the vector of adiabatic temperature rises, as explained in Equation (91). This result applies equally well to stirred tank reactors as to tubular reactors with axial dispersion [23,34,36].

However, the stability of the reactor is highly affected by the transition from isothermal to adiabatic operation, even if the mathematical extension is trivial. The only place the temperature effect is entering the stability problem, is in the matrix $\mathbf{R}$. Introducing the Jacobian of the reaction rate with respect to temperature gives the simple modification of the isothermal system matrix for stability:

$$\begin{aligned} \mathbf{A} &= (\mathbf{R}_a - s_i \mathbf{I}); \\ \mathbf{R}_a &= \mathbf{R}_i + \left(\tfrac{\partial \mathbf{r}}{\partial T}\right) \nu^{\mathrm{T}} \mathbf{B}_1^{-1} \end{aligned} \tag{92}$$

The Jacobian matrix $\mathbf{J}_2$ is augmented with the partial derivative: $\left(\tfrac{\partial \mathbf{r}}{\partial T}\right)^{\mathrm{T}}$.

Stirred Tank Reactor With Heat Exchange

Like in Equation (34), the heat exchange is included in the heat balance, and the dimension of the system matrix is extended to $\dim(\mathbf{r}) + 1$, since the temperature may no longer be transformed to a reaction invariant, i.e. Equation (91) is no longer applicable. This certainly has an effect on the stability of the uncontrolled reactor, as the system matrix now becomes:

$$\mathbf{A} = (\mathbf{R}_h - s_i \mathbf{I}) \tag{93}$$

The reaction matrix $\mathbf{R}_h$ contains the heat transfer parameter in Equation (34) defined as an equivalent number of heat transfer stages:

$$\beta = \frac{AU}{q_i \rho c_p} \tag{94}$$

The reaction matrix is partitioned and expanded as follows [34]:

$$\mathbf{R}_h = \left[\frac{\mathbf{R}_i}{\nu^{\mathrm{T}} B_1^{-1} \mathbf{R}_i} \right] \left[\frac{B_1 \left(\tfrac{\partial \mathbf{r}}{\partial T}\right)}{-\beta s_i + \nu^{\mathrm{T}} \left(\tfrac{\partial \mathbf{r}}{\partial T}\right)} \right] \tag{95}$$

Even this simple example makes it possible to extract some general expert advice to the modeling of the type of reactors considered, in particular the stirred tank type. With a fairly wide applicability, the following Expert Statements may be made:

- The concept of reaction invariants may be used for homogeneous reactors where the distribution of residence times and the expected life time through the reactor are the same for all components. For heterogeneous reactions, this is normally not true (due to the chromatographic effect), and the decomposition into reaction variants and invariants may not be generally possible for multiphase systems.

- The open loop stability of a stirred tank reactor is determined by two major factors, the space velocity through the reactor and the eigenvalues of the reaction matrix which is reaction specific.

- The mode of operation of the reaction has a strong effect on the reactor stability. The change from isothermal to adiabatic operation introduces a modification of the reaction matrix which now includes the sensitivity in the reaction rate to the temperature and the adiabatic temperature rise, a measure of the exothermicity of the reactions. The effect of heat transfer is to stabilize the reactor as compared to adiabatic operation.

10   Conclusion

A general systems engineering approach to process modeling and to chemical engineering in general, seems to have great potentials. It is seen that new outlooks to the process technology may disclose effects that where ignored or missed in a traditional approach. It is also felt that the role of the development of nonlinear models and their simulation is somewhat under-emphasized in process control, while the linear theory is given a unfairly large part of the academic teaching and research. More emphasis should be given to nonlinear process phenomena and even nonlinear control actions. Typically, the handling of constraints appears only in scattered publications in the chemical engineering literature before the publication of the QDMC approach to constrained optimal predictive control [37].

The industrial facts are that realistic nonlinear models should be preferred over linearized ones, because the limitations of the linear theory that often ignores mean value shifts during fluctuations and constraints in process variables and control manipulators. Modern process control problems should not be solved analytically with inappropriate simplifications of the models to make them feasible for linearized theory. Rather, they should be solved with as realistic as possible models for the process. This inevitably involves inequality constraints and process nonlinearities. This again requires efficient and user-friendly program packages for extensive computer simulations [9].

Regardless approach and algorithm for the solution of process control problems, it seems mandatory to master the mathematical modeling to a high degree of perfection, but also to as low a cost as possible. This calls for improved and reliable model development and extensive simulation for learning, because the manpower cost of the modeling expertise within a company is escalating. Therefore, a systematic development and expansion of the process knowledge base for modeling and cause and effect analysis of the process technology, seems to be a right and rational long term planning. In this effort, it also seems natural to concentrate on methods for improving the efficiency of the learning process by Knowledge Engineering, Artificial Intelligence and Symbolics Manipulation. The major role of the Academia in this effort is clearly in the area of fundamental principles and not in the industrial application.

However, the fact that process control from a chemical engineering point of view should emphasize the applications in its evolution and natural learning process, puts the chemical engineering departments into a difficult position.

On the one hand, they do not have a comparable laboratory to supplement the industrial practice, and on the other hand, most graduate students in chemical engineering have a background in systems and control engineering which is often weak. To make it even more complicated for the rational education in chemical engineering, the systems engineering approach to the discipline as a whole is not very well developed. The undergraduate emphasis on a systems engineering approach to chemical engineering in general, should be improved and encouraged.

References

[1] Hengstenberg, J., B. Sturm and O. Winkler (1964): *Messen und Regeln in der Chemischen Technik.*, Springer Verlag, Berlin.

[2] Moe, T. (Ed.) (1978): *General Specification for Instrumentation.* Norsk Hydro a.s, Engineering Division.

[3] Asbjørnsen, O. A. (1976): "The Economical, Technical, Intellectual and Social Gap between Theory and Practice in Process Control," *AIChE Symp. Ser.*, no. 159, 72, 80.

[4] Asbjørnsen, O. A. (1982): "Theory at Work in Process Control and Plant Operation," *Kemia-Kemi*, 9, 653.

[5] Renfro, J. G. (1986): Computational Studies in the Optimization of Systems Described by Differential/Algebraic Equations., *Ph. D. Dissertation*, Chemical Engineering Department, University of Houston - University Park.

[6] Asbjørnsen, O. A. (1982): "Project Oriented Chemical Engineering Education," *Norsk Hydro a.s*, Oslo Norway.

[7] Åstrom, K. J. and B. Wittenmark (1984): *Computer Controlled Systems - Theory and Design.* Prentice-Hall, Inc. Englewood Cliffs, New Jersey.

[8] Foss, A. S. (1973): "Critique of Chemical Process Control," *AIChE J.*, 19, 209.

[9] Balchen, J. G., M. Fjeld and S. Slid (1983): "Significant Problems and Potential Solutions in Future Process Control," *Proc. of AIChE Diamond Jubilee Meeting*, Washington D.C.

[10] Asbjørnsen, O. A. (1967): *Process Control Laboratory Experiments.* University of the West Indies, Trinidad and Tobago.

[11] Shinskey, F. G. (1967): *Process Control Systems - Application, Design, Adjustment.* McGraw-Hill, New York.

[12] Asbjørnsen, O. A., L. Foyen and P. M. Solberg (1978): "Plant Reliability, Faults and Fault Detection," *Kemia-Kemi*, 5, 71.

[13] Asbjørnsen, O. A. (1979): "Plant Reliability and Safety," *Comp. & Chem. Engr.*, 3, 21.

[14] Bristol, E. H. (1966): "On a New Measure of Interaction for Multivariable Process Control.," *IEEE Trans. Auto. Contr.*, AC-11 (7), 133.

[15] Tung, L. S. and T. F. Edgar (1982): "Dynamic Interaction Analysis and its Application to Distillation Column Control," *Proc. IEEE Decision and Control Conf.*, 1, 107.

[16] Garcia, C. E. and M. Morari (1982): "Internal Model Control - A Unifying Review and Some New Results," *IEC Proc. Des. Dev.*, 21, 308.

[17] Lau, H., J. Alvarez and K. F. Jensen (1985): "Synthesis of Control Structures by Singular Value Analysis: Dynamic Measures of Sensitivity and Interaction," *AIChE J.*, 31, 427.

[18] Asbjørnsen, O. A. and M. Fjeld (1970): "Response Modes in Stirred Tank Reactors," *Chem. Engg. Sci.*, 25, 1627.

[19] Fjeld, M., O. A. Asbjørnsen and K. J. Åstrom (1974): "Reaction Invariants and their Importance in the Analysis of Eigenvectors, State Observability and Controllability of the Continuous Stirred Tank Reactor," *Chem. Engg. Sci.*, 29, 1917.

[20] Asbjørnsen, O. A. (1986): "Modeling Techniques: Theory and Practice," *Modeling Identification and Control*, 6, 105.

[21] Villadsen, J. V., O. A. Asbjørnsen, K. E. Harg, N. Heyerdahl and M. Harg (1977): "Simulation of an Industrial Ethylene Cracker, a Comparison between the Stirred Tank and the Differential Equation Approach," in *Computer Applications in the Analysis of Chemical Data and Plant.* pp 27-41, Science Press, Princeton.

[22] Johnson, C. D. (1986): "Disturbance Accommodating Control; An Overview," *ACC Proceedings*, 1, 526.

[23] Asbjørnsen, O. A. (1972): "Reaction Invariants in the Control of Chemical Reactors," *Chem. Engg. Sci.*, 27, 709.

[24] Ray, W. H. (1981): *Advanced Process Control.* McGraw-Hill, New York.

[25] Levenspiel, O. (1962): *Chemical Reaction Engineering.* Wiley, New York.

[26] Asbjørnsen, O. A. (1961): "The Distribution of Residence Times in a Falling Water Film.," *Chem. Engg. Sci.*, 14, 211.

[27] Luyben, W. L. (1973): *Process Modeling, Simulation, and Control for Chemical Engineers*. McGraw-Hill, New York.

[28] Douglas, J. M. (1972): *Process Dynamics and Control - Volume 1 - Analysis of Dynamic Systems*. Prentice-Hall,Inc., Englewood Cliffs, New Jersey.

[29] Mäkilä, P. M. and K. V. Waller (1981): "The Energy Balance in Modeling Gas-Phase Chemical Reactor Dynamics," *Chem. Engg. Sci.*, 36, 643.

[30] Asbjørnsen, O. A. and B. Wang (1971): "Heat Transfer and Diffusion in Fixed Beds," *Chem. Engg. Sci.*, 26, 585.

[31] Asbjørnsen, O. A. (1973): "The Stage Efficiency in Dynamic Models of Phase Separation Processes," *Chem. Engg. Sci.*, 28, 2223.

[32] Dente, M. E. Ranzi, G. Antolini and F. Losco (1970): "Study of a Theoretical Model for Simulating Thermal Cracking of Hydrocarbon Mixtures," in *The Use of Computers in the Studies Preceeding the Design of Chemical Plants,* Proceedings, 97th Event of the EFChE, Florence, Italy.

[33] Holland, C. D. (1963): *Multicomponent Distillation*. Prentice-Hall, Inc., Englewood Cliffs, New Jersey.

[34] Asbjørnsen, O. A. (1974): Process Dynamics., *Tapir*, University of Trondheim, Trondheim, Norway.

[35] Asbjørnsen, O. A. (1982): "Modeling and Simulation in Computer Aided Design and Operation of Chemical Plants," *Kemia-Kemi*, 9, 103.

[36] Asbjørnsen, O. A. (1972): "Reaction Invariants in Chemical Reactor Dynamics," *Kemian Teollisuus*, no. 10, 1972, 633.

[37] Garcia, C. E., and A. M. Morshedi (1984): "Quadratic Programming Solution of Dynamic Matrix Control (QDMC)." *Proceedings, Canadian Conference on Industrial Computer Systems.*

Modelling and Control of Dispersed Phase Systems

James B. Rawlings
Department of Chemical Engineering
The University of Texas
Austin, Texas 78712

December 15-18, 1986

Abstract

The modelling, measurement, and control problems associated with dispersed phase systems are discussed. Crystallization and emulsion polymerization are selected as representative systems of this class of problems, and the present understanding of their dynamic behavior and existing control strategies are reviewed.

1   Introduction

A dispersed phase system consists of some type of particles dispersed in a continuous phase. Examples of such dispersed phase systems are numerous, including crystallizers, microbial fermenters, emulsion polymerizers, liquid-liquid extractors, and fluidized bed reactors.

The unifying feature of these diverse processes is the importance of the distribution of particles in some general measure of particle size. For example, in the emulsion polymerization field, one is interested in knowing the number of polymer particles in the reactor as a function of their radius or age. For a biochemical reactor, the key distribution might be the number of cells of a given species as a function of their mass.

This discussion of dispersed phase systems is organized as follows. A general modelling approach, based on a population balance, is discussed first. In the next sections, two important industrial processes, crystallization and emulsion polymerization, are discussed in greater detail. The modelling equations specific to these two systems are presented, and the unusual dynamic behavior and existing control approaches are reviewed. The particle size distribution (PSD) measurement technology is reviewed next, emphasizing the technology that is available for rapid, on-line applications. Finally, conclusions from this discussion are drawn, and areas that require further study are highlighted.

2   Model Development

The objective of dispersed phase system modelling is to provide a description of the population of the particles dispersed in the continuous phase. The mathematical framework that has been developed to describe dispersed phase systems is known as population balance modelling. The concept of a population balance probably arose as a generalization of the residence time distribution. In early work, Rudd [1,2] demonstrated that one cannot always formulate a model based on particle age, and it is necessary to consider a generalized describing variable (catalyst particle activity in his example). Hulburt and Katz [3] and Randolph and Larson [4] extended this concept further to include an arbitrary number of "internal" coordinates to fully describe the state of a particle. Tsuchiya et al. [5], Eakman et al. [6], and Ramkrishna [7] applied this methodology to microbial populations. Friedlander [8] and Hidy [9] review population balance methods in aerosol science applications. The interested reader is directed to Ramkrishna's [10] recent review article for an extensive discussion of population balance modelling, applications, and solution methods.

The key property of the particulate system is some general particle size distribution (PSD). Let $f(\underline{z}, t)$ denote the concentration of particles in the system, which depends on the time, t, and other suitably chosen coordinates, $\underline{z}$. The $\underline{z}$ coordinates include the usual three spatial (external) coordinates as well as the additional internal coordinates which are required to completely specify the state of the particle.

Consider an arbitrary, simply connected, volume element, V, in the z coordinate space shown in Figure 1.

In the following, the element V is considered to be moving with the particles, so that however the particles' coordinates change in time, the element's coordinates change in the identical manner. V is also chosen large enough so that it contains a statistically significant number of particles. The surface bounding the volume, V, is denoted by S and the outwardly directed unit normal vector is $\underline{n}$. We can then make a simple statement accounting for the accumulation of particles in the region, V,

$$\frac{d}{dt} \int_V f(\underline{z}, t) d\tau = \int_V (B - D) d\tau \tag{1}$$

B is the rate of particle formation due to birth mechanisms, and D is the rate of particle loss due to death mechanisms. Since V moves with the particles, the left hand side of Equation (1) can be expressed as

$$\frac{d}{dt} \int_V f(\underline{z}, t) d\tau = \int_V \frac{\partial f(\underline{z}, t)}{\partial t} d\tau + \int_S f \left(\frac{\partial \underline{z}}{\partial t} \cdot \underline{n} \right) d\Omega \tag{2}$$

Using the divergency theorem, the last term in Equation (2) can be converted from a surface to a volume integral,

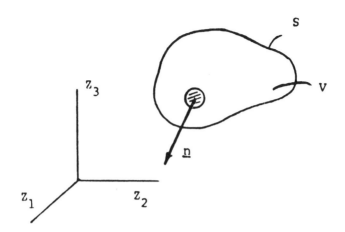

Figure 1: Microscopic Volume Element, V.

$$\int_S f \left(\frac{\partial \underline{z}}{\partial t} \cdot \underline{n} \right) d\Omega = \int_V \nabla \cdot \left(f \frac{\partial \underline{z}}{\partial t} \right) dt \qquad (3)$$

Combining Equations (1) to (3) leads to

$$\int_V \left[\frac{\partial f(\underline{z}, t)}{\partial t} + \nabla \cdot \left(f \frac{\partial \underline{z}}{\partial t} \right) - (B - D) \right] d\tau = 0 \qquad (4)$$

Since Equation (4) must apply for any volume element, the integrand must vanish identically leading to

$$\frac{\partial f(\underline{z}, t)}{\partial t} + \nabla \cdot \left(f \frac{\partial \underline{z}}{\partial t} \right) = B - D \qquad (5)$$

Equation (5) is the microscopic population balance and is simply an accounting of changes in the particle number distribution due to continuous changes in the coordinates $\underline{z}$ and any birth and death mechanisms.

   If we consider the ideally stirred reactor which has no spatial dependence, it is convenient to split the $\underline{z}$ coordinates into the external, $\underline{x}$, and internal, $\underline{y}$, parts,

$$\underline{z} = \left[\begin{array}{c} \underline{x} \\ \underline{y} \end{array} \right] \qquad (6)$$

Since neither f, B or D have any $\underline{x}$ dependence, the microscopic population balance can be integrated to give,

$$\frac{\partial f(\underline{y}, t)}{\partial t} + \nabla \cdot \left(f \frac{\partial \underline{y}}{\partial t} \right) = B - D + \left(\frac{Q_f}{V_R} \right) f_f - \left(\frac{Q}{V_R} \right) f \qquad (7)$$

in which Q_f and Q are the volumetric flowrates into and out of the reactor. The particulate system is still clearly a distributed parameter system regardless of whether it is spatially distributed as well.

To complete the mathematical description of the system, one has to choose the internal coordinates, $\underline{y}$, and specify the functional forms of $\partial \underline{y}/\partial t$, B, D, and supply the necessary initial and boundary conditions. Indeed, specifying these functions and conditions requires a detailed understanding of the transport and kinetic events taking place, and is the major hurdle to overcome in modelling any particulate system.

The key difference between the usual distributed parameter system and the particulate system is the need for the additional coordinates, $\underline{y}$, in the particulate system. The partial differential or difference equations describing the usual spatially distributed systems such as heat exchangers, fixed bed reactors, and separation columns are generally well known. On the other hand, the best choice of internal coordinates for many particulate systems and expressions for $\partial \underline{y}/\partial t$ are still subject to controversy.

In the next two sections, the general population balance equation is applied to the crystallization and emulsion polymerization processes.

3   Crystallization

Crystallization in the chemical industry is a unit operation of enormous economic importance, with applications to production of fertilizer chemicals, sucrose, pharmaceuticals, pesticides, catalysts, and proteins.

The crystal size distribution (CSD) is of overriding importance in determining the ease and efficiency of subsequent solid/liquid separation steps and suitability of the crystals for further processing or sales. In spite of its importance, the fundamental theory of CSD control is largely undeveloped, and industrial control practice is often unsuccessful resulting in large cycles of the CSD and slow damping of upsets.

Population balance techniques have been used extensively to analyze crystallizers' open loop dynamic behavior [3,4]. Although models are still sometimes formulated without explicitly considering the CSD [11], the dynamic population balance has become the dominant approach to modelling the crystallizer. The early adoption of the population balance description was undoubtedly due to the relatively simple birth and death terms required. The common starting point for crystallization studies is to consider the crystal to be completely described by a single internal coordinate, L, some characteristic crystal length. The population balance for the well-stirred crystallizer then takes the form,

$$\frac{\partial f(L,t)}{\partial t} + \frac{\partial \left(G(L)f \right)}{\partial L} + \frac{Q(L)}{V_R} f = 0 \qquad (8)$$

The crystal growth rate is in general denoted by $G(L)$ but is assumed a

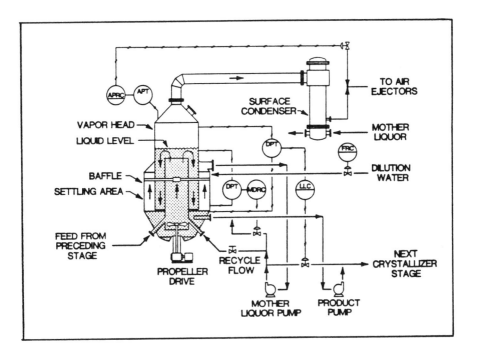

Figure 2: Cooling KC1 Crystallizer, after Bennett [28].

constant independent of L in many modelling efforts [11-19]. More detailed analyses consider the crystal growth to consist of consecutive diffusion and surface reaction steps and G is then size dependent [20-26].

The product removal rate is given by $Q(L)$. For a batch crystallizer, $Q = 0$ since no product is removed until the batch is finished. For a continuous crystallizer with a well-stirred product stream, $Q(L) = 1$ because product is removed at equal rates for all crystal sizes. This type of crystallizer is known as a mixed suspension mixed product removal (MSMPR) crystallizer. In usual industrial practice, however, particles above a desired size are preferentially removed from the reactor as product. This practice is known as product classification (cf. Fig. 2). An idealized or perfect classifier is described by the unit step function,

$$Q(L) = \begin{cases} 0, & L < L_{\max} \\ 1, & L \geq L_{\max} \end{cases} \tag{9}$$

Classification will generally result in a narrower product CSD, but causes operability problems and increases the tendency for the crystallizer to be unstable [27].

To reduce the number of fine crystals, often a fines stream is removed from the continuous crystallizer, the fine crystals are redissolved by heating, and are fed back into the bottom of the crystallizer [28,29]. The flowrate of the fines destruction stream is one of the important manipulated variables available for controlling the CSD of a continuous crystallizer. The removal and subsequent dissolution of the small crystals in the fines destruction loop is also described by an appropriate choice of $Q(L)$.

The driving force for crystal nucleation is the supersaturation, which can be achieved by evaporation, cooling, addition of a third component, or chemical reaction [27,30,31].

The nucleation rate enters the model in the boundary condition for $f(L,t)$ at $L = 0$. The following mechanisms are important in the nucleation rate:

- homogeneous nucleation; formation of nuclei as a result of supersaturation alone

- heterogeneous nucleation; formation of nuclei due to the presence of insoluble impurities

- secondary culeation; nuclei formation induced by the presence of other crystals

- attrition; nuclei formation by mechanical degradation of existing crystals.

The crystallizer design must ensure that there is sufficient recirculation in the body of the crystallizer to prevent high areas of local supersaturation which lead to excessive nucleation rates. The recirculation must also be gentle enough to avoid breading the crystals. Gootscholten et al. [32] showed that the impeller type has a large influence on the nucleation rate, and remarked that secondary nucleation is the dominant nucleation mechanism in industrial applications [33]. Accounting for the attrition mechanism would require an integral term in the population balance, but this mechanism is believed to be negligible in well-designed crystallizers.

Crystal agglomeration on the other hand can be an important problem, causing difficulties with subsequent solid-liquid separations, and a decline in product purity. Including crystal agglomeration in the model requires further progress in both the fundamental description of the agglomeration process, and solution methods for the resulting integro-differential equations.

3.1 Continuous Crystallizer Control

The need for large-tonnage fertilizers created the demand for continuous crystallizers that can produce large and uniform crystals, and operate stably for long periods of time [28,34,35]. Sustained oscillations in continuous crystallizers are an important industrial problem and can lead to off-specification products, overload of dewatering equipment, and increased equipment fouling. Figure 3 shows this oscillatory behavior for an industrial KC1 crystallizer.

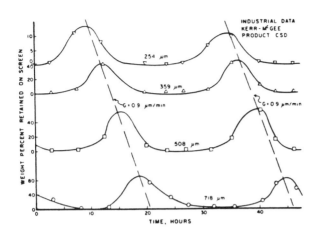

Figure 3: Unstable Behavior in an Industrial KC1 Crystallizer, after Randolph et al. [36].

Similar results have been observed in academic investigations with bench-scale crystallizers [31,36].

The stability of the open-loop, continuous crystallizer has received considerable attention [4,17,25,26,36-41]. The analytical techniques employed are either: Laplace transform of the linearized population balance [37], conversion to a set of ordinary differential equations by taking moments [4,17,26,41], or spectral analysis of the linear operator obtained from the population balance [25,38,39]. All of these analyses are obviously local, and no one has determined Lyapunov functions to ascertain the global stability. There have not been any simulations of the nonlinear equations, however, that indicate the global stability is markedly different from the local stability.

These stability analyses have shown that oscillations in continuous crystallizers can be caused by either: a high order kinetic relationship between the nucleation rate and the supersaturation, or nonrepresentative product removal. In a review of this topic, Randolph [42] concludes that oscillations in inorganic industrial crystallizers are most likely caused by combined fines destruction and classified product removal.

Numerous investigators have addressed the problem of stabilizing an oscillating crystallizer using feedback control. Many of the early studies proposed measuring some moment of the CSD on line and using a proportional feedback controller to adjust the flowrate of the fines destruction system [39,43-45]. After reviewing these studies, Rousseau and Howell [46] proposed using supersaturation as the measured variable, since on-line measurements of properties of the CSD are difficult to make. They demonstrated through simulation that a proportional feedback controller could stabilize a cycling crystallizer, although the sensitivity to measurement error was more severe.

Randolph and coworkers [40,42,47-51] proposed inferring the nuclei density from a measurement of the CSD in the fines stream in conjunction with a proportional feedback controller. Significantly improved disturbance rejection was demonstrated experimentally for the proposed control scheme. Recently, Rohani [52] reported preliminary tests on using a turbidimeter for an on-line measurement of slurry density in the fines loop. Again, stabilization and improved disturbance rejection were demonstrated by simulation using a proportional feedback controller.

There have been relatively few studies of multivariable controllers for continuous crystallizers. Hashemi and Epstein [53] applied singular value decomposition to a crystallizer model with four inputs and outputs and developed measures of the system observability and controllability. Finally, Johnson et al. [54] investigated multivariable, optimal control of an MSMPR crystallizer. A Kalman filter was used in conjunction with the linear quadratic optimal controller. On-line densitometry was proposed as a means of measuring the third moment of the CSD. Feed flowrate and heat exchanger duty were employed as additional manipulated variables in the MIMO studies.

3.2   Control of Batch Crystallizers

Batch crystallizers are used extensively in the chemical industry, often for small volume, high value-added specialty chemicals. Batch operation offers a flexible and simple processing step for plants with frequently changing recipes and product lines.

The control objectives for batch operation are often quite different from those of continuous crystallization. In the production of specialty organic materials, such as pharmaceuticals, product purity and CSD are of prime importance. If acceptable CSD and purity standards are not met, a batch of crystals has to undergo further processing steps, such as milling or recrystallization.

Most of the previous process control studies of batch crystallizers have dealt with finding the open-loop, supersaturation versus time trajectory that optimizes some performance criterion derived from the CSD. The "natural" cooling curve displays the greatest cooling in the initial period when the temperature difference between the hot crystallizer fluid and the surroundings is the greatest (cf. Fig. 4). This causes a large degree of supersaturation at early times which leads to excessive nucleation, smaller final crystals, and aggravated operability problems such as fouling.

Mullin and Nyvlt [55] showed that adjusting the supersaturation to maintain a constant level of nucleation would increase crystal size and decrease fouling of heat exchanger surfaces. Jones and Mullin [56] concluded that controlled cooling at constant nucleation rate significantly improved both the mean size and the variance of the CSD compared with natural or linear cooling.

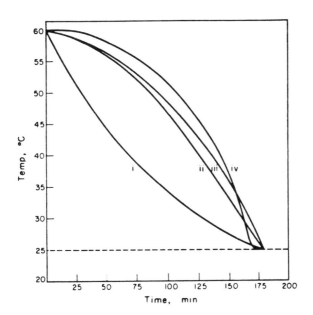

Figure 4: Cooling Curves for Batch Crystallization after Jones [57]: (i) Natural Cooling, (ii) Constant Nucleation rate, (iii) Metastable Control, and (iv) Size-Optimal Control.

Jones [57] applied optimal control theory and control vector iteration to find the cooling curve that maximized the final size of the seed crystals. The temperature of the size-optimal control profile decreases slowly at early times and then rapidly at the end of the batch time, in direct contrast to the natural cooling curve.

Chang and Epstein [58] also computed optimal temperature profiles for a variety of objective functions based on average size, volume, or variance of the final CSD. They used a gradient method to solve the two point boundary value problem for the optimal profile. Surprisingly, their results show an initial increase in supersaturation followed by a slow decrease, which is much closer to natural cooling than to Jone's result. Although these works employ slightly different objective functions and crystallization models, the source of the disagreement is not immediately clear. Jones [59] discusses this point in further detail and provides a general review of control of the batch crystallizer.

In another vein, Jones et al. [60] investigated the feasibility of using fines destruction in a batch crystallizer. Their experimental results indicate that fines destruction can remove the crystals formed by secondary nucleation and increase the average crystal size.

Another application of batch crystallization is in the sugar industry. Essentially all sugar production, both refined and raw, involves batch crystallization as one of the final processing steps. In a series of articles, Virtanen [61-63] discusses the automatic control of batch sucrose crystallization. The goals of automating the existing batch equipment were to shorten the crystallization time, produce a more uniform CSD, and reduce the labor and energy costs. In the final control scheme, a microcomputer adjusts t e steam flowrate to maintain constant supersaturation based on on-line density and refractive index measurements. The computer also regulates the timing of the feed addition, concentration, seeding, and growing steps for each batch. The control scheme was implemented at the Finnish Sugar Co. Ltd. and the biggest gains were achieved in decreasing batch to batch variations, decreasing the labor costs, and improving power plant economy.

4   Emulsion Polymerization

Emulsion polymerization is an important process for the production of adhesives, latex paints, paper and plastic coatings, ink pigments, floor coverings, etc., and annual worldwide sales of emulsion polymers exceeds one billion dollars. The polymerization is carried out in batch, semi-batch and continuous stirred-tank reactors (CSTR's). Batch and semi-batch reactors are well suited for low volume and specialty product lines. The continuous reactor offers strong economic advantages for high volume lines that require infrequent recipe changes. Unfortunately, continuous processes are used only for a limited range of products because of the perverse nonlinear behavior of the reactors. For a number of monomers, isothermal multiple steady states have been observed, while for practically all monomers, sustained periodic oscillations plague these reactors – often appearing and disappearing in an unpredictable fashion. For some 40 years, researchers have collected experimental data displaying these two phenomena (cf. Table 1) and have tried to understand the complex behavior through models.

Rather complete surveys of the previous modelling literature may be found in [64-67]. These models fall into two categories:

- simple particle models which ignore the particle size distribution (PSD) and consider the particles to be indistinguishable from each other, or

- models that explicitly account for the particle size distribution through population balance equations.

Nomura et al. [68] have shown that simple models that do not account for the PSD can be safely used to simulate the low conversion, steady-state behavior of rather ideal monomers such as styrene. Chu [69] and Schork et al. [70] have shown that these simple models fail, however, to describe reactor behavior at high conversion, or dynamic behavior of monomers with significant radical desorption, such as vinyl acetate. Most importantly, the simple models are incapable of simulating the sustained oscillations that have often been observed in experimental studies. For these reasons, this discussion will only be concerned with the population balance models. They are the only models which have realistic predictive capabilities over the full range of operating conditions and different monomers of interest.

Rawlings [66] and Rawlings and Ray [71,72] have recently reviewed the population balance models and presented a model that is capable of simulating all of the experimentally observed phenomena including steady-state multiplicity, sustained oscillations, ignition and extinction dynamics, and overshoot in conversion during reactor startup. This model will be presented briefly in the next section. Detailed bifurcation results are available to determine when the continuous reactor becomes unstable. Finally, the process control literature for emulsion polymerization is reviewed.

4.1   Model Development

Continuous emulsion polymerization reactors (cf. Fig. 5) typically have a feed consisting of monomer (M_f), water (V_w), surfactant (S_f), initiator (I_f), and other additives. For surfactant concentrations above the critical micelle concentration (CMC), micelles form the principle locus of particle initiation. New particles are nucleated by free radicals, in the aqueous phase, entering monomer-swollen micelles and initiating polymerization. Based on this simple physical picture, Omi et al. [73] proposed a qualitative explanation for the observed sustained oscillations:

Table 1: Nonlinear Phenomena Reported from Experimental Studies of Continuous Emulsion Polymerization Reactors.

Author	Year	Monomer	Observation
Owen et al.	1947	Styrene-butadiene	Oscillations
Jacobi	1952	Vinyl chloride	Oscillations
Gershberg and Longfield	1961	Styrene	Oscillations
Gerrens et al.	1971	Styrene	Multiple steady states
Ley and Gerrens	1974	Styrene	Multiple steady states Oscillations
Greene et al.	1976	Methylmethacrylate Vinyl Acetate	Oscillations Oscillations
Brooks et al.	1978	Styrene	Oscillations
Kiparissides et al.	1978	Vinyl Acetate	Oscillations
Nomura et al.	1980	Vinyl Acetate	Oscillations
Schork and Ray	1980	Methylmethacrylate	Multiple steady states Oscillations

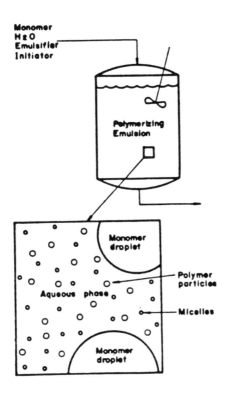

Figure 5: The Emulsion Polymerization Reactor.

Population balance equation in particle age, τ, and time, t:

$$\frac{\partial F}{\partial t} = -\frac{\partial F}{\partial \tau} - QF$$

$$F(0,t) = P_m V_w$$

$$P_m = C_5 Rm'H(m') \quad - \text{ nucleation from micelles}$$

Growth of polymer particle:

$$\frac{\partial V}{\partial t} = (1-C_1)C_6 g_p \frac{\phi}{1-\phi}\, \overline{I} + \frac{V}{1-\phi}\frac{d\phi}{dt}$$

$$V(0,t) = 1$$

$$r = \left(\frac{V}{4/3\pi}\right)^{1/3} \quad - \text{ particle radius}$$

Quasi-steady balances:

$$m' = C_7(S-S_{wc})V_w - C_8 \langle F,r^2\rangle \quad \text{(surfactant)}$$

$$\langle F,r^2\rangle \equiv \int_o^\infty F(\tau,t)r^2(\tau,t)d\tau \quad \begin{array}{l}\text{(total particle}\\ \text{surface area)}\end{array}$$

$$R = \frac{-b_R + \sqrt{b_R{}^2 - 4a_R c_R}}{2a_R} \quad \begin{array}{l}\text{(aqueous phase}\\ \text{free radicals)}\end{array}$$

$$a_R = C_{12}/V_w$$

$$b_R = C_{15}J + \frac{C_{10}\langle F,r^n\rangle}{V_w} + C_9\, m' + Q$$

$$c_R = -2fC_4 IV_w - C_{11}\langle F, \frac{\overline{1}}{\frac{C_{18}}{\phi} + r^2}\rangle g_{tr}$$

$$1-\phi + \ell n\phi + \psi(1-\phi)^2 = \ell n\left(\frac{M_w}{M_{sat}}\right) \quad \begin{array}{l}\text{(monomer concen-}\\ \text{tration in}\\ \text{polymer particle)}\end{array}$$

$$\overline{I}(V,t) = \frac{a}{4}\frac{I_b(a)}{I_{(b-1)}(a)} \quad \begin{array}{l}\text{(polymer particle}\\ \text{radical number)}\end{array}$$

$$a = C_{14}\, r^{\frac{3+n}{2}}\, \sqrt{R/g_t}$$

$$b = \frac{C_{13}}{g_t}\frac{r^3}{\frac{C_{18}}{\phi} + r^2}\, g_{tr}$$

Dynamic balances:

$$\frac{dM}{dt} = M_f Q_f - MQ - C_2 C_3 \langle F,\overline{I}\phi g_p\rangle \quad \text{(total monomer)}$$

$$\frac{d(IV_w)}{dt} = Q_f I_f V_{wf} - QIV_w - C_4 IV_w \quad \text{(total initiator)}$$

$$\frac{d(SV_w)}{dt} = Q_f V_{wf}S_f - QV_w S \quad \text{(total surfactant)}$$

$$\frac{dV_w}{dt} = Q_f V_{wf} - QV_w \quad \text{(aqueous phase)}$$

Table 2: Detailed Emulsion Polymerization Model

- Polymer particles are initiated by radical entry into micelles provided free emulsifier is available (emusilifier concentration is above the critical micelle concentration).

- The particles polymerize and grow, requiring more free emulsifier to stabilize their increased surface area. Eventually all the free emulsifier is absorbed on the particles and there are no micelles. Particle initiation then ceases.

- The large particles eventually wash out of the reactor and micelles reappear due to the emulsifier entering in the feed.

This qualitative explanation points out the essential feedback mechanism that allows uncontrolled emulsion polymerization (and continuous crystallization) reactors to oscillate. The rate of new particle initiation (boundary condition of the PSD) is influenced by the total area of existing particles (2nd moment of the PSD). The reverse is also true: the total particle area at future times will obviously be influenced by the rate of the new particle initiation at the present time.

The detailed model considering fundamental mechanistic steps consists of nonlinear, coupled integro-differential-algebraic equations. These equations, which are specific to the emulsion polymerization process, are listed in Table 2. This model was solved using orthogonal collocation on finite elements to reduce the two PDE's to sets of ODE's; and the resulting differential-algebraic system was integrated numerically. Physical parameters and kinetic rate constants were chosen from the literature, and model predictions were compared with experimental data for three monomer systems: styrene, methylmethacrylate, and vinyl acetate. In general, there was good qualitative agreement with the data without any parameter fitting and in some cases, excellent quantitative predictions. Figures 6-9 show sample comparisons to data displaying multiple steady states and sustained oscillations.

The model is able to describe steady-state multiplicity and to distinguish between runs which seem to oscillate and ones which do not. Obviously, in Figures 6-9 one could obtain closer quantitative agreement if allowed the option of tuning model parameters. However, the goal here is to demonstrate that the model is structurally correct and has *a priori* predictive capabilities with a single set of fundamental parameters.

4.2 CSTR Dynamics and Stability

Given a differential equation model,

$$\frac{dy}{dt} = \underline{f}(\underline{y}), \tag{10}$$

one can analyze the bifurcation structure by examining the eigenvalues of the Jacobian matrix, $\underline{L}$, evaluated at the steady state, $\underline{y}_s$.

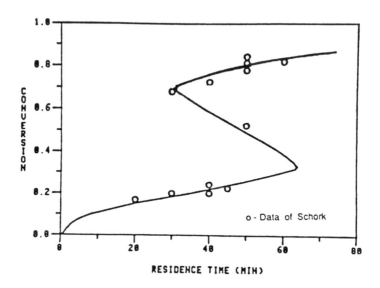

Figure 6: Steady-State Conversion for the Emulsion Polymerization of Methyl-methacrylate in a CSTR: o - data of Schork, ————— - Model of Rawlings and Ray.

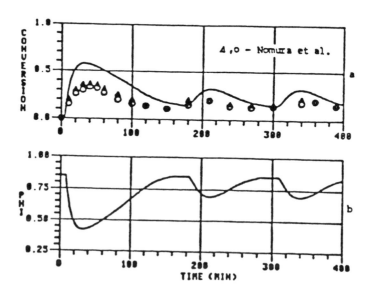

Figure 7: Model Simulation Compared with Vinyl Acetate Polymerization Data of Nomura et al. $S_f = 0.0069$ mol/L, $I_f = 0.0046$ mol/L, $V_{wf} = 0.82$, $\Theta = 20$ min, Temp = 50°C.

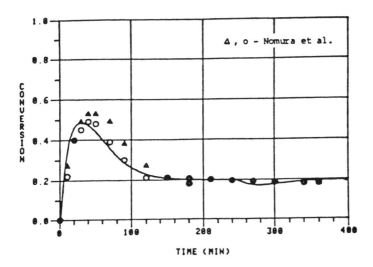

Figure 8: Model Simulation Compared with Vinyl Acetate Polymerization Data of Nomura et al. $S_f = 0.00243$ mol/L, $I_f = 0.00925$ mol/L, $V_{wf} = 0.82$, $\Theta = 20$ min, Temp $= 50°C$.

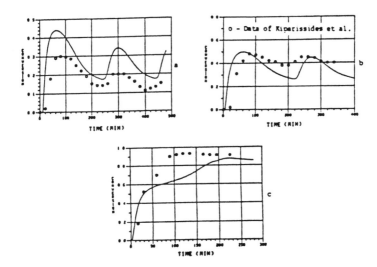

Figure 9: Model Simulation Compared with Vinyl Acetate Polymerization Data at Various Initiator Feed Concentrations. $S_f = 0.01$ mol/L, $V_{wf} = 0.70$, $\Theta = 30$ min, Temp $= 50°C$ (a) $I_f = 0.005$ mol/L, (b) $I_f = 0.01$ mol/L, (c) $I_f = 0.02$ mol/L.

$$\underline{0} = \underline{f}(\underline{y}_s) \tag{11}$$

$$\underline{L} = \frac{\partial \underline{f}}{\partial \underline{y}}\Big|_{\underline{y}=\underline{y}_s} \tag{12}$$

The system, Eq. (1), is stable to small perturbations if all of the eigenvalues of $\underline{L}$ are in the left half plane. If a pair of eigenvalues pass through the imaginary axis into the right half plane as a parameter of the model varies, then a periodic solution to Eq. (10) emerges at that value of the parameter. If a single eigenvalue passes through the origin, then one has a static bifurcation (i.e., ignition or extinction points of s-shaped steady-state curves).

One must extend these standard bifurcation results to differential/algebraic models in order to apply them to the emulsion polymerization model. Consider the model to be partitioned into a set of differential and a set of algebraic equations. Let vector $\underline{y}_1$ and $\underline{y}_1$ contain the differential and algebraic states, respectively.

The model can be written in the form,

$$\frac{d\underline{y}_1}{dt} = \underline{f}_1\left(\underline{y}_1, \underline{y}_2\right) \tag{13}$$

$$\underline{0} = \underline{f}_2\left(\underline{y}_1, \underline{y}_2\right) \tag{14}$$

The appropriate system Jacobian, $\underline{L}$, for bifurcation analysis is then

$$\underline{L} = \left[\frac{\partial \underline{f}_1}{\partial \underline{y}_1}\right]_{\underline{f}_2 = \underline{0}} \tag{15}$$

i.e., the rate of change of $\underline{f}_1$ with respect to $\underline{y}_1$ when the algebraic relations (14) are satisfied. The Jacobian of the full differential/ algebraic problem, $\underline{L}'$, can be partitioned as

$$\underline{L}' = \begin{bmatrix} \frac{\partial \underline{f}_1}{\partial \underline{y}_1} & \frac{\partial \underline{f}_1}{\partial \underline{y}_2} \\ \frac{\partial \underline{f}_2}{\partial \underline{y}_1} & \frac{\partial \underline{f}_2}{\partial \underline{y}_2} \end{bmatrix} \tag{16}$$

One can show from the chain rule for partial derivatives that

$$\underline{L} = \frac{\partial \underline{f}_1}{\partial \underline{y}_1} - \left(\frac{\partial \underline{f}_1}{\partial \underline{y}_2}\right)\left(\frac{\partial \underline{f}_2}{\partial \underline{y}_2}\right)^{-1}\left(\frac{\partial \underline{f}_2}{\partial \underline{y}_1}\right) \tag{17}$$

Note that $\partial \underline{f}_2/\partial \underline{y}_2$ must have full rank.

The technique can be extended to determine $\underline{L}$ for implicit differential/ algebraic systems. Consider a more general form of Eq. (13).

$$\underline{0} = \underline{f}_1\left(\dot{\underline{y}}_1, \underline{y}_1, \underline{y}_2\right) \tag{18}$$

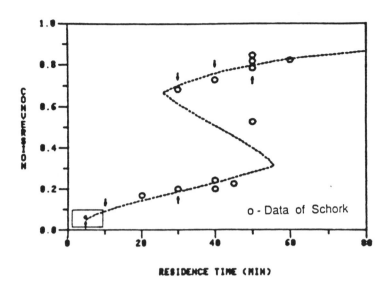

Figure 10: Steady-State Conversion versus Residence Time for Methyl
-methacrylate Polymerization. S_f = 0.03 mol/L, I_f = 0.03 mol/L, Temp
= 40°C, V_{wf} = 0.70. o - Data of Schork. o - Hopf bifurcation, ↑ - Dynamic
Simulation Point, ——— - Unstable Steady States.

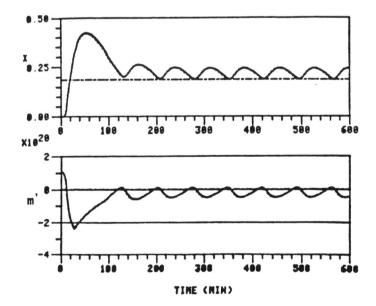

Figure 11: (a) Conversion and (b) Micelle Concentration (1/lit) Versus Time,
Lower Branch. S_f = 0.03, I_f = 0.03, Θ = 30 min, T = 40°C, V_{wf} = 0.70.

$$\underline{0} = \underline{f}_2\left(\underline{y}_1,\underline{y}_2\right) \tag{19}$$

Eq. (18) handles the case in which it is inconvenient or impossible to solve for the $\underline{y}_1$ time derivatives explicitly. This case can be important in the emulsion polymerization model since it is algebraically intractable to solve for $d\phi/dt$ explicitly. Application of the chain rule to this case gives

$$\underline{L} = \left[\frac{\partial \underline{f}_1}{\partial \underline{\dot{y}}_1}\right]^{-1}\frac{\partial \underline{f}_1}{\partial \underline{y}_1} - \frac{\partial \underline{f}_1}{\partial \underline{y}_2}\left[\frac{\partial \underline{f}_2}{\partial \underline{y}_2}\right]^{-1}\frac{\partial \underline{f}_2}{\partial \underline{y}_1} \tag{20}$$

where both $\partial \underline{f}_1/\partial \underline{\dot{y}}_1$ and $\partial \underline{f}_2/\partial \underline{y}_2$ must have full rank.

Eq. (17) and (20) offer substantial gains in computational efficiency since one does not have to explicitly solve the algebraic equations while computing $\underline{L}$ as in Eq. (15).

Rawlings and Ray [72] applied the bifurcation analysis methods of this section to the emulsion polymerization model in Table 2. Using methylmethacrylate polymerization as an example system, the bifurcation points with reaction residence time as the model parameter are shown in Figure 10. The reactor exhibits steady-state multiplicity for residence times between 25 and 60 minutes. Most strikingly, there is a bifurcation to limit cycles at about 5 minutes residence time (conversion = 5%). The steady states on most of the lower branch and all of the upper branch are predicted to be unstable and surrounded by limit cycles. Numerical simulations of the detailed model verify the existence of the limit cycles.

The monomer conversion and micelle concentration are plotted in Figure 11 at 30 minutes residence time. The period of the oscillation is 3 residence times, and the conversion amplitude is about 5%. The effect on the PSD is shown in Figure 12. The initial steady-state PSD is unstable and quickly develops into a series of 'waves' of particles that grow out into larger sizes while being washed out of the reactor.

Further discussion of the effects of start-up policy and other parameter variations are provided by Rawlings and Ray [72].

4.3   Emulsion Polymerization Control

There are only a small number of published control studies of the emulsion polymerization system. Dickinson [74] investigated the optimal start-up policy for a continuous reactor. He showed that one can successfully reduce the initial conversion overshoot (cf. Fig. 8, 11) and drive the system quickly to its steady state by controlling the rate of emulsifier addition. Kiparissides et al. [75] used an extended Kalman filter to reconstruct the PSD from turbidity measurements. They also solved the optimal start-up policy. The controller was generally unsuccessful, however, in stabilizing the oscillating reactor (Pollock, [76], Penlidis et al. [77]).

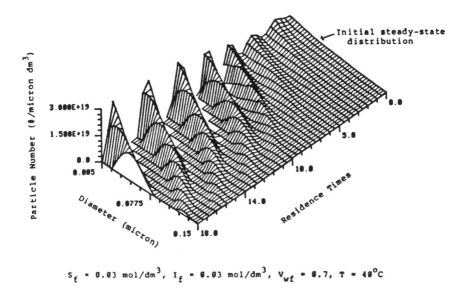

$S_f = 0.03 \text{ mol/dm}^3$, $I_f = 0.03 \text{ mol/dm}^3$, $V_{wf} = 0.7$, $T = 40^\circ C$

Figure 12: The Particle Size Distribution as a Function of Time, $\Theta = 10$ min., Unstable.

Leffew and Deshpande [78] applied dead time compensation with an analytical predictor suggested by Moore [79]. This controller could also only achieve a modest damping of the undesirable oscillation. Gordon and Weidner [80] studied the open-loop feed rate addition of surfactant to a semi-batch reactor, and its influence on the product PSD. They based the emulsifier addition on a reaction dependent variable such as monomer conversion, in order to get good reproducibility in spite of batch to batch variations in the monomer inhibitor levels.

Recently, Gugliotta and Meira [81] examined the feasibility of controlling PSD with forced oscillations in the feed rate to a continuous reactor. Theirs is also an open-loop scheme based on a steady-state model, so that it is only strictly valid in the limit of very low frequency oscillations. Louie et al. [82] proposed using oxygen injection as a means of controlling reactor temperature. Oxygen is an inhibitor for free-radical polymerization, and a feedback proportional controller was shown to successfully regulate the reaction rate without seriously lowering the product molecular weight.

Since the control schemes were generally unsuccessful in eliminating oscillations in the PSD, design changes in the process were investigated. Berrens [83], Green et al. [84], Nomura and Harada [85], and Pollock et al. [86] explored the idea of feeding a latex seed to the CSTR. A nonoscillating seed could be produced in a tubular, plug flow prereactor or a small CSTR. Particle nucleation was confined to the prereactor, and the CSTR or CSTR chain was used to grow the particles and achieve high conversion of monomer.

5   Measurement Technology

As discussed in the context of crystallization and emulsion polymerization processes, successful process control depends vitally on reliable, rapid measurement of the PSD. Other on-line measurements are also desirable, such as slurry density and surface tension in emulsions, but until recently, on-line PSD measurement was the major hurdle to be overcome.

Particle sizing technology is based on a wide variety of physical phenomena. Electrical properties, transport properties in the case of sedimentation and hydrodynamic chromatography, and optical properties including diffraction, blockage, and imaging techniques have all been exploited to measure particle size. The wide variety of commercially available instruments points out the great practical interest in measuring PSD. Several excellent review articles and books provide a thorough discussion of the various instruments and the advantages and disadvantages of each [87-89].

In process control applications, the emphasis is on rapid measurement and analysis techniques that are easily automated to provide a feedback signal to the controller. Several instruments based on laser light scattering would seem to fill this need [90]. As shown in Figure 13, a laser beam is directed through a sample cell containing the particles. The particles diffract the light which is collected and measured with a photodiode detector.

The PSD can then be calculated from the intensity of the light as a function of the scatter angle. The measurement and analysis time is on the order of ten seconds to a minute. A microcomputer, printed, and necessary software for calculation of the PSD are available with the optical instrument. Serial part communication allows direct transfer of data files from the instrument to a control computer without A/D interfaces.

Other advantages of this technique include: fairly wide size range (0.5 − 500 microns), nonintrusive measurement, no calibration requirements, and most importantly, no small orifices that can become plugged with particles. The biggest drawback is the need to have relatively dilute particulate concentrations to avoid multiple scatter. The concentration limits may necessitate an on-line dilution scheme upstream of the measuring device in industrial applications.

6 ̊ Conclusions

Dispersed phase system modelling and control is of interest to control engineers because of the large economic importance of these processes and the many unsolved control problems that they pose. Of the many industrial dispersed phase processes, this review has focused on crystallization and emulsion

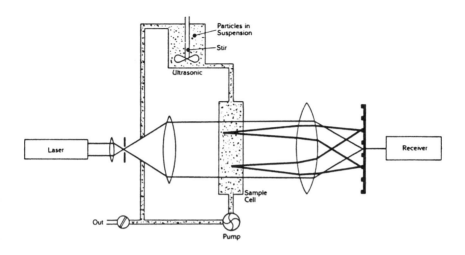

Figure 13: On-Line Particle Size Measurement with Laser Diffraction.

polymerization to illustrate the importance underlying characteristics of this class of processes.

Batch, semi-batch, and continuous modes of operation have found applications. Uncontrolled continuous processes are often unstable, resulting in limit cycle behavior and off-specification product. The instability is caused by the internal feedback and time delay mechanisms inherent to systems with a PSD that influences the rate of new particle formation. Simple proportional feedback controllers based on a measurement of some property of the PSD have successfully stabilized these reactors in some cases. Some attention has been focused on solving for optimal profiles for batch reactors. No consideration, however, has been given to on-line updating of the computed optimum to account for plant/model mismatch and disturbances.

The recent availability of complete, on-line particle sizing instruments has opened the door to rapid improvement in control methodology. There are several existing problems that must be solved before these gains can be realized. Further attention must be focused on reducing the distributed parameter model to a small set of ODE's. The common practice of model reduction via the method of moments is often unsuccessful. Controllers based on simultaneous optimization of an objective function and solution of the nonlinear lumped model, such as UDMC or nonlinear IMC, hold great potential for this class of problems [91-95].

The advances that can be realized for relatively simple dispersed phase systems, such as crystallization, will hopefully have applications in the much more difficult modelling and control problems associated with biochemical fermentations and aerosol technology.

Acknowledgements

The author would like to thank the NSF and NATO for postdoctoral support at Professor Gilles's Institute for System Dynamics and Process Control, where much of this paper was written. Helpful discussions with Professors Gilles, Zeitz, and Kohler, and Messrs. Marquardt, King, Epple and Buschulte are also gratefully acknowledged.

References

1. Rudd, D. F., *Chem. Eng. Sci.*, 12, 51 (1960).

2. Rudd, D. F., *Can. J. Chem. Eng.*, 40, 197 (1962).

3. Hulburt, H. M. and S. Katz, *Chem. Eng. Sci.*), 19, 555 (1964).

4. Randolph, A. D. and M. A. Larson, *Theory of Particulate Processes*, Academic Press, 1971.

5. Tsuchiya, H. M., A. G. Fredrickson, and R. Aris, in *Advances in Chemical Engineering*, Vol. 6, Academic Press, New York, 1966.

6. Eakman, J. M., A. G. Fredrickson, and H. M. Tsuchiya, *CEP Symp. Ser.* No. 69, 62, 37 (1966).

7. Ramkrishna, D., in *Advances in Biochemical Engineering*, Ghose, T. K., A. Fiechter, and N. Blakebrough (eds.), 11, 1 (1979).

8. Friedlander, S. K., *Smoke, Dust, and Haze*, Wiley, New York, 1977.

9. Hidy, G. M., *Aerosols*, Academic Press, Orlando, 1984.

10. Ramkrishna, D., *Reviews in Chem Eng.*, 3, 49 (1985).

11. Nyvlt, J. and J. W. Mullin, *Chem. Eng. Sci.*, 25, 131 (1970).

12. Randolph, A. D. and M. A. Larson, *AIChE J.*, 8, 639 (1962).

13. Randolph, A. D., *AIChE J.*, 11, 424 (1965).

14. Shadman, F. and A. D. Randolph, *AIChE J.*, 24, 782 (1978).

15. Han, C. D. and R. Shinnar, *AIChE J.*, 14, 612 (1968).

16. Han, C. D., *Chem. Eng. Sci.*, 22, 611 (1967).

17. Epstein, M. A. F. and L. Sowul, in *Design, Control, and Analysis of Crystallization Processes,* A. D. Randolph (ed.), AIChE Symp. Ser. 193, New York, 1980.

18. Timm, D. C. and M. A. Larson, *AIChE J.,* 14, 452 (1968).

19. Yu, K. M. and J. M. Douglas, *AIChE J.,* 21, 917 (1975).

20. Moyers, C. G. and A. D. Randolph, *AIChE J.,* 19, 1089 (1973).

21. O'dell, F. P. and R. W. Rousseau, *AIChE J.,* 24, 738 (1978).

22. Bourne, J. R. and M. Zabelka, *Chem. Eng. Sci.,* 35, 533 (1980).

23. Karpinski, P., *Chem. Eng. Sci.,* 35, 2321 (1980).

24. Wey, J. S. and R. Jagannathan, *AIChE J.,* 28, 697 (1982).

25. Ishii, T. and A. D. Randolph, *AIChE J.,* 26, 507 (1980).

26. Sherwin, M. B., R. Shinnar, and S. Katz, *AIChE J.,* 13, 1141 (1967).

27. Garside, J., *Chem. Eng. Sci.,* 40, 3 (1985).

28. Bennett, R. C., *Chem. Eng. Prog.,* 80, 89 (1984).

29. Saeman, W. C., *AIChE J.,* 2, 107 (1956).

30. Murray, D. C. and M. A. Larson, *AIChE J.,* 11, 728 (1965).

31. Song, Y. H. and J. M. Douglas, *AIChE J.,* 21, 917 (1975).

32. Grootscholten, P.A.M., B.G.M. DeLeer, E. J. de Jong, and C. J. Asselbergs, *AIChE J.,* 28, 728 (1982).

33. Clontz, N. A. and W. L. McCabe, *Chem. Eng. Symp. Ser.,* 110, 6 1971.

34. Miller, P. and W. C. Saeman, *Chem. Eng. Prog.,* 43, 667 (1947).

35. Robinson, J. N. and J. E. Roberts, *Can. J. Chem. Eng.,* 35, 105 (1957).

36. Randolph, A. D., J. R. Beckman, and Z. I. Kraljevich, *AIChE J.,* 23, 500 (1977).

37. Anshus, B. E. and E. Ruchenstein, *Chem. Eng. Sci.,* 28, 501 (1973).

38. Randolph, A. D., G. L. Beer, and J. P. Keener, *AIChE J.,* 19, 1140 (1973).

39. Lei, S., R. Shinnar, and S. Katz, *AIChE J.,* 17, 1459 (1971).

40. Beckman, J. R. and A. D. Randolph, *AIChE J.,* 23, 510 (1977).

41. Jerauld, G. R., Vasatis, Y., and M. F. Doherty, *Chem. Eng. Sci.*, 38, 1675 (1983).

42. Randolph, A. D., in *Design, Control, and Analysis of Crystallization Processes,* A. D. Randolph (ed.), AIChE Symp. Ser. 193, New York, 1980.

43. Han, C. D., *Ind. Eng. Chem. Process Des. Dev.*, 8, 150 (1969).

44. Lei, S. J., R. Shinnar, and S. Katz, *Chem. Eng. Prog. Symp. Ser.*, 67, 129 (1971).

45. Gupta, G. and D. C. Timm, *Chem. Eng. Prog. Symp. Ser.*, 67, 121 (1971).

46. Rousseau, R. W. and T. R. Howell, *Ind. Eng. Chem. Process Des. Dev.*, 21, 606 (1982).

47. Rovang, R. D. and A. D. Randolph, in *Design, Control, and Analysis of Crystallization Processes,* A. D. Randolph (ed.), AIChE Symp. Ser. 193, New York, 1980.

48. Randolph, A. D., E. T. White, and C. C. Low, *Ind. Eng. Chem. Process Des. Dev.*, 20, 496 (1981).

49. Randolph, A. D. and C. G. Low, in *Industrial Crystallization* 81, S. J. Jancic and E. J. de Jong (eds.), North-Holland Pub. Co., 1982.

50. Bennett, R. C. and A. D. Randolph, in *Potash 83*, R. M. McKercher (ed.), Pergamon Press, Willowdale, Ontario, 1983.

51. Randolph, A. D., L. Chen, and A. Tavana, AIChE Meeting, New Orleans, 1986.

52. Rohani, S., *Can. J. Chem. Eng.*, 64, 112 (1986).

53. Hashemi, R. and M. A. Epstein, *AIChE Symp. Ser. No. 215*, 78, 81 (1982).

54. Johnson, J. L., F. J. Schork, and A. S. Myerson, AIChE Meeting, Chicago, 1985.

55. Mullin, J. W. and J. Nyvlt, *Chem. Eng. Sci.*, 26, 369 (1972).

56. Jones, A. G. and J. W. Mullin, *Chem. Eng. Sci.*, 29, 105 (1974).

57. Jones, A. G., *Chem. Eng. Sci.*, 29, 1075 (1974).

58. Chang, C. and M. A. Epstein, *AIChE Symp. Ser. No. 215*, 78, 68 (1982).

59. Jones, A. G., in *Tailoring of Crystal Growth*, J. Garside (ed.), No. 2, I. Chem. E. N. W. Brance Symp. Ser., 1982.

60. Jones, A. G., A. Chianese, and J. W. Mullin, in *Industrial Crystallization 84*, S. J. Jancic and E. J. de Jong (eds.), Elsevier Science Publishers, Amsterdam, 1984.

61. Virtanen, J., in *Industrial Crystallization* 84, S. J. Jancic and E. J. de Jong (eds.), Elsevier Science Publishers, Amsterdam, 1984.

62. Virtanen, J., *Int. Sugar. J.*, 86, 175 (1984).

63. Virtanen, J., *Sugar J.*, 8 (Feb. 1986).

64. Ray, W. H., *ACS Symposium Series*, No. 226, 1983.

65. Poehlein, G., in *Emulsion Polymerization*, Acad. Press, 1982.

66. Rawlings, J. B., Ph.D. Thesis, University of Wisconsin-Madison, 1985.

67. Min, K. W. and W. H. Ray, *J. Macro. Sci.-Revs.*, C11, 177 (1974).

68. Nomura, M., H. Kojima, M. Harada, W. Eguchi, and S. Nagata, *J. Appl. Poly. Sci.*, 15, 675 (1971).

69. Chu, G. C., Master's Thesis, University of Wisconson-Madison, 1980.

70. Schork, F. J., G. C. Chu, and W. H. Ray, 73rd AIChE Meeting, Chicago, 1980.

71. Rawlings, J. B. and W. H. Ray, "Stability of Continuous Emulsion Polymerization Reactors - A Simplified Model Analysis," *AIChE J.*, (submitted for publication).

72. Rawlings, J. B. and W. H. Ray, "Stability of Continuous Emulsion Polymerization Reactors - A Detailed Model Analysis," *Chem. Eng. Sci.*, (submitted for publication).

73. Omi, S., T. Ueda, and H. Kubota, *J. Chem. Eng. of Japan*, 2, 193 (1969).

74. Dickinson, R. F., Ph.D. Thesis, University of Waterloo, Waterloo, Ontario, 1976.

75. Kiparissides, C., J. F. MacGregor, and A. E. Hamielec, *AIChE J.*, 27, 13 (1981).

76. Pollock, M. J., Ph.D. Thesis, McMaster University, Hamilton, Ontario, 1984.

77. Penlidis, A., J. F. MacGregor, and A. E. Hamielec, *AIChE J.*, 31, 881 (1985).

78. Leffew, K. W. and P. B. Deshpande, in *Emulsion Polymers and Emulsion Polymerization,* D. R. Bossett and A. E. Hamielec (eds.), ACS Symp. Ser., 165, Washington, D.C., 1981.

79. Moore, C. F., Ph.D. Thesis, Louisiana State University, 1969.

80. Gordon, D. L. and K. R. Weidner, in *Emulsion Polymers and Emulsion Polymerization,* D. R. Bossett and A. E. Hamielec (eds.), ACS Symp. Ser., 165, Washington, D. C., 1981.

81. Gugliotta, L. M. and G. R. Meira, *Macromol. Chem.,* Macromol. Symp., 2, 209 (1986).

82. Louie, B. M., W. Y. Chiu, and D. S. Soong, *J. Appl. Poly. Sci.*, 30, 3189 (1985).

83. Berrens, A. R., *J. Appl. Poly. Sci.*, 18, 2379 (1974).

84. Green, R. K., R. A. Gonzalez, and G. W. Poehlein, in *Emulsion Polymerization,* I. Piirma and J. L. Gardon (eds.), ACS Symp. Ser. 24, Washington, D. C., 1976.

85. Nomura, M. and M. Harada, in *Emulsion Polymers and Emulsion Polymerization,* D. R. Bossett and A. E. Hamielec (eds.), ACS Symp. Ser., 165, Washington, D. C., 1981.

86. Pollock, M. J., J. F. MacGregor, and A. E. Hamielec, *ACS Symp. Ser.,* 197, 209 (1981).

87. Barth, H. G., (ed.), *Modern Methods of Particle Size Analysis,* Wiley, New York, 1984.

88. Lieberman, A., *AIChE Symp. Ser.,* 78, 76 (1982).

89. Malihi, F. B., T. Provder, and M. E. Koehler, *J. Coatings Tech.,* 55, 41 (1983).

90. Weiner, B. B., in *Modern Methods of Particle Size Analysis,* Wiley, New York, 1984.

91. Biegler, L. T., *Comp. Chem. Eng.,* 8, 243 (1984).

92. Cuthrell, J. E. and L. T. Biegler, in *Process Systems Engineering PSE 85,* Inst. Chem. Engr. No. 92, 1985.

93. Economou, C. G. and M. Morari, Conf. on Decision and Control, Ft. Lauderdale, 1985.

94. Economou, C. G., M. Morari, and B. Palsson, *Ind. Eng. Chem. Process Des. Dev.*, 25, 403 (1986).

95. Morshedi, A. M., in *Chemical Process Control - CPC III*, M. Morari and T. J. McAvoy (eds.), Elsevier, Amsterdam, 1986.

Robust Control
An Overview and Some New Directions

Vasilios Manousiouthakis
Department of Chemical Engineering
UCLA
Los Angeles, CA 90024

October, 1986

I Introduction

In the last five years, robustness has become a central control research theme since it holds the premise of making control theory realistic and readily applicable. Most robustness studies have concentrated on the class of finite-dimensional linear time-invariant systems (FDLTIS). Both the system models and the allowable perturbations are assumed to belong to this class. The purpose of this paper is dual. First, to provide an overview of the current state of robustness studies for FDLTIS and then to present some new ideas for linear time-invariant system control (LTIS) and finite-dimensional linear time-varying system control (FDLTVS).

The structure of the rest of the paper is as follows: In section II, background information on fractional representation theory is provided. In Section III, a general control scheme is proposed which contains all other control schemes as special cases. A linear parameterization of all stabilizing controllers for this scheme is also provided. In sections IV and V, LTIS and FDVLTVS robustness is discussed. Finally, in section IV, conclusions are drawn and future research directions are outlined and in sections VII and VIII references and figures respectively are given.

II Mathematical Background

Fractional representation theory for linear systems is an input-output approach based upon the premise that single-input single-output linear systems can be considered as elements of an algebraic structure, namely a ring. The theory was pioneered by Youla, Bongiorno, and Lu [8] and Desoer and Chen [2] and is presented in a comprehensive book by Vidyasagar [7]. It has provided a plethora of results on feedback stabilization and H_∞ optimal controller syntheses. Most of the work has concentrated on the standard feedback configuration

213

while more recently the two degree of freedom compensator has been investigated [3]. A characterization of all such stabilizing compensators has been achieved. In the sequel a brief exposition of the theory is presented. Let

G: commutative ring with unity: SISO systems considered (proper).

H: commutative subring of **G** with the same unity: proper stable SISO systems considered.

I: monoid, subset of **H** containing units of G: proper stable SISO systems with inverse.

J: group, units of **H**: proper stable SISO systems with proper stable inverse.

Q: $\{xy^{-1} \mid x \in \mathbf{H}, y \in \mathbf{I}\}$: commutative subring of **G** isomorphic to the ring of quotients of **H** by **I** (for the rest of this work **Q** is identified with **G**).

M(A): set of matrices with entries in **A**.

G(H)-unimodular: square matrices in **M(H)** with determinant in **I(J)**.

Let also **N, D, Ñ, Ñ** $\in$ **M(H)**, **S, S′** $\in$ **M(G)**. Then the notions of left and right coprimeness (l.c., r.c.), left and right fractional representations (l.f.r., r.f.r.), left and right coprime fractional representations (l.c.f.r., r.c.f.r.) and left and right containment ($\subset, \supset$) are defined as follows:

$$(\tilde{\mathbf{D}}, \tilde{\mathbf{N}}) \text{ is l.c. } \Leftrightarrow \exists \tilde{\mathbf{U}}, \tilde{\mathbf{V}} \in \mathbf{M(H)} \text{ such that } \tilde{\mathbf{N}}\tilde{\mathbf{U}} + \tilde{\mathbf{D}}\tilde{\mathbf{V}} = \mathbf{I}$$

$$(\mathbf{N}, \mathbf{D}) \text{ is r.c. } \Leftrightarrow \exists \mathbf{U}, \mathbf{V} \in \mathbf{M(H)} \text{ such that } \mathbf{U}\mathbf{N} + \mathbf{V}\mathbf{D} = \mathbf{I}$$

$$(\tilde{\mathbf{D}}, \tilde{\mathbf{N}}) \text{ is an l.f.r. of } \mathbf{S} \Leftrightarrow \begin{bmatrix} \det \tilde{\mathbf{D}} \in \mathbf{I} \\ \mathbf{S} = \tilde{\mathbf{D}}^{-1}\tilde{\mathbf{N}} \end{bmatrix}$$

$$(\mathbf{N}, \mathbf{D}) \text{ is an r.f.r. of } \mathbf{S} \Leftrightarrow \begin{bmatrix} \det \mathbf{D} \in \mathbf{I} \\ \mathbf{S} = \mathbf{N}\mathbf{D}^{-1} \end{bmatrix}$$

$$(\tilde{\mathbf{D}}, \tilde{\mathbf{N}}) \text{ is an l.c.f.r. of } \mathbf{S} \Leftrightarrow \begin{bmatrix} \exists \tilde{\mathbf{U}}, \tilde{\mathbf{V}} \in \mathbf{M(H)} \text{ such that } \tilde{\mathbf{N}}\tilde{\mathbf{U}} + \tilde{\mathbf{D}}\tilde{\mathbf{V}} = \mathbf{I} \\ \det \tilde{\mathbf{D}} \in \mathbf{I} \\ \mathbf{S} = \tilde{\mathbf{D}}^{-1}\tilde{\mathbf{N}} \end{bmatrix}$$

$$
(\mathbf{N}, \mathbf{D}) \text{ is an r.c.f.r. of } \mathbf{S} \Leftrightarrow \left[\begin{array}{c} \exists \mathbf{U}, \mathbf{V} \in \mathbf{M}(\mathbf{H}) \text{ such that } \mathbf{UN} + \mathbf{VD} = \mathbf{I} \\ \det \mathbf{D} \in \mathbf{I} \\ \mathbf{S} = \mathbf{ND}^{-1} \end{array} \right]
$$

$$
\mathbf{S} \subset \mathbf{S}' \Leftrightarrow \exists \mathbf{X}, \mathbf{Y} \in \mathbf{M}(\mathbf{H}) \text{ such that } \mathbf{S} = \mathbf{X} + \mathbf{S}'\mathbf{Y}
$$

$$
\mathbf{S}' \supset \mathbf{S} \Leftrightarrow \exists \mathbf{X}, \mathbf{Y} \in \mathbf{M}(\mathbf{H}) \text{ such that } \mathbf{S} = \mathbf{X} + \mathbf{Y}\mathbf{S}'.
$$

III   The (F,P,L,C) Configuration and its Stability Properties

The notion of control is intimately related to the concepts of measurement and manipulation. In addition, a control scheme is simply a way to interconnect two systems, the plant and the controller. Thus, in order to identify the most general control scheme one need only classify the inputs (outputs) of each system in three classes: one that contains all inputs (outputs), one that contains all manipulated inputs, and one that contains all measured inputs outputs. Then the most general control scheme results from using the measured plant inputs and outputs as plant manipulated inputs. Since, however, a system's measured inputs can be considered as part of the measured outputs of an appropriately augmented system, the most general compensation scheme can be represented by configuration (I). To illustrate that configuration (II) (see Figure 1) can also depict the most general way to interconnect two systems, we show next that (I) is equivalent to (II).

$$
\bar{\mathbf{y}}_2 \equiv \left[\begin{array}{c} \mathbf{z} \\ \mathbf{y} \end{array} \right] = \begin{array}{c} n_1 \\ n_2 \end{array} \left[\begin{array}{cc} \overset{n_3}{\mathbf{P}_{11}} & \overset{n_4}{\mathbf{P}_{12}} \\ \mathbf{P}_{21} & \mathbf{P}_{22} \end{array} \right] \left[\begin{array}{c} \mathbf{w} \\ \mathbf{u} \end{array} \right] \equiv \mathbf{P}\bar{\mathbf{e}}_2, \quad \bar{\mathbf{u}}_2 = \left[\begin{array}{c} \mathbf{w} \\ \eta' \end{array} \right] \quad (1)
$$

$$
\bar{\mathbf{y}}_1 \equiv \left[\begin{array}{c} \mathbf{z}' \\ \mathbf{y}' \end{array} \right] = \begin{array}{c} n_5 \\ n_4 \end{array} \left[\begin{array}{cc} \overset{n_6}{\mathbf{C}_{11}} & \overset{n_2}{\mathbf{C}_{12}} \\ \mathbf{C}_{21} & \mathbf{C}_{22} \end{array} \right] \left[\begin{array}{c} \mathbf{w}' \\ \mathbf{u}' \end{array} \right] \equiv \mathbf{C}\bar{\mathbf{e}}_1, \quad \bar{\mathbf{u}}_1 = \left[\begin{array}{c} \mathbf{w}' \\ \eta \end{array} \right] \quad (2)
$$

$$
\bar{\mathbf{w}}_1 \equiv \left[\begin{array}{c} \mathbf{0} \\ \mathbf{y}' \end{array} \right] = \begin{array}{c} n_3 \\ n_4 \end{array} \left[\begin{array}{cc} \overset{n_5}{\mathbf{0}} & \overset{n_4}{\mathbf{0}} \\ \mathbf{0} & \mathbf{I} \end{array} \right] \left[\begin{array}{c} \mathbf{z}' \\ \mathbf{y}' \end{array} \right] \equiv \mathbf{L}\bar{\mathbf{y}}_1 \quad (3)
$$

$$
\bar{\mathbf{w}}_2 \equiv \left[\begin{array}{c} \mathbf{0} \\ \mathbf{y} \end{array} \right] = \begin{array}{c} n_6 \\ n_2 \end{array} \left[\begin{array}{cc} \overset{n_1}{\mathbf{0}} & \overset{n_2}{\mathbf{0}} \\ \mathbf{0} & \mathbf{I} \end{array} \right] \left[\begin{array}{c} \mathbf{z} \\ \mathbf{y} \end{array} \right] \equiv \mathbf{F}\bar{\mathbf{y}}_2 \quad (4)
$$

Based on this set of defintions the equivalence of (I) and (II) becomes apparent. In the sequel we will only be concerned with configuration (II).

It holds that

$$\bar{e} \equiv \begin{bmatrix} \bar{e}_1 \\ \bar{e}_2 \end{bmatrix} = \begin{bmatrix} \overset{n_6+n_2}{(I+FPLC)^{-1}} & \overset{n_3+n_4}{-FP\,(I+LCFP)^{-1}} \\ LC\,(I+FPLC)^{-1} & (I+LCFP)^{-1} \end{bmatrix} \begin{bmatrix} \bar{u}_1 \\ \bar{u}_2 \end{bmatrix} \equiv E\bar{u} \tag{5}$$

$$\bar{y} \equiv \begin{bmatrix} \bar{y}_1 \\ \bar{y}_2 \end{bmatrix} = \begin{bmatrix} C & 0 \\ 0 & P \end{bmatrix} E\bar{u} \equiv Y\bar{u} \tag{6}$$

$$\bar{w} \equiv \begin{bmatrix} LC & 0 \\ 0 & FP \end{bmatrix} E\bar{u} \equiv W\bar{u} \tag{7}$$

Then it is easy to verify that

$$W = B_1(E - I) = B_2 Y \tag{8}$$

$$E = I - B_1 W \tag{9}$$

where

$$B_1 \equiv \begin{bmatrix} \overset{n_6+n_2}{0} & \overset{n_3+n_4}{I} \\ -I & 0 \end{bmatrix} \qquad B_2 \equiv \begin{matrix} n_3 + n_4 \\ n_6 + n_2 \end{matrix} \begin{bmatrix} \overset{n_5+n_4}{L} & \overset{n_1+n_2}{0} \\ 0 & F \end{bmatrix} \tag{10}$$

Theorem

Let $(N_{P_{22}}, D_{P_{22}})$, $(\tilde{D}_{P_{22}}, \tilde{N}_{P_{22}})$ be any r.c.f.r. and l.c.f.r. of P_{22} and select $X_{22}, Y_{22}, \tilde{X}_{22}, \tilde{Y}_{22} \in M(H)$ so that

$$X_{22}N_{P_{22}} + Y_{22}D_{P_{22}} = I$$

$$\tilde{N}_{P_{22}}\tilde{X}_{22} + \tilde{D}_{P_{22}}\tilde{Y}_{22} = I$$

In addition, let

$$P_{21} \subset P_{22}$$

$$[P_{21}P_{22}] \supset [P_{11}P_{12}].$$

Define

$$C_{P_1} : \mathcal{D}(C_{P_1}) \to M(G)$$

$$C_{P_2} : \mathcal{D}(C_{P_2}) \to M(G), \text{where}$$

$$\mathcal{D}(C_{P_1}) = \{(R_1, R_2, R_3, R_4) \in M(H) \times M(H) \times M(H) \times M(H) :$$
$$\det(\tilde{Y}_{22} - N_{P_{22}} R_4) \in I\}$$

$$\mathcal{D}(C_{P_2}) = \{(Q_1, Q_2, Q_3, Q_4) \in M(H) \times M(H) \times M(H) \times M(H) :$$
$$\det(Y_{22} - Q_4 \tilde{N}_{P_{22}}) \in I\}$$

$$C_{P_1}(R_1, R_2, R_3, R_4) =$$
$$\begin{bmatrix} R_1 & R_2 \\ D_{P_{22}} R_3 & \tilde{X}_{22} + D_{P_{22}} R_4 \end{bmatrix} \begin{bmatrix} I & 0 \\ -N_{P_{22}} R_3 & \tilde{Y}_{22} - N_{P_{22}} R_4 \end{bmatrix}^{-1}$$

$$C_{P_2}(Q_1, Q_2, Q_3, Q_4) =$$
$$\begin{bmatrix} I & -Q_3 \tilde{N}_{P_{22}} \\ 0 & Y_{22} - Q_4 \tilde{N}_{P_{22}} \end{bmatrix}^{-1} \begin{bmatrix} Q_1 & Q_3 \tilde{D}_{P_{22}} \\ Q_2 & X_{22} + Q_4 \tilde{D}_{P_{22}} \end{bmatrix}$$

Then

1. C_{P_1} is a 1:1 map of $\mathcal{D}(C_{P_1})$ onto $S(P)$

2. C_{P_2} is a 1:1 map of $\mathcal{D}(C_{P_2})$ onto $S(P)$

When the open loop plant P is stable we have:

$$C_{P_1}(R_1, R_2, R_3, R_4) =$$
$$\begin{bmatrix} R_1 & R_2 \\ R_3 & R_4 \end{bmatrix} \begin{bmatrix} I & 0 \\ -P_{22} R_3 & I - P_{22} R_4 \end{bmatrix}^{-1}$$

$$C_{P_2}(Q_1, Q_2, Q_3, Q_4) =$$
$$\begin{bmatrix} I & -Q_3 P_{22} \\ 0 & I - Q_4 P_{22} \end{bmatrix}^{-1} \begin{bmatrix} Q_1 & Q_3 \\ Q_2 & Q_4 \end{bmatrix}$$

It becomes apparent that the above controller has **4 degrees of freedom**. The **1-degree of freedom** controller commonly used corresponds to $R_1 = 0$, $R_2 = 0$, $R_3 = 0$, or equivalently $Q_1 = 0$, $Q_2 = 0$, $Q_3 = 0$. The **2-degree of freedom** controller corresponds to $R_1 = 0$, $R_2 = 0$, or equivalently $Q_1 = 0$, $Q_3 = 0$.

IV   Robustness for Linear Time Invariant Systems

The results presented in this section are applicable to linear time- invariant systems (both finite and infinite dimensional). The rings $\mathbf{G}, \mathbf{H}$ used in Section II are identified with the algebras $\mathbf{B}, \hat{\mathbf{A}}_o$ respectively, proposed by Callier and Desoer [1]. More details on the implications of this correspondence are given in a forthcoming paper [5]. To study the robustness issue we must make these algebras topological rings. For the ring $\hat{\mathbf{A}}_o$ this can be achieved through the $\mathbf{H}_\infty$ norm defined as:

$$\parallel \mathbf{P} \parallel_\infty = \operatorname*{ess\,sup}_{\omega \in \mathbf{R}} \max_i \sigma_i \left[\mathbf{P}(j\omega) \right]$$

where $\mathbf{P} \in \hat{\mathbf{A}}_o$. Elements of the algebra $\mathbf{B}$ can only have a finite number of singularities in the closed right half plane and these are finite order poles. Therefore the Nyquist theorem (as extended in [4]) can be used to provide robustness results. Using additive and multiplicative types of perturbations and assuming that the number of RHP poles is the same for all perturbed systems, we have for the **1-degree of freedom feedback scheme.**

Additive Uncertainty:

$$\parallel \mathbf{P}(j\omega) - \mathbf{P}_o(j\omega) \parallel < |\delta(j\omega)| \ \forall \omega \in \mathbf{R}.$$

Then

$$(\mathbf{P}, \mathbf{C}) \text{ stable for all such } \mathbf{P} \Leftrightarrow$$

$$\parallel \mathbf{C} \left(\mathbf{I} + \mathbf{P}_o \mathbf{C} \right)^{-1} (j\omega) \parallel \cdot |\delta(j\omega)| \le 1 \ \forall \omega \in \mathbf{R}.$$

Multiplicative Uncertainty:

$$\mathbf{P}(s) = \left[\mathbf{I} + \mathbf{M}(s) \right] \mathbf{P}_o(s),$$

$$\parallel \mathbf{M}(j\omega) \parallel < |\delta(j\omega)| \ \forall \omega \in \mathbf{R}.$$

Then

$$(\mathbf{P}, \mathbf{C}) \text{ stable for all such } \mathbf{P} \Leftrightarrow$$

$$\parallel \mathbf{P}_o \mathbf{C} \left(\mathbf{I} + \mathbf{P}_o \mathbf{C} \right)^{-1} (j\omega) \parallel \cdot |\delta(j\omega)| \le 1 \ \forall \omega \in \mathbf{R}.$$

The above conditions can also be cast in an $\mathbf{H}_\infty$−norm form:

$$\parallel \mathbf{C} \left(\mathbf{I} + \mathbf{P}_o \mathbf{C} \right)^{-1} \delta \parallel_\infty \le 1$$

and

$$\parallel \mathbf{P}_o \mathbf{C} \left(\mathbf{I} + \mathbf{P}_o \mathbf{C} \right)^{-1} \delta \parallel_\infty \le 1.$$

To alleviate the constant number of RHP poles assumption, one can utilize feedback uncertainty representations.

Proposition 1

Feedback Uncertainty:

$$\mathbf{P} = \mathbf{P}_o - \mathbf{I} + (\mathbf{I} - \Delta)^{-1}, \ \Delta \text{ stable, } \| \Delta(j\omega) \| < \delta(j\omega) \ \forall \omega \in \mathbf{R}.$$

Then

$$(\mathbf{P}, \mathbf{C}) \text{ stable for all such } \mathbf{P} \Leftrightarrow$$

$$\left\| \left(\mathbf{I} - (\mathbf{I} + \mathbf{CP}_o)^{-1} \mathbf{C} \right) \delta \right\|_\infty \leq 1.$$

Proof

Define $\mathbf{w} = (\mathbf{I} - \Delta)^{-1} \Delta \mathbf{e}_2$, $\mathbf{z} = \mathbf{w} + \mathbf{e}_2$.
Then it holds that:

$$\mathbf{y}_2 = \mathbf{P}\mathbf{e}_2 = \left[\mathbf{P}_o - \mathbf{I} + (\mathbf{I} - \Delta)^{-1} \right] \mathbf{e}_2 = \mathbf{P}_o \mathbf{e}_2 + \mathbf{w}$$

$$\mathbf{e}_2 = \mathbf{u}_2 + \mathbf{C}\mathbf{e}_1 = \mathbf{u}_2 + \mathbf{C}\mathbf{u}_1 - \mathbf{C}\mathbf{P}_o \mathbf{e}_2 - \mathbf{C}\mathbf{w} \Rightarrow$$

$$\mathbf{e}_2 = (\mathbf{I} + \mathbf{CP}_o)^{-1} \mathbf{C}\mathbf{u}_1 + (\mathbf{I} + \mathbf{CP}_o)^{-1} \mathbf{u}_2 - (\mathbf{I} + \mathbf{CP}_o)^{-1} \mathbf{C}\mathbf{w}$$

$$\mathbf{e}_1 = \mathbf{u}_1 - \mathbf{P}_o \mathbf{e}_2 - \mathbf{w} = \left[\mathbf{I} - \mathbf{P}_o (\mathbf{I} + \mathbf{CP}_o)^{-1} \mathbf{C} \right] \mathbf{u}_1$$

$$-\mathbf{P}_o (\mathbf{I} + \mathbf{CP}_o)^{-1} \mathbf{u}_2 + \left[\mathbf{P}_o (\mathbf{I} + \mathbf{CP}_o)^{-1} \mathbf{C} - \mathbf{I} \right] \mathbf{w}.$$

$$\mathbf{y}_1 = \mathbf{C}\mathbf{e}_1.$$

It then holds that:

$$\begin{bmatrix} \mathbf{e}_1 \\ \mathbf{e}_2 \\ \mathbf{z} \end{bmatrix} =$$

$$\begin{bmatrix} \mathbf{I} - \mathbf{P}_o(\mathbf{I} + \mathbf{CP}_o)^{-1}\mathbf{C} & -\mathbf{P}_o(\mathbf{I} + \mathbf{CP}_o)^{-1} & \mathbf{P}_o(\mathbf{I} + \mathbf{CP}_o)^{-1}\mathbf{C} - \mathbf{I} \\ (\mathbf{I} + \mathbf{CP}_o)^{-1}\mathbf{C} & (\mathbf{I} + \mathbf{CP}_o)^{-1} & -(\mathbf{I} + \mathbf{CP}_o)^{-1}\mathbf{C} \\ (\mathbf{I} + \mathbf{CP}_o)^{-1}\mathbf{C} & (\mathbf{I} + \mathbf{CP}_o)^{-1} & \mathbf{I} - (\mathbf{I} + \mathbf{CP}_o)^{-1}\mathbf{C} \end{bmatrix} \begin{bmatrix} \mathbf{u}_1 \\ \mathbf{u}_2 \\ \mathbf{w} \end{bmatrix}$$

$$\mathbf{w} = \Delta \mathbf{z}$$

Let

$$\begin{bmatrix} \mathbf{e}_1 \\ \mathbf{e}_2 \end{bmatrix} \equiv \mathbf{e}, \ \begin{bmatrix} \mathbf{u}_1 \\ \mathbf{u}_2 \end{bmatrix} \equiv \mathbf{u}$$

and partition the above matrix as indicated below:

$$\begin{bmatrix} \mathbf{e} \\ \mathbf{z} \end{bmatrix} = \begin{bmatrix} \mathbf{R} & \mathbf{S} \\ \mathbf{T} & \mathbf{U} \end{bmatrix} \begin{bmatrix} \mathbf{u} \\ \mathbf{w} \end{bmatrix}$$

By substituting $\mathbf{w} = \Delta \mathbf{z}$ we eventually have:

$$\mathbf{e} = \left[\mathbf{R} + \mathbf{S}\Delta \left(\mathbf{I} - \mathbf{U}\Delta \right)^{-1} \mathbf{T} \right] \mathbf{u}, \; \mathbf{z} = \left(\mathbf{I} - \mathbf{U}\Delta \right)^{-1} \mathbf{T}\mathbf{u}$$

where $\mathbf{R}, \mathbf{S}, \mathbf{T}, \mathbf{U}$ are stable (since the nominal closed loop system is stable) and Δ is chosen to be stable. The signals $\mathbf{e}, \mathbf{z}$ will be stable iff $(\mathbf{I} - \mathbf{U}\Delta)^{-1}\mathbf{T}$ is stable. $\Leftrightarrow$

$$\left[\mathbf{I} - \left(\mathbf{I} - (\mathbf{I} + \mathbf{CP}_o)^{-1}\mathbf{C} \right) \Delta \right]^{-1} \left[\; (\mathbf{I} + \mathbf{CP}_o)^{-1}\mathbf{C} \quad (\mathbf{I} + \mathbf{CP}_o)^{-1} \; \right]$$

is stable. Since this must hold for all Δ satisfying $\| \, \Delta(j\omega) \, \| < |\delta(j\omega)|$ we have that stability is guaranteed iff $(\mathbf{I} - \mathbf{U}\Delta)^{-1}$ is stable for all these Δ, which in turn is equivalent to

$$\| \left(\mathbf{I} - (\mathbf{I} + \mathbf{CP}_o)^{-1}\mathbf{C} \right) \delta \|_{\infty} \leq 1 \quad \text{Q.E.D.}$$

Due to space limitations we do not endeavor into robust controller synthesis at this point. We instead concentrate our efforts on time-varying systems.

V Robustness for Linear Time-Varying Systems

The fractional representational theory has been presented in Section II for the case of commutative rings. We are interested in the set of operators whose states space representation is

$$\dot{\mathbf{x}}(t) = \mathbf{A}(t)\mathbf{x}(t) + \mathbf{B}(t)\mathbf{u}(t)$$

$$\mathbf{y}(t) = \mathbf{C}(t)\mathbf{x}(t) + \mathbf{D}(t)\mathbf{u}(t)$$

This type of system description arises often in chemical process control as the result of a nonlinear system's linearization around a time-dependent trajectory. However the set of SISO systems of this form, when equipped with map composition and addition, forms a **noncommutative** rather than **commutative** ring. The implications of this fact are discussed in a forthcoming paper [6]. In this work we try to quantify the robustness issue. Again we must make the rings $\mathbf{G}, \mathbf{H}$ topological and therefore we use for the ring $\mathbf{H}$ the $\mathbf{L}_{\infty}$-norm defined as

$$\| \, \mathbf{P} \, \|_{\infty} = \sup_{t \in \mathbf{R}} \int_{-\infty}^{\infty} \| \, \mathbf{C}(t)\mathbf{Q}(t,\tau)\mathbf{B}(\tau) + \mathbf{D}(t)\delta(t - \tau) \, \| \, d\tau$$

where $\| \cdot \|$ indicates norm on $\mathbf{R}^{m \times m}$ and $\delta(\cdot)$ is a delta function. Then the following robustness result holds:

Proposition 2

Feedback Uncertainty:

$$P = P_o - I + (I - \Delta)^{-1}, \; \| \Delta \| < \delta.$$

Then

$$(P, C) \text{ stable for all such } P \Leftrightarrow$$

$$\| \left(I - (I + CP_o)^{-1} C \right) \delta \|_\infty \leq 1.$$

Proof

Exactly as in the proof of Proposition 1 we have that P, C stable $\Leftrightarrow$

$$\left[I - \left(I - (I + CP_o)^{-1} C \right) \Delta \right]^{-1} \text{ stable } \forall \Delta \text{ s.t. } \| \Delta \|_\infty < \delta.$$

From the small gain property of the algebra of time-varying systems, we have that

$$\| \left(I - (I + CP_o)^{-1} C \right) \delta \|_\infty \leq 1 \Leftrightarrow \| \left(I - (I + CP_o)^{-1} C \right) \|_\infty \cdot \delta \leq 1$$

$$\Rightarrow \| \left(I - (I + CP_o)^{-1} C \right) \|_\infty \cdot \| \Delta \|_\infty \leq 1 \Rightarrow$$

$$\left[I - \left(I - (I + CP_o)^{-1} C \right) \Delta \right]^{-1} \text{ stable.}$$

VI   Discussion-Conclusions

The importance of Proposition 1 is that it makes obsolete the constant of RHP poles assumption. The important of Proposition 2 is that it provides tools for robustness analysis of time-varying systems. Some significant research areas open to investigation are:

1. Establishment of necessity in proposition 2.

2. Controller synthesis using $H_\infty(L_\infty)$ optimization.

3. Use of the four degrees of freedom in the F, P, L, C scheme to improve overall system properties.

Acknowledgments

Discussions with M. Morari, E. Zafiriou, S. Skogestad and financial assistance from the Shell Corporation are gratefully acknowledged.

References

[1] Callier, F. M. and Desoer, C. A., "An algebra of transfer functions of distributed linear time-invariant systems," *IEEE Trans. Circ. Sys.,* CAS-25, pp. 651-662, 1978.

[2] Desoer, C. A. and Chan, W. S., "The Feedback Interconnection of Lumped Linear Time-Invariant Systems," *J. Franklin Inst.,* Vol. 30, pp. 335-351, 1975.

[3] Desoer, C. A. and Gustafson, C. L., "Algebraic theory of linear multivariable feedback systems," *IEEE Trans. Auto. Control,* AC-29, pp. 909-917, 1984.

[4] Desoer, C. A. and Wang, Y. T., "On the Generalized Nyquist Stability Criterior," *IEEE Trans. Auto. Control,* Vol. AC-25, No. 1, 1980.

[5] Manousiouthakis, V., "4-degree of freedom control for distributed parameter systems," in preparation.

[6] Manousiouthakis, V., "Time Varying Control: An alternative to the fictitious 'fixed versus adaptive' dilemma," in preparation.

[7] Vidyasagar, M., *Control System Synthesis: A Factorization Approach,* MIT Press, 1985.

[8] Youla, D., J. Bongiorno, and C. Lu, "Single-loop Feedback Stabilization of Linear Multivariable Plants," *Automatica,* Vol. 10, pp. 159-173, 1974.

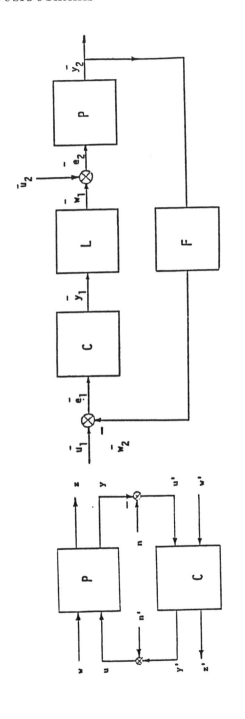

I.   (P,C) general configuration   II.   (F,P,L,C) general configuration

Figure 1.

An Overview of Nonlinear Geometrical Methods for Process Control *

Jeffrey C. Kantor
Department of Chemical Engineering
University of Notre Dame
Notre Dame, IN 46556

December 15-18, 1986

Abstract

Differential geometry offers a wide range of tools for analyzing non-linear dynamical systems. Recent developments have included conditions for nonlinear controllability and reachability, stability, the development of exact finite dimensional nonlinear filters, and the feedback transformation of nonlinear systems to linear systems. Generally speaking, the practical application of these results are often limited to restricted classes of nonlinear systems. Nevertheless, many problems in chemical process control may be amenable to these approaches. This paper presents a overview of some areas of application in process control, particularly with regard to nonlinear controllability, linearizing feedback transformations, and the construction of stable inverses. The notion of a 'distinguished process output' is used to provide a unifying viewpoint. Some areas of opportunity as well as potential limitations are described.

1 Introduction

It is a trite comment that one of the outstanding problems in control theory is the development of analytical tools for nonlinear systems. Nowhere is this more evident then in chemical process control where nonlinearities are an intrinsic part of essentially every device and model. While it is probably pointless to attempt a catalog, one should at least mention nonlinearities deriving from such physical phenomena as thermodynamics, reaction kinetics, and transport processes. In fact, a large class of process control problems involve the manipulation of extensive flows to regulate intensive quantities such as temperature, pressure, or concentration. A dimensional argument reveals a common underlying source of nonlinearities in such problems.

*This work was supported by the National Science Foundation through an NSF-PYI award CBT 84-51592, and by matching support from the Shell Development Company and the Shell Companies Foundation.

Nonlinear process control is important when the assumptions regarding the application of linear control are not valid. This most likely to occur when a process is subject to perturbations from steady state that are large with respect to the 'magnitude' of the nonlinearities present, or else when a process in not operated near a steady state or is operated over a wide range of conditions. These situations often occur in batch processing and in the control of reaction systems.

The last decade has seen intense research focussed on nonlinear dynamics. In part, this is due to the spectacular results of bifurcation and stability theory. In the last year, even the *New York Times* and the *Chicago Tribune* have run articles in their Sunday magazines on chaos and fractals. Though not nearly as well developed, nonlinear control theory is also undergoing rapid development, particularly with regard to the application of geometrical methods [1].

Geometrical methods are a basis for understanding the structural charac- teristics of nonlinear models. The word geometry carries two connotations. One is that the results obtained are independent of a coordinate system, i.e., the manner in which model variables are identified. In the spirit of Euclid, notions like orthogonality and integrability do not depend on the coordinates used to present a particular model. The familiar control concepts of stabil- ity, observability, controllability and reachability are examples of geometric properties. Completely coordinate-free analyses are possible, but, however, are also quite abstract and removed from normal engineering practice. Fortu- nately, this level of abstraction is not needed in most applications.

The second connotation of geometry is the absence of a metric. Results are often obtained without regard to a norm or an underlying topology. Thus geometric results are sometimes difficult to apply in practice. For example, the vitally important question of robustness is unanswered in many instances. This cautionary note must be carefully observed. In this regard, the reader should consult the recent thesis of C. Economu, or the review of Desoer and Wang, for examples of topological approaches to nonlinear control synthesis [11,12].

The purpose of this paper is present an overview of some selected results regarding the transformation of nonlinear dynamical systems through static state feedback. Discussed is the geometric construction of nonlinear inverses, stable and unstable inversion, exact linearization, and the significance of these results in some process control problems. It will be quite clear that a compre- hensive strategy for nonlinear control synthesis does not exist. But it should also be evident that geometrical methods provide key insights into the nature of the nonlinear control problem, and to the structure of specific process mod- els. Typical types of problems that can be addressed are the determination of appropriate measurements, selection of control variables, and calculation of nonlinear static feedback control laws.

The plan of the paper is to first introduce the notion of feedback lineariza-

---

[1]Casti's recent review is an accessible introduction to the literature [10].

tion for a very simple example. Subsequently, linear closed-loop models for the transformed systems are discussed, and it is pointed out that the companion form is a suitable target for feedback linearization. This is in contrast with the developments in the literature where the Brunovsky form is typically used. The companion form is much more suited to applications by providing a clean interpretation of a transformed input as a setpoint for a 'distinguished process output.' For example, steady-state tracking is demonstrated.

The Hirschorn construction of a nonlinear inverse is introduced as a preliminary to the development of a linearizing feedback by Su's approach. The notion of a 'distinguished process output' is useful. Hirschorn's inverse defines a relative order for a given output, whereas Su's linearization method establishes conditions for the existence of an output with relative order equal to the order of the process.

Partial linearization is probably all that can be accomplished for most models of chemical processes. By relaxing the requirement for linearization with respect to all states, one can greatly broaden the class of systems that can be treated. The key is to introduce an additional constraint that assures closed loop stability. Moreover, it may be possible to introduce integrable constraints on measurable state variables. Some preliminary results are described in Section 5.

The paper concludes with a summary of several process control applications in which geometrical methods may be appropriate. Considered are some issues regarding the involutivity constraints for exact linearization and examples of measurement and control selection. The examples illustrate some of the practical issues regarding static feedback transformations for nonlinear process systems.

2   Feedback Linearization

The continuous dynamical models encountered in process systems typically have a *control linear* structure. In process systems, the manipulable control variable is typically an extensive flowrate, whereas the state variables include intensive quantities such as temperature, pressure, or concentration. A standard representation for a single-input, control-linear system is of the form

$$\frac{dx}{dt} = f(x) + ug(x) \tag{1}$$

where x is a vector of state variables, and u is a manipulable input. Where necessary, it will be assumed that $f(x)$ and $g(x)$ are smooth [2] functions of the state.

In many examples, $g(x)$ has the same units as x. Control variable u then plays the role of a variable rate coefficient with units of inverse time. Typically u will be a ratio of an extensive flowrate to a characteristic process volume.

[2] i.e., infinitely differentiable, C^∞ functions $f : \Re^n \mapsto \Re^n$ and $g : \Re^n \mapsto \Re^n$

A control linear static state feedback is written as

$$u = p(x) + vq(x) \tag{2}$$

where $p(x)$ and $q(x)$ are smooth scalar functions of the state variables, and v is an external reference input. In linear systems, linear state feedback yields a closed-loop linear system. Similarly, control-linear feedback for Equation 1 yields a control linear closed-loop system

$$\frac{dx}{dt} = [f(x) + p(x)g(x)] + v[q(x)g(x)] \tag{3}$$

The set of control-linear models is closed under the operation of control linear feedback. This is not a property shared by all classes of system models. Bilinear systems, for example, are not closed under either linear or bilinear feedback.

For linear systems, the problem is to design a linear feedback control law to achieve desirable linear closed-loop properties. The nonlinear case is more challenging since the control might be used to alter the algebraic structure of the closed-loop model as well as the dynamics. In particular, the closed-loop model of Equation 3 can be made linear under certain restrictive conditions on $f(x)$ and $g(x)$.

Feedback linearization is, in fact, a standard tool in many process control situations. The following example illustrates this point by computing is feedback linearization for level control in a tank. The result is familiar to control practioners as a gain-scheduling solution. A careful reading of Shinsky's books is a good source for many other examples of feedback linearization that are based on intuition or experience [23,24]. Georgakis has recently described an approach he calls 'extensive variable control' that may be regarded as an informal or empirical approach to the same problem[13]. What is presented in the sequel is the analytical generalization of this idea.

2.1   A Simple Example — Level Control

Consider the problem of synthesizing a controller that is to regulate the level of a fluid in a tank of constant cross-sectional area.[22] The inlet flow to the tank is assumed to be unregulated. Control is done by manipulating a valve located in the exit stream. A model for the liquid level is then

$$\frac{dh}{dt} = F_{in} - \frac{k}{A} u\sqrt{h} \tag{4}$$

where h is the liquid height, A is the cross-sectional area, and k is a valve constant. The control variable u corresponds to the valve stem position.

A static control law can be constructed that linearizes the response of h to a control input. That is, we can define a rule for computing u which brings Equation 4 to the form

$$\frac{dh}{dt} = F_{in} - ah + v \tag{5}$$

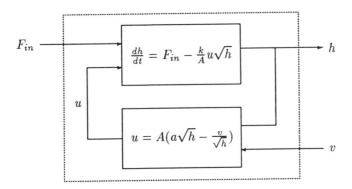

Figure 1: An example of feedback linearization for the level control of a tank. The static feedback control linearizes the response of h to changes in disturbance input F_{in} and to the external input v. The closed loop dynamics behave according to $\frac{dh}{dt} = F_{in} - ah + v$

where a is a chosen time constant, and where v is an external control input. Comparison of Equations 4 and 5 yield the required control law

$$u = A(a\sqrt{h} - \frac{v}{\sqrt{h}}). \tag{6}$$

Note that this control, like the model Equation 4, is not well-defined for $h = 0$.

This example illustrates two features that are characteristic of most solutions to the exact linearization problem. The physical control signal u is computed as function of the measured process state which, in this case, is simply the liquid level. State feedback of this type is a feature of the existing approaches to exact linearization. In more difficult examples, the dependence on process state variables can become algebraically complex. Moreover, on-line estimation of unmeasured process variables may be required to implement the control solution.

A second feature is a new control input, v, that has been defined with a simple dynamical relationship to the process state. Here v specifies the rate of change of liquid level. The input v represents a transformation of the physical control variable v that would serve as a basis for a secondary cascade control that would couple the tank to other process systems, and that would be suitable for use over wide range of operating conditions.

2.2 Closed-Loop Models

The conditions for which a nonlinear system admits a feedback linearization are of two types. The first are a class of controllability or reachability conditions that assure a nonsingular solution of the control law. The nonlinear controllability conditions are analogous, at least locally, to the controllability conditions for a linear system. A second type of condition is geometric in character, distinguishing between those models that are similar to a linear system and those which are not. In this sense, the phrase *similar to a linear system* means that the nonlinear model can be made linear by a nonlinear static state feedback and nonsingular transformations of the state and control variables. While the controllability conditions assure that the required control law is nonsingular, it is the second class of criteria which establishes the essential qualities of a nonlinear model that determine linearizability.

Consider a single-input model in the control linear form of Equation 1. A 'distinguished process output' $y = h(x)$ is defined whose dynamics are to be linearized with respect to an external input. That is, we seek to construct a scalar input v, and a vector ξ of α state variables such that

$$\frac{d\xi}{dt} = A\xi + bv \tag{7}$$
$$y = cz \tag{8}$$

where matrices A, b, c describe a controllable linear system of order α.

A standard result is that single-input controllable linear systems can be brought to the *Brunovsky canonical form*

$$\frac{d\xi}{dt} = \begin{bmatrix} 0 & 1 & 0 & \dots & 0 \\ 0 & 0 & 1 & \dots & 0 \\ \vdots & \vdots & \vdots & \dots & \vdots \\ 0 & 0 & 0 & \dots & 1 \\ 0 & 0 & 0 & \dots & 0 \end{bmatrix} \xi + \begin{bmatrix} 0 \\ 0 \\ \vdots \\ 0 \\ 1 \end{bmatrix} v \tag{9}$$

$$y = \begin{bmatrix} 1 & 0 & 0 & \dots & 0 \end{bmatrix} \xi \tag{10}$$

through a linear change of coordinates [21]. Thus if one can bring a nonlinear system to the Brunovsky form through static feedback and coordinate changes, then a subsequent linear coordinate change and linear static feedback can bring that nonlinear system to any controllable linear form. The converse also holds. Therefore, a given nonlinear system that is feedback equivalent to the Brunovsky canonical form is feedback equivalent to any controllable linear system [20,25].

The Brunovsky canonical form is essentially a chain of α integrators for which v is a transformed input. The distinguished output y satisfies the differential equation

$$\frac{d^\alpha y}{dt^\alpha} = v. \tag{11}$$

Because of its simplicity, this form has been the commonly used basis for establishing basic linearization results.

The subsequent development will show that there is no loss of simplicity if the companion form [21] of a controllable single-input linear system is used instead of the Brunovsky form. In this case, the target for the closed-loop linear model is given as

$$\frac{d\xi}{dt} = \begin{bmatrix} 0 & 1 & 0 & \cdots & 0 \\ 0 & 0 & 1 & \cdots & 0 \\ \vdots & \vdots & \vdots & \ddots & \vdots \\ 0 & 0 & 0 & \cdots & 1 \\ k_1 & k_2 & k_3 & \cdots & k_\alpha \end{bmatrix} \xi + \begin{bmatrix} 0 \\ 0 \\ \vdots \\ 0 \\ k_1 \end{bmatrix} r \qquad (12)$$

$$y = \begin{bmatrix} 1 & 0 & 0 & \cdots & 0 \end{bmatrix} \xi \qquad (13)$$

where the constants $k_1, k_2, \ldots k_\alpha$ are used to specify the closed-loop dynamics. The distinguished output y satisfies the differential equation

$$\frac{d^\alpha y}{dt^\alpha} + k_\alpha \frac{d^{\alpha-1} y}{dt^{\alpha-1}} + k_{\alpha-1} \frac{d^{\alpha-2} y}{dt^{\alpha-2}} + \cdots + k_2 \frac{dy}{dt} + k_1 y = k_1 r. \qquad (14)$$

There are two advantages to the companion form as a basis for feedback linearization. At steady state, $y = r$, so that the external input r serves as a setpoint for the distinguished output y. r parameterizes the steady state directly in terms of the 'distinguished output.' The Brunovsky form does not provide for a straightforward parameterization of the steady-state in terms of y, rather it is indirectly specified as an α Dirac delta function. Secondly, the implementation of the resulting static control law can be done in one step through choice of the coefficients $k_1, k_2, \ldots k_\alpha$. The Brunovsky form requires a two step implementation in which linear state feedback is attempted as a secondary control about a lower level nonlinear control.

2.3   The Distinguished Process Output

It is convenient in this discussion to talk about a distinguished scalar output $y = h(x)$. In fact, y may be a particular process output of either economic or operational interest, perhaps a key temperature or composition variable. In this context, one asks whether there exists a feasible feedback control for which that output responds in a linear fashion. This is a familiar problem in output feedback.

However, a broader class of problems presents itself if one asks what output variables can be defined which admit a linearizing feedback. This is an important question since it identifies those combinations of process variables for which the closed-loop dynamics are particularly simple. One is asking, in other words, what outputs should be measured in order to establish effective control over a wide range of operating conditions

The problem of selecting appropriate measurement and control variables is perhaps the key to solving many problems in process control. Multivariable control problems, for example, are often simplified by chosing appropriate measurements and controls. This is a basis for such familiar strategies as ratio

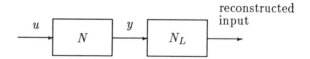

Figure 2: Schematic definition of a left inverse.

Figure 3: Schematic definition of a right inverse.

control for furnaces, inventory or energy controls for distillation columns, or using derived quantities like conversion, selectivity, or separation as setpoint variables in other applications. The utility of the derived variables is that they may respond in a simpler, noninteracting manner, thereby simplifying the multivariable control problem.

In the context of the nonlinear problems of this paper, we should expect the analytical development to provide a guide for systematically selecting those process variables for which a process presents itself in a dynamically simple manner. Simple means linear. Linear models are simple in the sense that a global analysis is the same as a local analysis. The goal of the subsequent analysis, therefore, is to identify those properties of a particular measurement which admit dynamical simplification as a linear closed-loop system.

3   Linearization by Inversion

An inverse of a nonlinear single-input, single-output system can be defined in two ways. Let N refer to a given nonlinear system. The *left inverse* is a dynamical system N_L, which given the history of outputs of N, reconstructs the input history. Conversely, the *right inverse* N_R produces the input history required to effect a particular output for N. These situations are shown schematically in Figures 2 and 3. The *left* and *right* terminology stem from the order in which the inverse operators would appear in composition with N, as opposed to their relative positions in the block diagram.

An exact inverse cannot be constructed using causal elements. The best that can be done is for the composition of a system and its inverse to respond as a series of integrations. The least possible number of integrators is defined as the relative order of the nonlinear system.

As in the case for linear systems, the nonlinear inverse is primarily a conceptual tool. It may be used to define a notion analogous to non-minimum phase linear systems. In the following discussion, feedback linearization is presented as an extreme case of nonlinear inversion in which outputs are constructed such that relative order is equal to the system order. The practical application of inverses, however, is severely constrained by closed-loop sensitivity and stability issues.

3.1 The Lie Derivative

It is convenient to introduce a special notation for the directional derivative of a scalar function $h(x)$ with respect to a vector field $f(x)$:

$$L_f h = \sum_{i=1}^{n} f_i(x) \frac{\partial h}{\partial x_i} \qquad (15)$$

The utility of this notation is that $L_f h$ is linear with respect to the scalar function h and to the vector field f. That is, $L_f h$ satisfies the identity

$$L_f(c_1 h_1 + c_2 h_2) = c_1 L_f h_1 + c_2 L_f h_2 \qquad (16)$$

where c_1 and c_2 are scalar constants, and

$$L_{a_1(x)f_1 + a_2(x)f_2} h = a_1(x) L_{f_1} h + a_2(x) L_{f_2} h \qquad (17)$$

where $a_1(x)$ and $a_2(x)$ are scalar functions of the variables $x_1, x_2, \ldots, x_n$ [3]. The notation L_f refers to an operator that acts on a scalar function. $L_f h$ is referred to as the *Lie derivative* of h.

3.2 Hirshorn's Construction of a Nonlinear Inverse

The constraints on the existence of local input-output linearization are found by applying a variation of Hirschorn's method for constructing a nonlinear inverse [15,16] for the distinguished output $y = h(x)$.

For a process model in the control-linear form of Equation 1, the rate of change of a distinguished output $y = h(x)$ is computed as

$$\frac{dy}{dt} = L_f h + u L_g h \qquad (18)$$

If $L_g h \not\equiv 0$, then $\alpha = 1$ and the control rule which linearizes the input-output response to order 1 is obtained by solving

$$L_f h + u L_g h = k_1(r - h(x)) \qquad (19)$$

for u.

If, on the other hand, $L_g h \equiv 0$, then a second derivative y is computed as

$$\begin{aligned} \frac{d^2 y}{dt^2} &= L_f L_f h + u L_g L_f h \\ &= L_f^2 h + u L_g L_f h \end{aligned} \qquad (20)$$

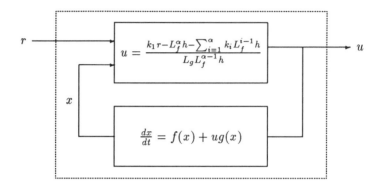

Figure 4: An implementation of a nonlinear inverse using Equation 22.

Repeating the process, if $L_g L_f h \not\equiv 0$, then $\alpha = 2$, and the linearizing control rule is given by the solution of

$$L_f^2 h + u L_g L_f h = k_1(r - h(x)) - k_2 L_f h \qquad (21)$$

If $L_g L_f h \equiv 0$, this program of differentiation is repeated until $L_g L_f^{\alpha-1} h \not\equiv 0$ for some $\alpha \leq n - 1$. The required control is given by

$$u = \frac{k_1 r - L_f^\alpha h - \sum_{i=1}^{\alpha} k_i L_f^{i-1} h}{L_g L_f^{\alpha-1} h} \qquad (22)$$

Comparing to Equation 81, this control law puts the closed-loop system into companion form where the variables $y_1, y_2, \ldots, y_\alpha$ are defined as

$$
\begin{aligned}
y_1 &= h \\
y_2 &= L_f h \\
&\;\;\vdots \\
y_\alpha &= L_f^{\alpha-1} h
\end{aligned}
\qquad (23)
$$

and where the distinguished output satisfies the α order differential Equation 14.

A nonlinear right or left inverse can be implemented by using Equation 22 as state feedback control applied to a process model. This is shown schematically in Figure 4. While the inverse is conceptually easy to construct, there are a number of reasons why it cannot be used directly for feedback control.

3.3   Stability of the Inverse and the 'Zero Dynamics'

By analogy with standard results for linear systems, Byrnes and Isidori suggested that the nonlinear inverse could be used to define the zero dynamics of a nonlinear system [8,9]. The right inverse describes an ideal feedforward controller. If a given nonlinear system has a stable inverse, then an approximation of the inverse might be used as part of a model reference feedback controller with potentially good results.

The zero dynamics are defined as the right inverse with the reference input set to zero. Using the above derivations, this is given by

$$\frac{dx}{dt} = f(x) - \frac{L_f^\alpha h + \sum_{i=1}^\alpha k_i L_f^{i-1} h}{L_f L_f^{\alpha-1} h} g \qquad (24)$$

By design, this n^{th} order system has an $n - \alpha$ order invariant manifold [14] defined by $h = L_f h = L_f^2 h = \cdots = L_f^{\alpha-1} h = 0$.

The Center Manifold theorem shows that stability in the neighborhood of the steady-state is determined by stability on the invariant manifold. The stability of the zero dynamics are determined by the stability of the nth order system

$$\frac{dx}{dt} = f(x) - \frac{L_f^\alpha h}{L_f L_f^{\alpha-1} h} g \qquad (25)$$

as it evolves on an $n - \alpha$ manifold embedded in $\Re^n$. As a consequence, inverse stability is an intrinsic system property that is independent of the particular control constants $k_1, k_2, \ldots, k_\alpha$.

4 Exact Local Linearization

4.1 Su's Construction of a Maximal Order Inverse

A special case is obtained if the relative order is equal to the order of the nonlinear system, that is $\alpha = n$. Equation 23 then describes a set of n coordinate transformations defining a new set of the state variables $y_1, y_2, \ldots, y_n$ in terms of the original set $x_1, x_2, \ldots, x_n$ according to the rule $y_i = L_f^{i-1} h$. Equation 22 completes the transformation by defining a new input, r, in terms of u and the original state variables. The effect of implementing the input transformation as a feedback control law is to cause the original nonlinear system to behave linearly with respect to the transformed input and state variables. It is clearly necessary that the state and input transformations be nonsingular for this to be a successful approach.

Thus we wish to find the class of distinguished outputs for which the relative order is n. The contribution Su made in 1982 was to provide local conditions for the existence of a scalar function $h(x)$ defining such outputs. The conditions are expressed as differential relations and are therefore local in nature [25]. Hunt, Su, and Meyer latter extended the results to find global criteria [20]. The question of global versus local criteria has been a topic of continuing discussion. Below is a version of Su's original approach, with some additional comments regarding the global extensions.

Examining the development of Equation 22, it is evident that the relative order is n only if

$$
\begin{aligned}
L_g h &= 0 \\
L_g L_f h &= 0 \\
&\vdots \\
L_g L_f^{n-2} h &= 0.
\end{aligned}
\tag{26}
$$

These equations form a set of $n - 1$ constraints on the class of distinguished outputs for which the relative order is n. In addition, the nonsingularity of the input transformation requires $L_g L_f^{n-1} \neq 0$.

4.2   The Lie Bracket

The Lie derivative $L_f h$ is the result of a first-order differential operator acting on the scalar function $h(x)$, where L_f is defined as

$$
L_f = \sum_{i=1}^{n} f_i(x) \frac{\partial}{\partial x_i}.
\tag{27}
$$

The composition of two operators, such as $L_g L_f$, is then second order. A calculation gives the explicit formula

$$
L_g L_f = \sum_{j=1}^{n} \sum_{i=1}^{n} \left(g_j f_i \frac{\partial^2}{\partial x_j \partial x_i} + g_j \frac{\partial f_i}{\partial x_j} \frac{\partial}{\partial x_i} \right).
\tag{28}
$$

Due to the second term inside the summation, the second order operator does not commute with respect to f and g, that is $L_g L_f \neq L_f L_g$. The difference, however, is a first-order operator and is denoted by the special notation

$$
\begin{aligned}
L_{[f,g]} &= L_f L_g - L_g L_f \tag{29} \\
&= \sum_{i=1}^{n} \left(\sum_{j=1}^{n} \left(f_j \frac{\partial g_i}{\partial x_j} - g_j \frac{\partial f_i}{\partial x_j} \right) \right) \frac{\partial}{\partial x_i}. \tag{30}
\end{aligned}
$$

$L_{[f,g]}$ describes the operation of taken a derivative in the direction of the vector $[f, g]$. The vector quantity $[f, g]$ is described as the *Lie bracket* of f and g. [3] The ith component is given explicitly by the formula

$$
([f,g])_i = \sum_{j=1}^{n} \left(f_j \frac{\partial g_i}{\partial x_j} - g_j \frac{\partial f_i}{\partial x_j} \right)
\tag{31}
$$

Letting f_x and g_x denote the Jacobians of f and g, a more succinct expression is $[f, g] = g_x f - f_x g$. [4]

---

[3] Alternatively, the *Poisson bracket* or the *commutator*.

[4] Some authors occasionally perform the differentiations in opposite order with a consequent change of sign. Beware!

The Lie bracket has several useful algebraic properties. More or less obvious is the linearity with respect to either f or g. For the constants c_1 and c_2,

$$[c_1 f_1 + c_2 f_2, g] = c_1 [f_1, g] + c_2 [f_2, g] \tag{32}$$
$$[f, c_1 g_1 + c_2 g_2] = c_1 [f, g_1] + c_2 [f, g_2]. \tag{33}$$

The Lie bracket is also *anti-symmetric* with $[f, g] = -[g, f]$.

A less obvious relationship is *Jacobi's identity*. Given three vector functions f_1, f_2 and f_3,

$$[[f_1, f_2], f_3] + [[f_2, f_3], f_1] + [[f_3, f_1], f_2] = 0. \tag{34}$$

Jacobi's identity is used in the feedback linearization problem to interchange the order in which certain derivatives are taken.

Finally, we introduce a recursive notation for iterated Lie brackets. Define $ad_f^k g = [f, ad_f^{k-1} g]$ where $ad_f^0 g = g$. Then

$$\begin{aligned}
ad_f^0 g &= g \\
ad_f^1 g &= [f, g] \\
ad_f^2 g &= [f, [f, g]] \\
ad_f^3 g &= [f, [f, [f, g]]]
\end{aligned} \tag{35}$$

$$\vdots \qquad \vdots$$

4.3   Local Linearization Conditions

In examining Equations 26, the task is to reduce this system of high-order differential equations to a set of first-order equations. The first equation, $L_g h = 0$ is already first-order, so consider the second equation. Then

$$\begin{aligned}
0 &= L_g L_f h \\
&= L_f L_g h - L_{[f,g]} h
\end{aligned} \tag{36}$$

by definition of the Lie bracket. $L_g h$, however, is identically zero by the prior constraint. Consequently, $L_g L_f h = 0$ leaving just the first-order equation $L_{[f,g]} h = 0$.

From the third constraint,

$$\begin{aligned}
0 &= L_g L_f^2 h \\
&= L_f L_g L_f h - L_{[f,g]} L_f h
\end{aligned} \tag{37}$$

where $L_f L_g L_f h = 0$ from the prior equation, giving

$$\begin{aligned}
0 &= -L_f L_{[f,g]} h + L_{[f,[f,g]]} h \\
&= L_{ad_f^2 g} h
\end{aligned} \tag{38}$$

again leaving a first-order constraint.

This procedure can be carried out to arbitrary order. The result is a set of $n-1$ first-order partial differential equations for $h(x)$ that are written as

$$
\begin{aligned}
L_{ad_f^0 g} h &= 0 \\
L_{ad_f^1 g} h &= 0 \\
&\vdots = \vdots \\
L_{ad_f^{n-2} g} h &= 0 \\
L_{ad_f^{n-1} g} h &\neq 0
\end{aligned} \tag{39}
$$

The last condition is required for the transformation of the control u to be nonsingular.

This is a type of local controllability result.

In addition to linear independence, it is necessary that the first-order partial differential equations be *integrable*. Not all systems of first-order PDE's can be integrated.

4.4   Integrability, Involutivity, and the Frobenius Theorem

A key question regarding systems of linear homogeneous first-order PDE's of the type shown in Equation 39 is whether or not a solution can exist. For example, the pair of PDE's

$$
0 = \frac{\partial h}{\partial x_2} \tag{40}
$$

$$
0 = \frac{\partial h}{\partial x_1} + x_2 \frac{\partial h}{\partial x_3} \tag{41}
$$

has only the trivial solution $h = 0$. The first constraint says that h is not a function of x_2, while second requires the ratio of derivatives of h to be a function of x_2. This contradiction means that no nonzero solution can exist.

To illustrate the basic notion of integrability, consider the solution $h(x)$ to a pair of first-order homogeneous linear PDE's in 3 variables

$$
0 = L_p h = p_1(x)\frac{\partial h}{\partial x_1} + p_2(x)\frac{\partial h}{\partial x_2} + p_3(x)\frac{\partial h}{\partial x_3} \tag{42}
$$

$$
0 = L_q h = q_1(x)\frac{\partial h}{\partial x_1} + q_2(x)\frac{\partial h}{\partial x_2} + q_3(x)\frac{\partial h}{\partial x_3} \tag{43}
$$

where p and q are vector functions. If a nontrivial solution exists, then $L_{[p,q]} h = L_p(L_q h) - L_q(L_p h) = L_p(0) - L_q(0) = 0$. One may impose constraints on p and q so that this is assured to happen. In particular, if $[p, q]$ can be written as a weighted sum

$$
[p, q] = a(x)p + b(x)q \tag{44}
$$

then $L_{[p,q]} h = 0$ from Equation 17. [5] The essence of the Frobenius Theorem is that constraints of this type turn out to be necessary and sufficient conditions for the existence of solutions to systems of linear PDE's.

---

[5]The coefficients a and b may be functions of x.

Theorem 1 (Frobenius [4]) *Consider a set of first order homogeneous linear partial differential equations in the form*

$$
\begin{aligned}
L_{f_1} h &= 0 \\
L_{f_2} h &= 0 \\
&\ \vdots \qquad \vdots \\
L_{f_\alpha} h &= 0
\end{aligned}
\tag{45}
$$

where $\alpha \leq n - 1$, and where $f_1, f_2, \ldots, f_\alpha$ are C^∞ on a C^∞ manifold M for which the span of $f_1, f_2, \ldots, f_\alpha$ is of constant dimension. There is a C^∞ solution $h(x)$ in a neighborhood of a point $x_0 \in M$ if and only if the Lie bracket of each pair f_i and f_j can be written as

$$
[f_i, f_j] = \sum_{k=1}^{\alpha} \gamma_{ijk}(x) f_k.
\tag{46}
$$

where the $\gamma_{ijk}(x)$ are scalar functions of x.

A set of vector functions which satisfies Equation 46 is said be *involutive*.[6] Involutivity, in fact, is a rather exceptional condition for a set of vector fields indicative of some underlying structure or symmetry. It is remarkable that some models of real chemical processes can be made involutive by a proper indentification of the input variable. Subsequent examples will demonstrate that integrability may be used as a guide for the selection of control variables.

4.5   Su's Theorem

Given the Frobenius theorem, the conditions for exact local linearization are then fairly clear. To satisfy the constraints posed by Equation 39, it is necessary and sufficient that the n constraints be linearly independent, and that the first $n - 1$ constraints constitute an integrable set of homogeneous linear first-order PDE's. We now cite a version of Su's theorem.

Theorem 2 (Su [25]) *A linear-analytic system in the form*

$$
\frac{dx}{dt} = f(x) + u g(x)
\tag{47}
$$

is feedback equivalent to a linear system in a neighborhood of a point x_0 if and only if

(i).   the vectors $g, ad_f^1 g, \ldots, ad_f^{n-1} g$ span $\Re^n$ about x_0, and

(ii).   the set of vector fields $g, ad_f^1 g, \ldots, ad_f^{n-2} g$ is involutive.

---

[6] Occasionally spelled as *involuative* [5].

Applications of this theorem to problems in process control are described in a subsequent section. Of particular note is that the existence of a linearizing feedback is a system property, not dependent on the controller to be implemented. An interesting line of research is to systematically determine choice of control variables for a given process that will result in a linearizable description.

5 Partial Linearization

The conditions of Su's theorem are generally quite restrictive. Involutivity is a rigorous constraint that limits the class of process systems admitting a linearizing feedback. A means for broadening the class of systems is to forego exact linearization, and instead search for incomplete or partial linearizations of order α. The linearization is then implemented as shown in Figure 4. The following subsections describe some goals for current research involving partial linearization.

5.1 Stability

A simple corollary of Su's theorem is that there exists a distinguished output $h(x)$ admitting an order α feedback linearization in a neighborhood of x_0 if and only if the vectors $g, ad_f^1 g, \ldots, ad_f^{\alpha-1} g$ are linearly independent about x_0, and the set $g, ad_f^1 g, \ldots, ad_f^{\alpha-2} g$ is involutive. If these conditions hold, then distinguished output is found as a solution to the equations

$$
\begin{aligned}
L_{ad_f^0 g} h &= 0 \\
L_{ad_f^1 g} h &= 0 \\
&\vdots = \vdots \\
L_{ad_f^{n-2} g} h &= 0 \\
L_{ad_f^{n-1} g} h &\neq 0
\end{aligned}
\tag{48}
$$

and the state feedback control implemented as shown in Figure 4.

In addition to the existence of $h(x)$, it necessary to require closed loop stability. Following the discussion of Section 4.3, stability for $r = 0$ is obtained when the zero dynamics given by the dynamical system

$$
\frac{dx}{dt} = f(x) - \frac{L_f^\alpha h}{L_f L_f^{\alpha-1} h} g
\tag{49}
$$

are stable.

A Lyapunov function can be introduced establish stability. Let $V(x)$ be a Lyapunov function such that $V(x_0) = 0$, and $V(x) > 0$ for $x \neq x_0$. A sufficient condition for asymptotic stability is that $\frac{dV}{dt} < 0 \; \forall x \neq x_0$. Computing,

$$\frac{dV}{dt} = L_f V - \frac{L_f^\alpha h}{L_f L_f^{\alpha-1} h} L_g V \tag{50}$$

so that a condition for asymptotic stability is

$$L_f L_f^{\alpha-1} h L_f V - L_f^\alpha h L_g V = -L_f L_f^{\alpha-1} \phi(x) \tag{51}$$

where $\phi(x)$ is a positive-definite function of x. Without loss of generality it can been assumed that $L_f L_f^{\alpha-1} \geq 0$.

Continuing research is attempting to exploit this stability result by using a Liebnitz-type formula to cast stability as an additional constraint on the definition of a distinguished output. A technical difficulty is that $V(x)$ and $\phi(x)$ must introduced into the problem such that the resulting constraints are integrable.

5.2 Measurement Constraints

Partial linearization provides a mechanism for introducing additional constraints into the determination of the distinguished output, and consequently, the feedback control. Consider, for example, a set vector of functions $p_1(x)$, $p_2(x)$, ..., $p_m(x)$ that describe a set of unmeasured state variables. An entry in $p_i(x)$ is nonzero if the corresponding state cannot be measured.

The control law resulting from partial linearizationis independent of the unmeasured state variables if

$$0 = L_{p_i} L_f^\alpha h \tag{52}$$
$$0 = L_{p_i} L_g L_f^{\alpha-1} h \tag{53}$$

and

$$0 = L_{p_i} h$$
$$0 = L_{p_i} L_f h \tag{54}$$
$$\vdots \quad \vdots$$
$$0 = L_{p_i} L_f^{\alpha-1} h$$

$\forall i = 1, \ldots, m$. Jacobi's identity then allows one to cast the measurement constraints as additional first-order constraints on the distinguished output.

6 Applications

6.1 Exothermic Stirred Tank Reactor – An Involutive System of Order 3

Involutivity appears to be a very strong constraint on the class of nonlinear models for which exact linearization is possible. Nevertheless, applications show that a surprising number of models are involutive. This seems to be because of the manner in which control variables are introduced to regulate significant process states. Second order systems trivially satisfy the involutivity criterion since condition (ii) of Su's theorem contains only the single vector g. For single-input systems of order 3, however, it is necessary to test for the involutivity of the pair $g, [f, g]$.

Hoo and Kantor [17] previously demonstrated feedback linearization for an exothermic stirred tank reactor in which the model contained two state variables for temperature and conversion. It was unrealistically assumed that it was possible to directly manipulate the heat flux to the reactor. This example is reconsidered below in which a third state variable is introduced to denote the cooling jacket temperature. The control variable is then the cooling jacket flowrate. It is demonstrated the three state model is involutive with respect to this choice of control variable, so that feedback linearization is possible. It is shown that any nonsingular of the reactor conversion will serve as a 'distinguished output' in a feedback linearization.

The classic model for a single exothermic, irreversible reaction in a constant volume stirred tank reactor is given in dimensionless form as [27]

$$\frac{dx}{dt} = -x + Da R(x, y) \tag{55}$$

$$\frac{dy}{dt} = -y + B Da R(x, y) - \beta(y - z) \tag{56}$$

$$\frac{dz}{dt} = \frac{\beta}{C}(y - z) + u(z_f - z) \tag{57}$$

where the variable x, y, z denote dimensionless conversion, reactor temperature, and cooling jacket temperature, respectively. The parameters are defined in their conventional manner. In particular, z_f denotes the dimensionless cooling jacket feed temperature, and C is the relative heat capacity of fluid in the cooling jacket. The dimensionless control variable u corresponds to the manipulation of the cooling jacket flowrate. $R(x, y)$ is a temperature-dependent reaction rate expression, for example

$$R(x, y) = (1 - x_1) \exp\left(\frac{x_2}{1 + \frac{x_2}{\gamma}}\right) \tag{58}$$

though the subsequent development is for general $R(x, y)$.

The Lie bracket $[f, g] = g_x f - f_x g$ is computed to be

$$[f, g] = \begin{bmatrix} 0 \\ -\beta(z_f - z) \\ \frac{\beta}{C}(z_f - y) \end{bmatrix} \tag{59}$$

The involutivity condition requires that the vector function $[g, [f, g]]$ be expressed as a weighted sum of g and $[f,g]$. For this example,

$$[g,[f,g]] = \begin{bmatrix} 0 \\ \beta(z_f - z) \\ \frac{\beta}{C}(z_f - y) \end{bmatrix} \tag{60}$$

$$= \frac{2\beta}{C} \frac{(z_f - y)}{(z_f - z)} g - [f,g] \tag{61}$$

as required. Note that the scalar coefficient of g is not constant, but is a function of the reactor and cooling jacket temperatures.

The integrable constraints on the distinguished output h are given as

$$0 = L_g h \tag{62}$$
$$0 = L_{[f,g]} h \tag{63}$$

which yield the pair of first-order PDE's

$$0 = (z_f - z)\frac{\partial h}{\partial z} \tag{64}$$

$$0 = -\beta(z_f - z)\frac{\partial h}{\partial y} + \frac{\beta}{C}(z_f - z)\frac{\partial h}{\partial z} \tag{65}$$

from which it is evident that h must be function of conversion alone. Any nonsingular function of conversion will suffice and there seems to be no criteria for selection. Experience indicates that, in some instances, using the logarithm of a state variable may introduce some algebraic simplifications. That does not appear to be the case in this example, however.

Choosing $h(x) = x$ yields a set of transformed state variables $\xi_i = L_f^{i-1}h$. A computation shows

$$\xi_1 = x \tag{66}$$
$$\xi_2 = -x + DaR(x,y) \tag{67}$$
$$\xi_3 = (-1 + Da\frac{\partial R}{\partial x})(-x + DaR(x,y)) +$$
$$Da\frac{\partial R}{\partial y}(-y - \beta(y - z) + BDaR(x,y)) \tag{68}$$

The required control law for exact linearization is found by solving the relation

$$L_f^3 h + uL_g L_f^2 h = k_1(r - \xi_1) - k_2\xi_2 - k_3\xi_3 \tag{69}$$

for u, where k_1, k_2, k_3 are control constants. For this example, an analytical expression for u can be derived but is omitted here due to its length and complexity. In practice u would be computed numerically.

With the control in place, the closed loop process responds linearly to changes in the new input v. When cast in terms of the distinguished output $x, \xi_1 = h(x)$, one obtains the closed-loop dynamics

$$\frac{d^3 x}{dt} + k_3\frac{d^2 x}{dt} + k_2\frac{dx}{dt} + k_1 x = k_1 r \tag{70}$$

The constants k_1, k_2, k_3 are chosen to achieve an acceptable closed-loop response to changes in r. Note that r represents a setpoint for conversion.

6.2 Measurement Selection for an Unstable Biological Reactor

A feature of the approach outlined above for feedback linearization is the definition of distinguished process outputs. While the transformed variables have been treated only as functions of the states, in several examples it happens that transformed states may be directly measured. In such cases, the transformed states may then incorporated directly into the feedback control. The example cited below was originally presented by Hoo and Kantor [18].

A simple model for the growth of a single microorganism on a single rate-limiting substrate in a well stirred tank is given by

$$\frac{dx_1}{dt} = \mu(x_2)x_1 - Dx_1 \tag{71}$$

$$\frac{dx_2}{dt} = -\sigma(x_2)x_1 + D(x_{2,f} - x_2) \tag{72}$$

where x_1 and x_2 are the cell density and substrate concentration, respectively. The feed is sterile with substrate concentration of $x_{2,f}$. The dilution rate D is the physical control variable. The specific growth rate $\mu(x_2)$ and the specific substrate consumption rate $\sigma(x_2)$ are experimentally measured functions of the substrate concentration.

Involutivity is trivially satisfied for a second order model. The distinguished output for exact linearization satisfies the constraint

$$0 = L_g h = -x_1 \frac{\partial h}{\partial x_1} + (x_{2,f} - x_2) \frac{\partial h}{\partial x_2} \tag{73}$$

The solution to this equation is any nonsingular function $h\left(x_1/(x_{2,f} - x_2)\right)$, an estimate of the net yield as measured at the reactor effluent.

It is algebraically convenient to choose the logarithm of the net yield as the distinguished output. The transformed state variables are then

$$\xi_1 = \ln\left(\frac{x_{2,f} - x_2}{x_1}\right) \tag{74}$$

The second transformed state $\xi_2 = L_f h$ is computed then

$$\xi_2 = -\mu(x_2) + \frac{\sigma(x_2)x_1}{x_{2,f} - x_2} \tag{75}$$

Variable ξ_2 is an instantaneous 'excess' yield. It is the difference between the specific growth rate and the growth that would be inferred from a constant yield model.

The resulting control law was used by Hoo and Kantor to demonstrate global stabilization of an unstable steady state. Simulation studies included the effects of modeling error, and the use of an extended Kalman filter to estimate the substrate concentration from a measurement of the cell density.

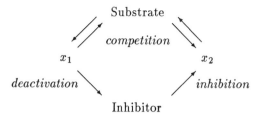

Figure 5: The interaction of cell populations, substrate, and inhibitor in a mixed-culture bioreactor.

6.3   Input Selection for a Mixed Culture Fermentation

The involutivity condition can serve as means for defining control variables which admit a linearizing feedback. The following example of a continuous, mixed-culture bioreactor illustrates this point. The control problem is to stabilize the simultaneous growth of two cell populations that compete for a single rate-limiting substrate. The coexistence steady-state is stabilized by controlled addition of an external inhibitor and control of the dilution rate. In contrast to the other problems treated in this paper, this problem is a two-input, three-state model. This study was originally presented by Hoo and Kantor [19] as a prototype for control studies of continuous fermentations of recombinant microorganisms.

The accompanying figure illustrates the interactions of the two cell populations. The cell densities are denoted by x_1 and x_2, respectively. Cell population one is a plasmid bearing organism that produces a valuable product. Population two is a plasmid free offspring that produces no product, and has a higher intrinsic growth rate. To provide a mechanism for stabilizing the cell populations, the plasmid bearing organism also produces an enzyme that can digest an externally supplied growth inhibitor to which the second population is sensitive. Without inhibitor present, the coexistence steady-state is unstable, and the faster growing, plasmid free population will dominate. The inhibitor levels the playing field so that plasmid bearing cell population can coexist.

An elementary model for the simulataneous growth of the two populations in a continuous fermentation is given as

$$\frac{dx_1}{dt} = (\mu_1(S) - D)x_1 \tag{76}$$

$$\frac{dx_2}{dt} = (\mu_2(S,I) - D)x_2 \tag{77}$$

$$\frac{dI}{dt} = -px_1I + D(I_f - I) \tag{78}$$

$$\frac{dS}{dt} = -\mu_1(S)\frac{x_1}{Y_1} - \mu_2(S,I)\frac{x_2}{Y_2} + D(S_f - S) \tag{79}$$

where S and I are the substrate and inhibitor concentrations, respectively. D denotes the dilution rate q/V, and Y_1, Y_2 are the yields of species 1 and 2 per unit mass of substrate. The term $-px_1I$ in Equation 78 represents the rate at which the first cell population consumes the inhibitor. $\mu_1(S)$ and $\mu_2(S, I)$ are the specific growth rates of respective cell populations given by Monod-like expressions. Note that $\mu_2(S, I)$ is a function of I.

The above equations reduce to a set of three independent equations after several residence times have passed. Summing the first three, one finds the operating constraint

$$S = S_f - \frac{x_1}{Y_1} - \frac{x_2}{Y_2}. \tag{80}$$

In contrast with the examples discussed above, in this case there is more than one candidate for the physical control variables. Among them are the dilution rate D, substrate addition rate DS_f, the inhibitor addition rate DI_f, the substrate and inhibitor feed concentrations S_f, I_f. One may choose any independent subset that can be constructed from these quantities. Some control sets will lead to integrable solutions for a linearizing feedback, while others will not. Integrability, therefore, is a criterion for screening candidate control variables.

Consider the problem of selecting two independent controls. The target closed-model is a linear system with two inputs. There are several possibilities, considered here is a multivariable linear model in companion form with controllability indices $\kappa_1 = 1, \kappa_2 = 2$ [21].

$$\frac{d\xi}{dt} = \begin{bmatrix} k_{11} & k_{12} & k_{13} \\ 0 & 0 & 1 \\ k_{21} & k_{22} & k_{23} \end{bmatrix} \xi + \begin{bmatrix} k_{11} & 0 \\ 0 & 0 \\ 0 & k_{22} \end{bmatrix} \begin{bmatrix} r_1 \\ r_2 \end{bmatrix} \tag{81}$$

Two distinguished outputs are required to fit this closed-loop model; these are denoted $y_1 = h_1(x)$ and $y_2 = h_2(x)$.

Control variables are to be selected so the reactor model is the two input form

$$\frac{dx}{dt} = f(x) + u_1 g_1(x) + u_2 g_2(x). \tag{82}$$

Comparing to the closed-loop model, the distinguished outputs must satisfy

$$\frac{dy_1}{dt} = L_f h_1 + u_1 L_{g_1} h_1 + u_2 L_{g_2} h_1 \tag{83}$$

$$= k_{11}(r_1 - y_1) - k_{12}y_2 - k_{13}y_3 \tag{84}$$

$$\frac{dy_2}{dt} = L_f h_2 + u_1 L_{g_1} h_2 + u_2 L_{g_2} h_2 \tag{85}$$

$$= y_3 \tag{86}$$

$$\frac{dy_3}{dt} = L_f^2 h_2 + u_1 L_{g_1} L_f h_2 + u_2 L_{g_2} L_f h_2 \tag{87}$$

$$= -k_{21}y_1 + k_{22}(r_2 - y_2) - k_{23}y_3 \tag{88}$$

where the notation $y_3 = L_f h_2$ has been introduced to simplify the notation. Fitting the closed-loop model requires that the coefficients of u_1 and u_2 be identically zero in Equation 85. This yields the two linearization constraints

$$0 = L_{g_1} h_2 \qquad (89)$$
$$0 = L_{g_2} h_2 \qquad (90)$$

For these to be integrable it is required that g_1 and g_2 be involutive.

The involutivity of g_1 and g_2 constrains the possible choices of control variables. One solution suggested by Hoo and Kantor was the choice

$$u_1 = D \qquad (91)$$
$$u_2 = DI_f. \qquad (92)$$

The control variables are the dilution rate and the total feedrate of the external inhibitor. Other choices, such as the inhibitor feed concentration, lead to nonintegrable constraints on the distinguished outputs of a linearizing control.

For this choice of control variables, a solution for the distinguished outputs is calculated to be

$$y_1 = x_1 \qquad (93)$$
$$y_2 = \ln\left(\frac{x_1}{x_2}\right) \qquad (94)$$
$$y_3 = \mu_1 - \mu_2 \qquad (95)$$

Hoo and Kantor present simulation results for the resulting linearizing state feedback. They demonstrate the global stabilization of the coexistence steady-state, and the use of an extended Kalman filter to estimate unmeasured states when only a measurement of the total cell density is available.

7 Acknowledgements

I would also like to acknowledge the aid of M. Keenan and L. Limqueco in checking the example calculations. Several of the above examples were originally prepared by Karlene Hoo in a Ph.D. thesis.

8 Notation

A	=	Cross-sectional area, matrix
a	=	time constant
b	=	column vector
c	=	row vector
Da	=	Damkohler number
F_{in}	=	volumetric inlet flowrate
h	=	liquid level

k	=	valve constant
$L_f h$	=	derivative of h in the direction of f, $< f, \nabla h >$
N	=	A nonlinear system
N_L	=	left inverse
N_R	=	right inverse
r	=	reference or transformed control variable
$R(x,y)$	=	reaction rate
u	=	control variable
v	=	transformed control variable
x	=	conversion
y	=	dimensionless reactor temperature, distinguished output
z	=	dimensionless cooling jacket temperature

Greek Symbols

α	=	relative order
β	=	heat transport coefficient
γ	=	dimensionless activation energy
ξ_i	=	transformed state variables

References

[1] Abraham, Ralph, Jerrold E. Marsden, Tudor Ratiu, *Manifolds, Tensor Analysis, and Applications.* Addison-Wesley, 1983.

[2] Arnold, V. I., *Ordinary Differential Equations.* MIT Press, 1973.

[3] Arnold, V. I., *Geometrical Methods in the Theory of Ordinary Differential Equations.* Springer-Verlag, 1983.

[4] Boothby, William M., *An Introduction to Differentiable Manifolds and Riemannian Geometry.* Academic Press, 1975.

[5] Brockett, Roger W., "Nonlinear Systems and Differential Geometry," *Proceedings of the IEEE,* **64**, 61–72, 1976.

[6] Burke, William L., *Applied Differential Geometry.* Cambridge University Press, 1985.

[7] Byrnes, Christopher I., "Remarks on nonlinear planar control systems which are linearizable by feedback," *Systems and Control Letters,* **5**, 363–367, 1985.

[8] Byrnes, Christopher I., and Alberto Isidori, "A Frequency Domain Philosophy for Nonlinear Systems, with Applications to Stabilization and to Adaptive Systems," Proceedings of the 23rd Conference on Decision and Control, Las Vegas, 1569–1573, 1984.

[9] Byrnes, Christopher I., and Alberto Isidori, "Global Feedback Stabilization of Nonlinear Systems," Proceedings of the 24th Conference on Decision and Control, Ft. Lauderdale, 1031–1037, 1985.

[10] Casti, John L., *Nonlinear System Theory*. Academic Press, 1985.

[11] Desoer, Charles A., and Yung-Terng Wang, "Foundations of Feedbck Theory for Nonlinear Dynamical Systems," *IEEE Trans. on Circuits and Systems*, **CAS-27**, 104–123, 1980.

[12] Economou, Constantin G., *An Operator Theory Approach to Nonlinear Controller Design*. Ph.D. Thesis, California Institute of Technology, 1985.

[13] Georgakis, Christos, "On the Use of Extensive Variables in Process Dynamics and Control," *Chemical Engineering Science*, **41**, 1471-1484, 1986.

[14] Guckenheimer, John, and Philip Holmes, *Nonlinear Oscillations, Dynamical Systems, and Bifurcations of Vector Fields*. Springer-Verlag, 1983.

[15] Hirschorn, R. M., "Invertibility of Nonlinear Control Systems," *SIAM J. Control and Optimization*, **17**, 289–297, 1979.

[16] Hirschorn, Ronald M., "Invertibility of Multivariable Nonlinear Control Systems," *IEEE Trans. Automat. Control*, **AC-24**, 855–865, 1979.

[17] Hoo, Karlene A., and Jeffrey C. Kantor, "An Exothermic Continuous Stirred Tank Reactor is Feedback Equivalent to a Linear System," *Chemical Eng. Commun.*, **37**, pp. 1–10, 1985.

[18] Hoo, Karlene A., and Jeffrey C. Kantor, "Linear Feedback Equivalence and Control of an Unstable Biological Reactor," *Chemical Eng. Commun.*, **46**, pp. 385–399, 1986.

[19] Hoo, Karlene A., and Jeffrey C. Kantor, "Global Linearization and Control of a Mixed Culture Bioreactor with Competition and External Inhibition," *Mathematical Biosciences*, (to appear), 1986.

[20] Hunt, L. R., Renjeng Su, and George Meyer, "Global Transformations of Nonlinear Systems," *IEEE Trans. Automat. Control*, **AC-28**, 24–31, 1983.

[21] Kailath, Thomas. *Linear Systems*. Prentice-Hall, 1980.

[22] Ogunnaike, Babatunde A., "Some results on the design of nonlinear controllers for nonlinear process systems," *Proceedings 1985 American Control Conference*, 657, 1985.

[23] Shinsky, F. G., *Controlling Multivariable Processes*. Instrument Society of America, 1981.

[24] Shinsky, F. G., *Distillation Control, 2nd Edition*. McGraw-Hill, 1984.

[25] Su, Renjeng, "On the linear equivalents of nonlinear systems," *Systems and Control Letters*, **2**, 48–52, 1982.

[26] Su, Renjeng, and L. R. Hunt, "A Canonical Expansion for Nonlinear Systems," *IEEE Trans. Automat. Control*, **AC-31**, 670–673, 1986.

[27] Uppal, A., W.H. Ray, and A. B. Poore, "On the dynamic behavior of continuous stirred tank reactors," *Chemical Eng. Sci.*, **29**, 967, 1974.

Recent Advances in the Use of the Internal Model Control Structure for the Synthesis of Robust Multivariable Controllers[*]

Evanghelos Zafiriou
Department of Chemical and Nuclear Engineering
and Systems Research Center
University of Maryland
College Park, MD 20742

December 15-18, 1986

Abstract

This paper presents the following recent theoretical developments in the IMC methodology:

- Multivariable controller design for the minimization of the Integral Squared error (ISE) for every input direction in a set and their linear combinations.

- Treatment of open-loop unstable plants; use of the two-degree-of-freedom controller.

- Minimization of the Structured Singular Value (SSV) for robust performance over the IMC Filter parameters; unconstrained problem; analytic computation of the gradients.

- Computation of the worst (over all possible plants) ISE for a particular setpoint or disturbance input.

  The paper deals with comtinuous systems. Extension to sampled-data systems is straightforward but not included here for lack of space.

1 Preliminaries

1.1 Internal Model Control

The Internal Model Control (IMC) structure, introduced by Garcia and Morari (1982), has been widely recognized as very useful in clarifying the issues related to the mismatch between the model used for controller design and the actual process. The IMC structure (Fig.1a), is mathematically equivalent to the

---

[*]This paper is based on the PhD research of the author. The work was carried out at the California Institute of Technology with Prof. M. Morari as PhD advisor.

classical feedback structure (Fig.1b). The IMC controller Q and the feedback C are related through

$$Q = C(I + \tilde{P}C)^{-1} \qquad\qquad (1.1.1)$$

$$C = Q(I - \tilde{P}Q)^{-1} \qquad\qquad (1.1.2)$$

where $\tilde{P}$ is the process model.

$\underline{P = \tilde{P}.}$ In this case the overall transfer function connecting the set-points r and disturbances d to the errors $e = y - r$, where y are the process outputs, is

$$e = y - r = (I - PQ)(d - r) \overset{\text{def}}{=} \tilde{E} \quad (d - r) \qquad (1.1.3)$$

Hence the IMC stucture becomes effectively open-loop (Fig.2a) and the design of Q is simplified. Note that the IMC controller is identical to the parameter of the Q-parametrization (Zames, 1981). Also the addition of a diagonal filter F by writing

$$Q = \tilde{Q}F \qquad\qquad (1.1.4)$$

introduces parameters (the filter time constants) which can be used for adjusting on-line the speed of response for each process output.

$\underline{P \neq \tilde{P}.}$ The model-plant mismatch generates a feedback signal in the IMC stucture which can cause performance deterioration or even instability. Since the relative modeling error is larger at higher frequencies, the addition of the low-pass filter F (Fig.2b) adds robustness characteristics into the control system. In this case the closed-loop transfer function is

$$e = y - r = (I - P\tilde{Q}F)(I - (P - \tilde{P})\tilde{Q}F)^{-1}(d - r) \overset{\text{def}}{=} E \quad (d - r) \qquad (1.1.5)$$

Hence the IMC structure gives rise rather naturally to a two step design procedure:

Step 1 Design $\tilde{Q}$, assuming $P = \tilde{P}$.

Step 2 Design F so that the closed-loop characteristics that $\tilde{Q}$ produces in Step 1, are preserved in the presence of model-plant mismatch ($P \neq \tilde{P}$).

1.2   Internal Stability

A linear time invariant control system is internally stable if the transfer functions between any two points of the control system are stable. A more detailed discussion of the concept of internal stability can be found in the literature (e.g Morari et al., 1987).

Examination of the feedback structure of Fig. 1b results in the requirement that all elements in the matrix $IS1$ in (1.2.1) are stable.

$$IS1 = (\, C(I + PC)^{-1} \quad PC(I + PC)^{-1} \quad CP(I + CP)^{-1} \quad (I + PC)^{-1}P \,)$$
$$\qquad\qquad (1.2.1)$$

For the remainder of this section we shall assume that $P = \tilde{P}$. The additional requirements to take care of modeling error are discussed in section 3.3. Use of (1.1.1) or (1.1.2) in (1.2.1) yields

$$IS1 = (Q \quad PQ \quad QP \quad (I - PQ)P) \tag{1.2.2}$$

Note that stability of each element in (1.2.2) implies internal stability when the control system is implemented as the feedback structure in Fig. 1b, where C is obtained from the Q used in (1.2.2) through (1.1.2).

In order for the control system to be stable when implemented in the IMC stucture of Fig.1a, internal stability arguments (Morari et al.,1987) lead to the requirement that all elements of IS2 are stable.

$$IS2 = (Q \quad PQ \quad QP \quad (I - PQ)P \quad PQP \quad P) \tag{1.2.3}$$

Hence if the process P is open-loop unstable, $IS2$ will also be unstable and the control system has to be imlemented in the feedback stucture of Fig.1b. Still, the two step IMC design procedure can be used for the design of Q, as described in the following sections. C can then be obtained from (1.1.2) and the structure in Fig.1b implemented.

Note that when the process is open-loop stable, it follows from (1.2.2) that the only requirement for internal stability is that Q is stable.

2   Step 1: Design of $\tilde{Q}$

Throughout this section the assumption is made that $P = \tilde{P}$.

2.1   Objective

The performance objective adopted in this paper is to minimize the Integral Squared Error (ISE) for the error signal e given by (1.1.3). This is an H_2−type objective. Other objectives like an H_∞−type can be used (Zafiriou and Morari, 1986) but they will not be discussed here.

For a specified external system input v ($v = d$ for $r = 0$; $v = -r$ for $d = 0$), the ISE is given by the square of the L_2−norm of e:

$$\Phi(v) \stackrel{\text{def}}{=} ||e||_2^2 = \frac{1}{2\pi} \int_{-\infty}^{+\infty} e^*(i\omega)e(i\omega) \quad d\omega \tag{2.1.1}$$

From (1.1.3) we get

$$\Phi(v) = ||e||_2^2 = ||\tilde{E}v||_2^2 = ||(I - P\tilde{Q})v||_2^2 \tag{2.1.2}$$

Hence one objective could be

$$\min_{\tilde{Q}} \Phi(v) \tag{2.1.3}$$

for a particular input $v = \begin{pmatrix} v_1 & v_2 & ... & v_n \end{pmatrix}^T$, where $\tilde{Q}$ satisfies the internal stability requirements of section 1.2.

Minimizing the ISE just for one vector v however is not very meaningful, because of the different directions in which the disturbances enter the process or the setpoints are changed. What is desirable is to find a $\tilde{Q}$, that minimizes $\Phi(v)$ for every single v in a set of external inputs v of interest for the particular process. This set can be defined as

$$\mathcal{V} = \{v(s)|v(s) = diag(v_1(s), ..., v_n(s))x, \quad x \in \mathbf{R}^n\} \qquad (2.1.4)$$

where $v_1(s), ..., v_n(s)$, describe the frequency content of the external system inputs, e.g. steps, ramps or other types of inputs.

The objective can then be written as

$$\min_{\tilde{Q}} \Phi(v) \qquad \forall v \in \mathcal{V} \qquad (2.1.5)$$

under the constraint that $\tilde{Q}$ satisfies the internal stability requirements. It should be noted however that a linear time invariant $\tilde{Q}$ that solves (2.1.5) does not necessarily exist. In section 2.3, it will be shown that this is the case for some $\mathcal{V}$'s.

2.2   Parametrization of all stabilizing $\tilde{Q}$'s.

The process P can in general be open-loop unstable. The following assumption simplifies the solution of the optimization problem:

Assumption A.1 *If π is a pole of the model $\tilde{P}$ in the* <u>open</u> *RHP, then:*

    a Its order is equal to 1.

    b $\tilde{P}$ has no zeros at π.

    c The residual matrix corresponding to π is full rank.

Assumption A.1.a, is made to simplify the notation and it is the usual case. The results can be extended to higher order poles. A.1.b is always true for SISO systems. MIMO systems however can have zeros at the location of a pole (Kailath, 1980). This requires an exact cancellation in $det[\tilde{P}(s)]$ and therefore the assumption that this does not happen is not restrictive because such a cancellation will usually not happen anymore when a slight perturbation in the coefficients of $\tilde{P}$ is introduced. A.1.c is also always true for SISO systems, but it can be quite restrictive for MIMO systems. Instead of A.1.c however, an additional assumption can be made on the input for which the optimal controller is designed. This is discussed in section 2.3.

Assumption A.1 is not made for poles at the origin because more than one such poles may appear in an element of $\tilde{P}$, introduced by capacitances that are present in the process. The following assumption true for all practical process control problems is made:

Assumption A.2 *Any poles of $\tilde{P}$ or P on the imaginary axis are at $s = 0$. Also $\tilde{P}$ has no finite zeros on the imaginary axis.*

Let $\pi_1, ..., \pi_q$ be the poles of $\tilde{P}$ in the open RHP. Define the allpass

$$b_p(s) = \prod_{i=1}^{q} \frac{-s + \pi_i}{s + \pi_i^*} \tag{2.2.1}$$

where the superscript * denotes complex conjugate (and transpose when applied to a matrix).

If A.1.c does not hold then define

$$b_p(s) = 1 \tag{2.2.2}$$

The following Theorem holds:

Theorem 2.2.1 *Assume that $Q_0(s)$ satisfies the internal stability requirements of section 1.2, i.e. it produces a matrix $IS1$ with stable elements. Then* <u>all</u> *Q's that make $IS1$ stable are given by*

$$Q(s) = Q_0(s) + b_p(s)^2 Q_1(s) \tag{2.2.3}$$

where Q_1 is any stable transfer matrix such that

i.) If A.1.c holds, then PQ_1P has no poles at $s = 0$.

ii.) If A.1.c does not hold, then PQ_1P has no poles in the closed RHP.

Proof: See Appendix A.1.

2.3   Solution to (2.1.3)

This is the first step towards obtaining a solution to (2.1.5), if such a solution exists. In this section we only consider one particular input v. The plant P can be factored into an allpass portion P_A and a minimum phase (MP) portion P_M such that

$$P = P_A P_M \tag{2.3.1}$$

Hence P_A is stable and such that $P_A^*(i\omega)P_A(i\omega) = I$. Also P_M^{-1} is stable. This inner-outer coprime factorization can be accomplished through the spectral factorization of $P(-s)^T P(s)$, where 'T' denotes transpose. Details on these problems can be found in the literature (Anderson, 1967; Chu, 1985; Doyle et al., 1984).

Let $v_0(s)$ be the scalar allpass that includes the <u>common</u> RHP zeros of the elements of v. Factor v as follows:

$$v(s) = v_0(s) \left(\hat{v}_1(s) \;\;\; \cdots \;\;\; \hat{v}_n(s) \right)^T \stackrel{\text{def}}{=} v_0(s)\hat{v}(s) \tag{2.3.2}$$

Without loss of generality make the following assumption for the input v for which $\tilde{Q}$ is designed:

Assumption A.3   *a The poles of each nonzero element of v (or $\hat{v}$) in the open RHP (if any) are the first q' poles π_i of the plant in the open RHP.*

   b If A.1.c does not hold, then every nonzero element of v (or $\hat{v}$) includes all the open RHP poles of $\tilde{P}$ each with degree 1.

To simplify the arguments in the paper, we shall assume that if A.3.b is satisfied, then A.1.c is not. In this way the proper choices in the definitions and the proofs will be made on the basis of A.1.c. If both A.1.c and A.3.b hold, then the results that apply to the case where A.1.c does not hold but A.3.b does, are still correct.

Define

$$b_v(s) = \prod_{i=1}^{q'} \frac{-s + \pi_i}{s + \pi_i^*} \qquad (2.3.3)$$

If A.1.c does not hold define

$$b_v(s) = 1 \qquad (2.3.4)$$

An different assumption is made for the poles of v at $s = 0$:

Assumption A.4 *Let l_i be the maximum number of poles at $s = 0$ that an element of the ith row of P has. Then $v_i(s)$ has at least l_i poles at $s = 0$. Also v has no other poles or any zeros on the imaginary axis.*

The above assumptions are not restrictive in the case where v is a output disturbance d, because in a practical situation we want to design for an output disturbance produced by a disturbance that has passed through the process and therefore includes the unstable process poles (e.g., an output disturbance produced by a disturbance on the manipulated variables). Note that the control system will still reject with no steady-state offset, other disturbances with fewer unstable poles. The assumption is different for poles at $s = 0$, because their number in each row of $\tilde{P}$ can be different, since capacitances may be associated with only certain process outputs. Also the output disturbance may have more poles at $s = 0$ than the process (e.g., a persistent disturbance in the manipulated variables).

The assumptions might be restrictive in the case of setpoints though. However for setpoint tracking the use of the Two-Degree-of-Freedom structure, which will be discussed briefly in section 2.5, allows us to disregard the existence of any unstable poles of P and therefore this assumption need not be made for setpoints.

The following theorem holds:

Theorem 2.3.1 *The set of controllers $\tilde{Q}$ that solve (2.1.3) satisfy*

$$\tilde{Q}\hat{v} = b_p b_v^{-1} P_M^{-1} \{ b_p^{-1} b_v P_A^{-1} \hat{v} \}_* \qquad (2.3.5)$$

where the operator $\{.\}_*$ denotes that after a partial fraction expancion of the operand all terms involving the poles of P_A^{-1} are omitted. Furthermore, for

$n \geq 2$ *the number of stabilizing controllers that satisfy (2.3.4) is infinite. Guidelines for the construction of such a controller are given in the proof.*

Proof: See Appendix A.2

2.4 Solution to (2.1.5)

Write

$$V(s) \stackrel{\text{def}}{=} diag(v_1(s), ..., v_n(s)) \qquad (2.4.1)$$

$$\hat{V}(s) \stackrel{\text{def}}{=} diag(\hat{v}_1(s), ..., \hat{v}_n(s)) \qquad (2.4.2)$$

The following Theorem holds:

Theorem 2.4.1 *Step 1 If all the RHP zeros of V appear in every element of V with the same degree, then the controller $\tilde{Q}$ that solves (2.1.5) is given by*

$$\tilde{Q} = b_p b_v^{-1} P_M^{-1} \{b_p^{-1} b_v P_A^{-1} \hat{V}\}_* \hat{V}^{-1} \qquad (2.4.3)$$

Step 2 If an element of V has a RHP zero that does not appear in all the other elements with the same degree, then there exists no stabilizing $\tilde{Q}$ that solves (2.1.5), unless P is stable and minimum phase in which case $\tilde{Q} = P^{-1}$.

Proof: See Appendix A.3.

The case descibed by Thm.2.4.1.ii, where no optimal solution exists, is not necessarily rare. Since v can be an output disturbence d, the designer might want to specify it as some common input, e.g. a step, going through some transfer matrix. For such a v, its elements may very well include different RHP zeros. When this happens, a solution to an alternative problem exists. Factor each element v_i of V into a stable allpass part v_{Ai} and a minimum phase v_{Mi}:

$$v_i(s) = v_{Ai}(s) v_{Mi}(s) \qquad (2.4.4)$$

The following theorem holds.

Theorem 2.4.2 *The controller*

$$\tilde{Q} = b_p b_v^{-1} P_M^{-1} \{b_p^{-1} b_v P_A^{-1} V_M\}_* V_M^{-1} \qquad (2.4.5)$$

minimizes $\Phi(v)$ for the following n directions x:

$$x = \begin{pmatrix} 1 \\ 0 \\ \vdots \\ 0 \end{pmatrix}, \begin{pmatrix} 0 \\ 1 \\ \vdots \\ 0 \end{pmatrix}, \begin{pmatrix} 0 \\ 0 \\ \vdots \\ 1 \end{pmatrix} \qquad (2.4.6)$$

and their multiples, as well as for the linear combinations of those directions that correspond to elements of V with the same RHP zeros in the same degree.

Proof: See Appendix A.4.

2.5   Two-Degree-of-Freedom Structure

From the discussion of the Internal Stability requirements in section 1.2, it
follows that RHP poles in the plant limit the possible choices of Q and thus
the achievable performance. This however need not be so for setpoint tracking.
Consider the general feedback structure of Fig.5. For the disturbance behavior
it is irrelevant if the controller is implemented as one block C as in Fig. 1b,
or as two blocks as in Fig. 5. Hence the achievable disturbance rejection is
restricted both by the RHP zeros and poles of P as the quantitative results of
the previous sections indicate.

Let us now proceed from the point where a stabilizing $\tilde{Q}$ and the corresponding C have been found through the results of the previous sections,
which produce a satisfactory disturbance response. We can then split C
into two blocks C_1 and C_2 such that C_1 is minimum phase and C_2 is stable. Then one can easily see that the only RHP zeros of the stabilized system
$PC_1(I + PC_1C_2)^{-1}$ are those of the process P. Thus C_3 can be designed
without regard for the RHP poles of P and the achievable setpoint tracking is
restricted by the RHP zeros of P only.

In summary, the achievable disturbance response of a system is restricted
by the presence of the plant RHP zeros and poles regardless of how complicated
a controller is used. If the Two-Degree-of-Freedom controller shown in Fig.5
is employed, the achievable setpoint response is restricted by the RHP zeros
only. A more rigorous discussion can be found in Vidyasagar (1985).

3   Model Uncertainty

3.1   Structured Singular Value

Potential modeling errors, described as uncertainty associated with the process
model, can appear in different forms and places in a multivariable model. This
fact makes the derivation of non-conservative conditions that guarantee robustness with respect to model-plant mismatch difficult. The Structured Singular
Value (SSV), introduced by Doyle (1982), takes into account the structure of
the model uncertainty and it allows the non-conservative quantification of the
concept of robust performance.

For a constant complex matrix M the definition of the SSV $\mu_{\boldsymbol{\Delta}}(M)$ depends also on a certain set $\boldsymbol{\Delta}$. Each element Δ of $\boldsymbol{\Delta}$ is a block diagonal
complex matrix with a specified dimension for each block, i.e.

$$\boldsymbol{\Delta} = \{diag(\Delta_1, \Delta_2, ..., \Delta_n) | \Delta_j \in \mathbf{C}^{m_j \times m_j}\} \qquad (3.1.1)$$

Then

$$\frac{1}{\mu_{\boldsymbol{\Delta}}(M)} = \min_{\Delta \in \boldsymbol{\Delta}} \{\bar{\sigma}(\Delta) | det(I - M\Delta) = 0\} \qquad (3.1.2)$$

and $\mu_{\boldsymbol{\Delta}}(M) = 0$ if $det(I - M\Delta) \neq 0 \qquad \forall \Delta \in \boldsymbol{\Delta}$. Note that $\bar{\sigma}$ is the maximum
singular value of the corresponding matrix.

Details on how the SSV can be used for studying the robustness of a control system can be found in Doyle (1985), where a discussion of the computational problems is also given. For three or fewer blocks in each element of Δ, the SSV can be computed from

$$\mu_\Delta(M) = \inf_{D \in \mathbf{D}} \bar{\sigma}(DMD^{-1}) \qquad (3.1.3)$$

where

$$\mathbf{D} = \{diag(d_1 I_{m_1}, d_2 I_{m_2}, ..., d_n I_{m_n}) | d_j \in \mathbf{R}_+\} \qquad (3.1.4)$$

and I_{m_j} is the identity matrix of dimension $m_j \times m_j$. For more than three blocks, (3.1.3) still gives an upper bound for the SSV.

3.2   Block Structure

In order to effectively use the SSV for designing F, some rearrangement of the block structure is necessary. The IMC structure of Fig.1a can be written as that of Fig.3a, where $v = d - r$, $e = y - r$ and

$$G = \begin{pmatrix} 0 & 0 & \tilde{Q} \\ I & I & \tilde{P}\tilde{Q} \\ -I & -I & 0 \end{pmatrix} \qquad (3.2.1)$$

where the blocks 0 and I have appropriate dimensions.

The structure in Fig.3a can always be transformed into that in Fig.3b, where Δ is a block diagonal matrix with the additional property that

$$\bar{\sigma}(\Delta) \leq 1 \qquad \forall \omega \qquad (3.2.2)$$

The superscript u in G^u denotes the dependance of G^u not only on G but also on the specific uncertainty description available for the model $\tilde{P}$. Only some of the more common types will be covered here to demonstrate how this is done, but it is straightforward to apply the same concepts to other types of uncertainty descriptions, like parametric uncertainty.

Multivariable Additive Uncertainty.   The information on the model uncertainty is of the form

$$\bar{\sigma}(P - \tilde{P}) \leq l_a(\omega) \qquad (3.2.3)$$

where l_a is a known function of frequency. In this case we can easily write $P - \tilde{P} = l_a \Delta$ where $\bar{\sigma}(\Delta) \leq 1$ and so obtain

$$G^u = G^a = \begin{pmatrix} l_a I & 0 & 0 \\ 0 & I & 0 \\ 0 & 0 & I \end{pmatrix} G \qquad (3.2.4)$$

Multivariable Input Multiplicative Uncertainty.

$$\bar{\sigma}(\tilde{P}^{-1}(P - \tilde{P})) \leq l_i(\omega) \tag{3.2.5}$$

where l_i is known. Then

$$G^u = G^i = G \begin{pmatrix} l_i\tilde{P} & 0 & 0 \\ 0 & I & 0 \\ 0 & 0 & I \end{pmatrix} \tag{3.2.6}$$

Multivariable Output Multiplicative Uncertainty.

$$\bar{\sigma}((P - \tilde{P})\tilde{P}^{-1}) \leq l_o(\omega) \tag{3.2.7}$$

$$G^u = G^o = \begin{pmatrix} l_o\tilde{P} & 0 & 0 \\ 0 & I & 0 \\ 0 & 0 & I \end{pmatrix} G \tag{3.2.8}$$

Element by Element Additive Uncertainty. For each element p_{ij} of P we have

$$|p_{ij} - \tilde{p}_{ij}| \leq l_{ij}(\omega), \qquad i = 1, ..., n; \quad j = 1, ..., n \tag{3.2.9}$$

Then

$$P - \tilde{P} = J_1 \Delta L J_2 \tag{3.2.10}$$

where

$$L = diag(l_{11}, l_{12}, ..., l_{1n}, l_{21}, ..., l_{nn}) \tag{3.2.11}$$

$$J_1 = \begin{pmatrix} 1 & \cdots & 1 & 0 & \cdots & 0 & \cdots & \cdots & 0 & \cdots & 0 \\ 0 & \cdots & 0 & 1 & \cdots & 1 & \cdots & \cdots & 0 & \cdots & 0 \\ \vdots & \ddots & \vdots & \vdots & \ddots & \vdots & \ddots & \ddots & \vdots & \ddots & \vdots \\ 0 & \cdots & 0 & 0 & \cdots & 0 & \cdots & \cdots & 1 & \cdots & 1 \end{pmatrix} \tag{3.2.12}$$

$$J_2 = \begin{pmatrix} I_n \\ I_n \\ \vdots \\ I_n \end{pmatrix} \tag{3.2.13}$$

From (3.2.10) it follows that

$$G^u = G^{ebe} = \begin{pmatrix} LJ_2 & 0 & 0 \\ 0 & I & 0 \\ 0 & 0 & I \end{pmatrix} G \begin{pmatrix} J_1 & 0 & 0 \\ 0 & I & 0 \\ 0 & 0 & I \end{pmatrix} \tag{3.2.14}$$

Note that all the above relations yield a G^u already partitioned as

$$G^u = \begin{pmatrix} G^u_{11} & G^u_{12} & G^u_{13} \\ G^u_{21} & G^u_{22} & G^u_{23} \\ G^u_{31} & G^u_{32} & G^u_{33} \end{pmatrix} \tag{3.2.15}$$

Then Fig.3b can be written as Fig.4 with

$$
\begin{aligned}
G^F &= \begin{array}{cc} G_{11}^u & G_{12}^u \\ G_{21}^u & G_{22}^u \end{array} + \begin{array}{c} G_{13}^u \\ G_{23}^u \end{array} (I - FG_{33}^u)^{-1} F \begin{array}{cc} G_{31}^u & G_{32}^u \end{array} \\
&\overset{\text{def}}{=} \begin{array}{cc} G_{11}^F & G_{12}^F \\ G_{21}^F & G_{22}^F \end{array}
\end{aligned}
\tag{3.2.16}
$$

3.3 Robust Stability

We now require that the matrix $IS1$ as given by (1.2.1) is stable for all possible plants P. The design of $\tilde{Q}$ according to section 2 resulted in a stable $IS1$ for $P = \tilde{P}$. In order for $IS1$ to remain stable we need to satisfy the requirements that as we move in a "continuous" way from the model $\tilde{P}$ to the plant P, no closed-loop RHP poles cross the imaginary axis and no such poles suddenly appear in the RHP. The latter requirement is satisfied if we assume that the model and the plant have the same number of RHP poles. The SSV can be used to determine if any crossings of the imaginary axis occur. Then we can say that the system is stable for any of the plants in the set defined from the bounds on the model uncertainty and which have the same number of RHP poles as the model, if and only if (Doyle, 1985)

$$
\mu_\Delta(G_{11}^F) < 1 \qquad \forall \omega
\tag{3.3.1}
$$

3.4 Robust Performance

In the first step of the IMC design procedure a controller $\tilde{Q}$ is obtained, which produces satisfactory disturbance rejection and/or setpoint tracking. This response is described by the "sensitivity" function $\tilde{E}$ given by (1.1.3). Since $\tilde{E}$ connects the external inputs to the error e, a well-designed control system produces a relatively "small" $\tilde{E}$. A measure of the magnitude of the known $\tilde{E}$ is its maximum singular value. Let $b(\omega)$ be a frequency function such that

$$
\bar{\sigma}(\tilde{E}(i\omega)) < b(\omega) \qquad \forall \omega
\tag{3.4.1}
$$

When $P \neq \tilde{P}$, the sensitivity function E is described by (1.1.5). Note that $E = \tilde{E}$ when $P = \tilde{P}$. In order for the performance of the control system to remain robust with respect to model-plant mismatch we have to keep E small in spite of the modeling error. Hence we require that

$$
\sup_\omega \bar{\sigma}(b(\omega)^{-1} E(i\omega)) < 1 \qquad \forall \Delta \in \Delta
\tag{3.4.2}
$$

We can now use the properties of the SSV (Doyle,1985) to obtain

$$
\sup_\omega \bar{\sigma}(b(\omega)^{-1} E(i\omega)) < 1 \qquad \forall \Delta \in \Delta \iff \sup_\omega \mu_{\Delta^\circ}(G^b) < 1
\tag{3.4.3}
$$

where

$$
G^b = \begin{pmatrix} I & 0 \\ 0 & b^{-1} \end{pmatrix} G^F
\tag{3.4.4}
$$

$$\Delta^0 = \left\{ diag(\Delta, \Delta^0) | \Delta \in \Delta, \Delta^0 \in \mathbf{C}^{n \times n} \right\} \tag{3.4.5}$$

The worst possible ISE that any plant within the uncertainty bounds can produce for a particular input v is given by the following theorem.

Theorem 3.4.1 *For a specified v define*

$$G^x \stackrel{def}{=} \begin{pmatrix} I & 0 \\ 0 & x \end{pmatrix} G^F \begin{pmatrix} I & 0 \\ 0 & v \end{pmatrix} \tag{3.4.6}$$

where x is a scalar function of ω and the blocks 0 have the appropriate dimensions (in general non-square). Augment G^x, which is in general a "tall" matrix, to obtain a square matrix:

$$G^x_{full} = (\, G^x \quad 0 \,) \tag{3.4.7}$$

Then

$$\mu_{\Delta^0}(G^x_{full}(i\omega)) = 1 \iff x(\omega) = x_0(\omega) \qquad \forall \omega \tag{3.4.8}$$

defines a function x_0 of frequency and

$$\sup_{\Delta \in \Delta} ||Ev||_2 = ||x_0^{-1}||_2 \tag{3.4.9}$$

Proof: See Appendix B.1.

Note that as it turned out $x_0^{-1} = \sup_{\Delta \in \Delta} \bar{\sigma}(Ev)$, but the only way to compute it is through (3.4.8). Also without loss of generality x can be assumed to be positive since the value of $\mu_{\Delta^0}(G^x_{full})$ depends only on $|x|$. The following theorem simplifies the problem of computing x_0.

Theorem 3.4.2 *Let*

$$M^x = \begin{pmatrix} M_{11} & M_{12} \\ x\,M_{21} & x\,M_{22} \end{pmatrix} \tag{3.4.10}$$

where x a positive scalar.

Then $\inf_{D \in \mathbf{D}} \bar{\sigma}(DM^x D^{-1})$ is a non-decreasing function of x, where $\mathbf{D} = \{ diag(D_1, D_2) \}$.

Proof: See Appendix B.2.

Note that G^x_{full} is a special case of M in the Theorem and so Theorem 3.4.2 applies to (3.4.11).

4   Step 2: Design of F

4.1   Filter Structure

The filter parameters can now be computed so that the robustness conditions that were discussed in section 3 are satisfied. To do so, some structure will have to be assumed for F, which can be of any general type that the designer wishes.

However in order to keep the number of variables in the optimization problem small, a rather simple structure like a diagonal F with first or second order terms would be recommended. In most cases this is not restrictive because the potentially higher orders of the model $\tilde{P}$ have been included in the controller $\tilde{Q}$ that was designed in the first step of the IMC procedure and which is in general a full matrix. Some additional restrictions on the filter exist in the case of an open-loop unstable plant. Also the use of more complex filter structure may be necessary in cases of highly ill-conditioned systems $(\bar{\sigma}(\tilde{P})/\underline{\sigma}(\tilde{P})$ very large).

4.1.1   Open-loop unstable plants.

The IMC filter $F(s)$ is chosen to be a diagonal rational function that satisfies the following requirements.

a Pole-zero excess. The controller $Q = \tilde{Q}F$ must be proper. Assume that the designer has specified a pole-zero excess of m for the filter $F(s)$.

b Internal stability. $IS1$ in (1.2.2) must be stable.

c Asymptotic tracking of disturbances. $(I - \tilde{P}\tilde{Q}F)v$ must be stable.

Write
$$F(s) = diag(f_1(s), ..., f_n(s)) \tag{4.1.1}$$
Under assumptions A.1,2,3,4, (b),(c) are equivalent to the following conditions. Let π_i $(i = 1, q)$ be an open RHP pole of $\tilde{P}$ (with order 1 according to A.1.a) and $\pi_0 = 0$ and l_{vk} the multiplicity of such a pole in the kth element of V. Then the kth element , f_k of the filter F must satisfy:

$$f_k(\pi_i) = 1, \qquad i = 0, 1, ..., q \tag{4.1.2}$$

$$\frac{d^j}{ds^j} f_k(s)|_{s=\pi_0} = 0, \qquad j = 1, ..., l_{vk} - 1 \tag{4.1.3}$$

(4.1.2) clearly shows the limitation that RHP poles place on the robustness properties of a control system designed for an open-loop unstable plant. Since because of (4.1.2) one cannot reduce the nominal $(P = \tilde{P})$ closed-loop bandwidth of the system at frequencies corresponding to the RHP poles of the plant, one can only tolerate a relatively small model error at those frequencies. One can write for a filter element $f_k(s)$:

$$f_k(s) = \frac{a_{n_k-1,k}s^{n_k-1} + ... + a_{1,k}s + a_{0,k}}{(\lambda s + 1)^{m+n_k-1}} \tag{4.1.4}$$

where
$$n_k = l_{vk} + q \tag{4.1.5}$$
and then compute the numerator coefficients for a specific tuning parameter λ from (4.1.2), (4.1.3).

In the simple case where $l_{vk} = 1$, one can develop an explicit formula for a filter element $f(s)$:

$$f(s) = \frac{1}{(\lambda s + 1)^{m+q}} \sum_{j=0}^{q} (\lambda \pi_j + 1)^{m+q} \prod_{i=0, i \neq j}^{q} \frac{s - \pi_i}{\pi_j - \pi_i} \qquad (4.1.6)$$

Example 4.1.1 *Assume that we have a pole-zero excess of m and there is only one pole π. Then from (4.1.6)*

$$f(s) = \frac{(\lambda \pi + 1)^m}{(\lambda s + 1)^m} \qquad (4.1.7)$$

If $\pi = 0$, (4.1.7) reduces to the standard filter for stable systems $f(s) = (\lambda s + 1)^{-m}$.

Example 4.1.2 *Assume that $m = 2$ and the only pole is a double pole at $s = 0$. Then from (4.1.2), (4.1.3) for $q = 0$*

$$f(s) = \frac{3\lambda s + 1}{(\lambda s + 1)^3} \qquad (4.1.8)$$

4.1.2 Ill-conditioned plants.

The problems arise because the optimal controller $\tilde{Q}$ designed for $\tilde{P}$ tends to be an approximate inverse of $\tilde{P}$ and as a result $\tilde{Q}$ is ill-conditioned as well, which means that a lot of detuning action will be required in a diagonal F to guarantee robust stability. The result is that although stability is maintained, the response is very sluggish and therefore the robust performance condition is very difficult to satisfy. A way to address this problem is to try to use a filter that acts directly on the singular values of $\tilde{Q}$, at the frequency where the condition number of $\tilde{Q}$ is highest, say ω^*. Let

$$\tilde{Q}(i\omega^*) = U_Q \Sigma_Q V_Q^* \qquad (4.1.9)$$

be the SVD of $\tilde{Q}$ at ω^* and let R_u, R_v, be real matrices that solve the pseudo-diagonalization problems:

$$U_Q^* R_u \approx I \qquad (4.1.10)$$

$$V_Q^* R_v \approx I \qquad (4.1.11)$$

Then for the IMC controller Q that includes the filter, use the expression

$$Q(s) = R_u F_1(s) R_u^{-1} \tilde{Q}(s) F_2(s) \qquad (4.1.12)$$

or

$$Q(s) = \tilde{Q}(s) R_v F_1(s) R_v^{-1} F_2(s) \qquad (4.1.13)$$

where $F_1(s)$, $F_2(s)$ are diagonal filters, such that $F_1(0) = F_2(0) = I$. Nota that when $\tilde{P}$ has poles at $s = 0$, every element of $F_1(s)$ must satisfy (4.1.3) for $j = 1, ..., l_v$, where

$$l_v = \max_{k=1,...,q} l_{vk} \qquad (4.1.14)$$

It should be pointed out that the success of this approach depends on how good any of the pseudo-diagonalizations (4.1.10) or (4.1.11) is. The diagonalization will be perfect if U_Q or V_Q is real. This will happen if $\omega^* = 0$, which is the case when the problems arise because the plant is ill-conditioned at steady-state, as for example high purity distillation columns are.

One can put this control structure in the form of Fig.3, as follows. Define

$$F(s) = diag(F_1(s), F_2(s)) \qquad (4.1.15)$$

$$\tilde{Q}_A(s) = R_u \quad or \quad \tilde{Q}(s)R_v \qquad (4.1.16)$$

$$A(s) = R_u^{-1}\tilde{Q}(s) \quad or \quad R_v^{-1} \qquad (4.1.17)$$

depending on whether (4.1.12) or (4.1.13) is used. Obtain G^u by substituting $\tilde{Q}$ with $\tilde{Q}_A$ in (3.2.1). Then in Fig.3 use instead of G^u, $G^{u,ill}$, where

$$G^{u,ill} = \begin{pmatrix} G_{11}^u & G_{12}^u & G_{13}^u & 0 \\ G_{21}^u & G_{22}^u & G_{23}^u & 0 \\ 0 & 0 & 0 & A \\ G_{31}^u & G_{32}^u & G_{33}^u & 0 \end{pmatrix} \qquad (4.1.18)$$

4.2 Objective

We can write

$$F \stackrel{def}{=} F(s; \Lambda) \qquad (4.2.1)$$

where Λ is an array with the filter parameters.

The problem can now be formulated as a minimization problem over the elements of the array Λ. A constraint is that the part of Λ corresponding to denominator time constants should be such that F is a stable transfer function. However the problem can be turned into an unconstrained one by writing the denominator of each element of F as a product of polynomials of degree 2 and one of degree 1 if the order is odd, with the constant terms of the polynomials equal to 1. Then the stability requirement translates into the requirement that the coefficients (elements of Λ) are positive, which is a constraint that can be eliminated by writing λ_k^2 or $|\lambda_k|$ instead of λ_k for the corresponding filter parameters.

Our goal is to satisfy (3.4.3). The filter parameters can be obtained by solving

$$\min_{\Lambda} \sup_{\omega} \mu_{\Delta^0}(G^b) \qquad (4.2.2)$$

It may be however that the optimum values for (4.2.2), still do not manage to satisfy (3.4.3). The reason may be that an F with more parameters is required, but more often that the performance requirements set by the selection of $b(\omega)$ in (3.4.1) are too tight to satisfy in the presence of model-plant mismatch. In this case one should choose a less tight bound b and resolve (4.2.2). Note that satisfaction of the Robust Performance condition (3.4.3) implies satisfaction of the Robust Stability condition (3.3.1) as well.

A different objective can be set in the case where the ISE for a particular external input direction v is of special interest to the designer. The objective is then to minimize (3.4.9) for a specified v (set-point or disturbance). Hence the filter parameters are obtained by solving

$$\min_{\Lambda} \|x_0^{-1}\|_2 \tag{4.2.3}$$

It should be pointed that contrary to the problems addressed in section 2, where a minimization for a set of v's could be carried out, (4.2.3) cannot be solved for a set of v's. The reason is the presence of modeling error in the problem definition.

4.3   Computational Issues

4.3.1   Solution of (4.2.2).

The computation of μ in (4.2.2) is made through (3.1.3); details can be found in Doyle (1982). As it was pointed out in Doyle (1985), the minimization of the Frobenious norm instead of the maximum singular value yields D's which are very close to the optimal ones for (3.1.3). Note that the minimization of the Frobenious norm is a very simple task. In the computation of the supremum in (4.2.2) only a finite number of frequencies is considered. Hence (4.2.2) is transformed into

$$\min_{\Lambda} \max_{\omega \in \Omega} \inf_{D \in \mathbf{D}^0} \bar{\sigma}(DG^b D^{-1}) \tag{4.3.1}$$

where Ω is a set containing a finite number of frequencies and $\mathbf{D}^0$ is the set corresponding to $\mathbf{\Delta}^0$ according to (3.1.1) and (3.1.4). Define

$$\Phi_\infty(\Lambda) \stackrel{\text{def}}{=} \max_{\omega \in \Omega} \inf_{D \in \mathbf{D}^0} \bar{\sigma}(DG^b D^{-1}) \tag{4.3.2}$$

The analytic computation of the gradient of Φ_∞ with respect to Λ is in general possible. This is not the case when the two or more largest singular values of $DG^b D^{-1}$ are equal. However this is quite uncommon and although the computation of a generalized gradient is possible, experience has shown the use of a mean direction to be satisfactory. A similar problem appears when the $\max_{\omega \in \Omega}$ is attained at more than one frequencies, but again the use of a mean direction seems to be sufficient. We shall now proceed to obtain the expression for the gradient of $\Phi_\infty(\Lambda)$ in the general case.

Assume that for the value of Λ where the gradient of $\Phi_\infty(\Lambda)$ is computed, the $\max_{\omega \in \Omega}$ is attained at $\omega = \omega_0$ and that the $\inf_{D \in \mathbf{D}^0} \bar{\sigma}(DG^b(i\omega_0)D^{-1})$ is obtained at $D = D_0$, where only one singular value σ_1 is equal to $\bar{\sigma}$. Let the singular value decomposition (SVD) be

$$D_0 G^b(i\omega_0) D_0^{-1} = (\, u_1 \quad U \,) \begin{pmatrix} \sigma_1 & 0 \\ 0 & \Sigma \end{pmatrix} \begin{pmatrix} v_1^* \\ V^* \end{pmatrix} \tag{4.3.3}$$

Then for the element of the gradient vector corresponding to the filter parameter λ_k we have under the above assumptions:

$$\frac{\partial}{\partial \lambda_k} \Phi_\infty = \frac{\partial}{\partial \lambda_k} \sigma_1(D_0 G^b(i\omega_0) D_0^{-1}) \tag{4.3.4}$$

because $\nabla_{D_0}(\sigma_1) = 0$ since we are at an optimum with respect to the D's. To simplify the notation use

$$A = D_0 G^b(i\omega_0) D_0^{-1} = U_A \Sigma_A V_A^* \tag{4.3.5}$$

By using the properties of the SVD we obtain from (4.3.3)

$$\begin{aligned}
& AA^* = U_A \Sigma_A^2 U_A^* \Rightarrow u_1^* \frac{\partial}{\partial \lambda_k}(AA^*) u_1 = u_1^* U_A \frac{\partial}{\partial \lambda_k}(\Sigma_A^2) U_A^* u_1 \\
\Rightarrow\ & u_1^*(\frac{\partial}{\partial \lambda_k}(A)A^* + A\frac{\partial}{\partial \lambda_k}(A^*))u_1 = u_1^* U_A(2\Sigma_A \frac{\partial}{\partial \lambda_k}(\Sigma_A))U_A^* u_1 \\
\Rightarrow\ & u_1^* \frac{\partial}{\partial \lambda_k}(A)v_1\sigma_1 + \sigma_1 v_1^* \frac{\partial}{\partial \lambda_k}(A^*)u_1 = 2\sigma_1 \frac{\partial}{\partial \lambda_k}(\sigma_1) \\
\Rightarrow\ & \frac{\partial}{\partial \lambda_k}(\sigma_1) = \mathrm{Re}\left[u_1^* \frac{\partial}{\partial \lambda_k}(D_0 G^b(i\omega_0) D_0^{-1})v_1 \right]
\end{aligned} \tag{4.3.6}$$

Use of (4.3.4), (3.2.16), (3.4.4), (4.3.6), and of the property

$$\frac{d}{dz}(M(z)^{-1}) = -M(z)^{-1}\frac{d}{dz}(M(z))M(z)^{-1} \tag{4.3.7}$$

where M(z) is a matrix, yields after some algebra

$$\begin{aligned}
\frac{\partial}{\partial \lambda_k} \Phi_\infty = \mathrm{Re}\Big[\ & u_1^* D_0 \begin{pmatrix} G_{13}^u \\ b^{-1} w G_{23}^u \end{pmatrix} (I - FG_{33}^u)^{-1}\frac{\partial}{\partial \lambda_k}(F(i\omega_0)) \\
& (I - FG_{33}^u)^{-1}(\, G_{31}^u \quad G_{32}^u \,) D_0^{-1} v_1 \Big]
\end{aligned} \tag{4.3.8}$$

where F, G_{ij}^u, b, w are computed at $\omega = \omega_0$. The derivatives of F with respect to its parameters (elements of Λ) depend on the particular form that the designer selected and they can be easily computed.

4.3.2 Solution of (4.2.3).

The first issue in this case is the computation of x_0. Note that this computation has to be made at every frequency ω. In practice only a set Ω with a finite number of frequencies is used, from which $||x_0^{-1}||_2$ can be computed approximately. Theorem 2 indicates that any basic descent method should be sufficient. The fact that it is possible to obtain an analytic expression for the gradient of $\mu_{\Delta^\circ}(G_{full}^x(i\omega))$ with respect to x, simplifies the problem even further. This is possible when (1.2.3) is used for the computation of μ and the two largest singular values of $DG^x_{full}D^{-1}$ for the optimal D's at the value of x where the gradient is computed, are not equal to each other. If this not the case a mean direction can be used as mentioned in the H_∞ case above.

Let the $\inf_{D \in \mathbf{D}^\circ} \bar{\sigma}(DG^x_{full}(i\omega)D^{-1})$ be attained for $D_0 = D_0(\omega; x)$ and let σ_1 be the maximum singular value and u_1, v_1 the corresponding singular vectors. Then the same steps for obtaining (4.3.6) are valid. Hence by using (3.4.6) and (3.4.7) we get after some algebra

$$\frac{\partial}{\partial x}\left(\mu_{\Delta^\circ}(G^x_{full}(i\omega))\right) = \mathrm{Re}\left[u_1^* D_0 \begin{pmatrix} 0 & 0 & 0 \\ WG^F_{21} & WG^F_{22}v & 0 \end{pmatrix} D_0^{-1}v_1\right] \qquad (4.3.9)$$

The second computational issue is the solution of (4.2.3). To obtain the gradient of $\|x_0^{-1}\|_2$ with respect to the filter parameters, we need to compute the gradient of $x_0(\omega)$ with respect to these parameters for every frequency $\omega \in \Omega$. From the definition of x_0 in (3.4.8) we see that as some filter parameter λ_k changes, $x_0(\omega)$ will also change so that $\mu_{\Delta^\circ}(G^x_{full}(i\omega))$ remains constantly equal to 1. Hence we can write

$$\frac{\partial \mu}{\partial x_0}\frac{\partial x_0}{\partial \lambda_k} + \frac{\partial \mu}{\partial \lambda_k} = 0 \implies \frac{\partial x_0}{\partial \lambda_k} = -\frac{\partial \mu}{\partial \lambda_k}\bigg/\frac{\partial \mu}{x_0} \qquad (4.3.10)$$

where μ is computed through (3.1.3). The denominator in the right hand side of (4.3.10) is given from (4.3.9). As for the numerator, it can be computed in the same way as (4.3.6) and (4.3.8) but with G^x_{full} instead of G^b:

$$\frac{\partial}{\partial \lambda_k}\left(\mu_{\Delta^\circ}(G^x_{full}(i\omega))\right) = \mathrm{Re}\left[u_1^* D_0 \begin{pmatrix} G^u_{13} \\ xWG^u_{23} \end{pmatrix}(I - FG^u_{33})^{-1}\right.$$

$$\left.\frac{\partial}{\partial \lambda_k}(F(i\omega))(I - FG^u_{33})^{-1}\begin{pmatrix} G^u_{31} & G^u_{32}v & 0 \end{pmatrix} D_0^{-1}v_1\right] \qquad (4.3.11)$$

Hence $\partial x_0/\partial \lambda_k$ can be computed from (4.3.9), (4.3.10), (4.3.11).

Appendix A

A.1 Proof of Thm. 2.2.1.

We shall show that any Q given by (2.2.2) makes $IS1$ stable. From substitution of (2.2.2) into (1.2.2) it follows that all that is required is that $(Pb_p^2Q_1 \quad b_p^2Q_1P \quad Pb_p^2Q_1P)$ be stable, which is true because of Assumptions A.1, A.2 and the properties of Q_1.

Assume that Q makes $IS1$ stable. Then the difference matrix

$$IS1(Q) - IS1(Q_0) = (\,(Q - Q_0) \quad P(Q - Q_0) \quad (Q - Q_0)P \quad P(Q - Q_0)P\,)$$
$$(A.1.1)$$

is stable. The fact that P has no zeros at the location of the unstable poles makes the stability of the matrix in (A.1.1) equivalent to the stability of $(Q - Q_0)$, $P(Q - Q_0)P$. Then, when assumption A.1.c holds, we have $P = b_p\hat{P}$, where $\hat{P}$ has no zeros at the open RHP poles of P and its only only unstable poles are at $s = 0$, from which it follows that $(Q - Q_0) = b_p^2Q_1$ with Q_1 stable and such that PQ_1P has no poles at $s = 0$. If A.1.c does not hold, Q_1 should also have the property that it makes PQ_1P stable.

A.2 Proof of Thm. 2.3.1.

We shall assume that a Q_0 exists, which in addition to the properties mentioned in Thm. 2.1.1, it also produces a matrix $(I - PQ_0)V^0$ with no poles at $s = 0$, where V^0 is a diagonal matrix with l_v poles at $s = 0$ in every element with l_v the maximum number of such poles in any element of v. If assumption A.1.c does not hold, then each column of V^0 also satisfies A.3.b and Q_0 makes $(I - PQ_0)V^0$ stable. Its existence will be proven by finding an optimal solution that has such properties. Substitution of (2.2.2) into (2.1.2) and use of the fact that pre- or postmultiplication of a function with an allpass does not change its L_2-norm, yields:

$$\Phi(v) = \|b_p^{-1}b_v P_A^{-1}(I - PQ_0)\hat{v} - b_pb_v P_M Q_1\hat{v}\|_2^2$$
$$\stackrel{\text{def}}{=} \|f_1 - f_2 Q_1\hat{v}\|_2^2 \qquad (A.2.1)$$

L_2, the space of functions square integrable on the imaginary axis, can be decomposed into two subspaces, H_2 the subspace of functions analytic in the RHP (stable functions) and its orthogonal complement $H_2^\perp$ that includes any strictly unstable functions. Then f_1 can be uniquely decomposed into two orthogonal functions $\{f_1\}_- \in H_2$ and $\{f_1\}_+ \in H_2^\perp$:

$$f_1 = \{f_1\}_- + \{f_1\}_+ \qquad (A.2.2)$$

From (A.2.1) one can see that if improper Q's are allowed, then f_1 may not be an L_2 function. However, in order for $\Phi(v)$ to be finite, the optimal Q_1 has to make $f_1 - f_2 Q_1\hat{v}$ strictly proper. The assumption will be made that is the case and it will be verified at the solution has this property. Hence to proceed we shall use the convention that when a decomposition as in (A.2.2) of a function is obtained through a partial fraction expansion, all improper and the constant terms are included in $\{.\}_-$.

When A.1.c holds, inspection of (A.2.1) shows that $f_2 Q_1 \hat{v}$ can have no poles in the closed RHP except possibly for some poles at $s = 0$ introduced by $\hat{v}$. f_1 however has no poles at $s = 0$ because $(I - PQ_0)V^0$ has no such poles. Hence for $\Phi(v)$ to be finite, $f_2 Q_1 \hat{v}$ should have no poles at $s = 0$. Hence the optimal Q_1 has to be such that these poles are cancelled. When A.1.c does not hold, then the fact that $(I - PQ_0)V^0$ is stable impies that an acceptable Q_1 (and therefore the optimal Q_1 as well) makes PQ_1v stable and therefore the optimal Q_1 is such that $f_2 Q_1 v$ is stable. We shall assume that Q_1 has this property. It should be verified at the end however that the solution indeed has the property. We can then write

$$\Phi(v) = \|\{f_1\}_+\|_2^2 + \|\{f_1\}_- - f_2 Q_1 \hat{v}\|_2^2 \qquad (A.2.3)$$

The first term in the right hand side of (A.2.3) does not depend on Q_1. Hence for solving (2.1.3) we only have to look at the second term. The obvious solution is

$$Q_1 \hat{v} = f_2^{-1}\{f_1\}_- \qquad (A.2.4)$$

Clearly such a Q_1 produces a stable $f_2 Q_1 \hat{v}$ as it was assumed. Also $f_1 - f_2 Q_1 \hat{v} = \{f_1\}_+$, which has no improper or constant terms.

It should now be proved that Q_1's that satisfy the internal stability requirements exist among those described by (A.2.4) so that the obvious solution is a true solution. For $n = 1$, (A.2.4) yields a unique Q_1, which can be shown to satisfy the requirements by following the arguments in the Proof of Thm 2.4.1 in Appendix A.3. For $n \geq 2$ write

$$Q_1 \overset{\text{def}}{=} (\, q_1 \quad q_2 \,) \qquad (A.2.5)$$

$$\hat{V}_2 \overset{\text{def}}{=} (\, \hat{v}_2 \quad \ldots \quad \hat{v}_n \,)^T \qquad (A.2.6)$$

where q_1 is $n \times 1$ and q_2 is $n \times (n-1)$. Then from (A.2.4) it follows that

$$Q_1 = (\, \hat{v}_1^{-1}(f_2^{-1}\{f_1\}_- - q_2 \hat{V}_2) \quad q_2 \,) \qquad (A.2.7)$$

We now need to show that a stable q_2 exists such that Q_1 is stable and produces a PQ_1P with no poles at $s = 0$ (or in the closed RHP when A.1.c does not hold). Write

$$q_2 = s^{3l_v} \prod_{i=1}^{q}(s - \pi_i)^3 \hat{q}_2 \qquad (A.2.8)$$

where $\hat{q}_2$ is stable. Then from (A.2.7) it follows that in order for PQ_1P not to have any poles at $s = 0$ it is sufficient that $P\hat{v}_1^{-1}f_2^{-1}\{f_1\}_-\{P\}_{1strow}$ have no such poles. This holds because the poles in the P on the left cancel with the

P_M^{-1} in f_2^{-1} and v_1 has by assumption A.4 at least as many poles at $s = 0$ as the 1st row of P. When A.1.c does not hold, then the same type of argument and the fact that A.3.b holds, imply that PQ_1P has no poles in the open RHP either. Let us now examine the stability of Q_1. The only poles in the open RHP may come from $\hat{v}_1^{-1}$. Let α be such a pole (zero of v_1). Then for stability we need to find $\hat{q}_2$ such that

$$\hat{q}_2(\alpha)\hat{V}_2(\alpha) = \alpha^{-3l_v}\prod_{i=1}^{q}(\alpha - \pi_i)^{-3}f_2^{-1}(\alpha)\{f_1\}_-(\alpha) \qquad (A.2.9)$$

The above equation always has a solution because the vector $\hat{V}_2(\alpha)$ is not identically zero since any common RHP zeros in v were factored out in v_0.

We shall now proceed to obtain an expression for $Q\hat{v}$. (2.2.2) and (A.2.7) yield

$$\begin{aligned} Q\hat{v} &= b_pb_v^{-1}P_M^{-1}\left[b_p^{-1}b_vP_A^{-1}PQ_0\hat{v} - \{b_p^{-1}b_vP_A^{-1}PQ_0\hat{v}\}_- + \{b_p^{-1}b_vP_A^{-1}\hat{v}\}_-\right] \\ &= b_pb_v^{-1}P_M^{-1}\left[\{b_p^{-1}b_vP_A^{-1}PQ_0\hat{v}\}_{0+} + \{b_p^{-1}b_vP_A^{-1}\hat{v}\}_-\right] \qquad (A.2.10) \end{aligned}$$

where $\{.\}_{0+}$ indicates that in the partial fraction expansion all poles in the closed RHP are retained. For (A.2.10), these poles are the poles of $b_p^{-1}b_v\hat{v}$ in the closed RHP; $P_A^{-1}PQ_0 = P_MQ_0$ is strictly stable because Q_0 is a stabilizing controller. When A.1.c holds, the stability of $(I - PQ_0)P$ and the fact that the residues of P at the open RHP poles are full rank imply that at these poles $I - PQ_0 = 0$. Also the fact that $(I - PQ_0)V^0$ has no poles at $s = 0$ imply that $(I - PQ_0)$ and its derivatives up to the $(l_v - 1)^{th}$ are also equal to zero at $s = 0$. When A.1.c does not hold, the fact that $(I - PQ_0)V^0$ is stable and that the columns of the diagonal V^0 satisfy A.3.b, imply that $(I - PQ_0) = 0$ at $\pi_1, ..., \pi_q$. Thus (A.2.10) simplifies to (2.3.4).

We simply need to establish that a stabilizing controller Q_0 with the property that $(I - PQ_0)V^0$ has no unstable poles exists. The selection of a V^0 with the properties mentioned in the beginning of this section and no RHP zeros and its use instead of V in (2.4.3) yields such a controller as it follows from the proof of Thm. 2.4.1 in Appendix A.3.

A.3   Proof of Thm. 2.4.1.

A stabilizing controller that solves (2.1.5) has to solve (2.1.3) for all $x \in \mathbf{R}^n$. Hence it has to satisfy (2.3.4) for all $v = Vx$, $x \in \mathbf{R}^n$. For each of the n linearly independent directions $(2, 1, ..., 1)$, $(1, 2, ..., 1)$, $(1, 1, ..., 2)$, the factor $v_0(s)$ containing the common RHP zeros of its elements is the same as the one for the direction $(1, 1, ..., 1)$. Therefore for each of them we can substitute in (2.3.4) $\hat{v} = \hat{V}x$, where $\hat{V}$ is defined through (2.2.1),(2.3.2),(2.4.2). Then from Linear Algebra it follows that there is only one Q with this property:

$$Q = b_pb_v^{-1}P_M^{-1}\{b_p^{-1}b_vP_A^{-1}\hat{V}\}_*\hat{V}^{-1} \qquad (A.3.1)$$

This solution however is not necessary stabilizing because not every Q that satisfies (2.3.4) for some x, is. To start with, Q is not stable if $\hat{V}$ has RHP zeros (unless of course P is stable and minimum phase). This will be the case when there RHP zeros in V that are not present in every element of V. In this case, there exists no solution to (2.1.5), which is part (ii) of the Theorem. When $\hat{V}^{-1}$ is stable, we still have to establish that the internal stability matrix $IS1$ is stable. Careful inspection shows that both Q and PQ are stable. We also have

$$(I - PQ)P = b_p b_v^{-1} P_A \{b_p^{-1} b_v P_A^{-1} \hat{V}\}_A \hat{V}^{-1} P \qquad (A.3.2)$$

where $\{.\}_A$ indicates that after a partial fraction expansion, ony the terms corresponding to poles of P_A^{-1} are retained. These poles are cancelled in (A.3.1) by P_A. Then from assumptions A.3, A.4, it follows that $(I - PQ)P$ is stable.

A final, but very important point is to show that the above Q minimizes the ISE for any $x \in \mathbf{R}^n$, when of course all the RHP zeros of V appear in every element with the same degree. But then $v_0(s)$ is the same for any direction $x \in \mathbf{R}^n$ and therefore for any x it suffices that (2.3.4) is satisfied, a property which the above controller has.

A.4   Proof of Thm. 2.4.2.

The proof follows that of Thm. 2.4.1 in A.3, with V_M used instead of $\hat{V}$. V_M appears because the directions in (2.4.6) are used and as a result for each direction the corresponding v_0 includes all the RHP zeros of the corresponding element v_i of V.

Appendix B

B.1   Proof of Theorem 3.4.1.

For a matrix K partitioned as

$$K = \begin{pmatrix} K_{11} & K_{12} \\ K_{21} & K_{22} \end{pmatrix} \qquad (B.1.1)$$

define

$$R(K, \Delta) \overset{\text{def}}{=} K_{22} + K_{21} \Delta (I - K_{11}\Delta)^{-1} K_{12} \qquad (B.1.2)$$

Then the transfer function relating v to e in Fig.4 is $R(G^F, \Delta)$ and since Fig.1a and Fig.4 are equivalent, we get by using (1.1.5)

$$E = R(G^F, \Delta) \qquad (B.1.3)$$

The properties of the SSV and (3.4.8) imply (Doyle,1985) that

$$\sup_{\Delta \in \Delta} \bar{\sigma}(R(G^{x_0}_{full}, \Delta)) = 1 \qquad (B.1.4)$$

From (3.4.6), (3.4.7), (B.1.2), (B.1.3), it follows after some algebra that

$$R(G_{full}^{x_0}, \Delta) = (\, x_0 Ev \quad 0 \,) \tag{B.1.5}$$

Then from (B.1.4),(B.1.5) and the definition of the singular values, it follows, since $x_0 Ev$ is a vector:

$$\sup_{\Delta \in \Delta} (x_0^2 v^* E^* Ev) = 1 \qquad \forall \omega$$

$$\Longrightarrow \sup_{\Delta \in \Delta} \int_{-\infty}^{+\infty} v^* E^* Ev \; d\omega \;\; = \int_{-\infty}^{+\infty} x_0^{-2} \; d\omega$$

$$\Longleftrightarrow \sup_{\Delta \in \Delta} ||Ev||_2 = ||x_0^{-1}||_2 \tag{B.1.6}$$

B.2   Proof of theorem 3.4.2.

Let $0 < x_2 \leq x_1$. Then we can write $x_2 = x_1\beta$, where $0 < \beta \leq 1$. From (3.4.10) we have

$$DM^{x_2}D^{-1} = \begin{pmatrix} D_1 & 0 \\ 0 & D_2 \end{pmatrix} \begin{pmatrix} I & 0 \\ 0 & \beta I \end{pmatrix} M^{x_1}D^{-1}$$

$$= \begin{pmatrix} I & 0 \\ 0 & \beta I \end{pmatrix} DM^{x_1}D^{-1} \tag{B.2.1}$$

Then the properties of the singular values yield

$$(B.2.1) \; \Longrightarrow \bar{\sigma}(DM^{x_2}D^{-1}) \leq \bar{\sigma}\begin{pmatrix} I & 0 \\ 0 & \beta I \end{pmatrix} \bar{\sigma}(DM^{x_1}D^{-1}) \tag{B.2.2}$$

$$\Longrightarrow \bar{\sigma}(DM^{x_2}D^{-1}) \leq \bar{\sigma}(DM^{x_1}D^{-1}) \qquad \forall D \in \mathbf{D} \tag{B.2.3}$$

$$\Longrightarrow \inf_{D \in \mathbf{D}} \bar{\sigma}(DM^{x_2}D^{-1}) \leq \inf_{D \in \mathbf{D}} \bar{\sigma}(DM^{x_1}D^{-1}) \tag{B.2.4}$$

$$QED \tag{B.2.5}$$

C References

1. B. D. O. Anderson, "An Algebraic solution to the Spectral Factorization Problem", *I.E.E.E. Trans. Aut. Control*, AC-12, p. 410, 1967.

2. C. C. Chu, *H∞-Optimization and Robust Multivariable Control*, Ph.D. Thesis, Dept. of Electr. Eng., Univ. of Minnesota, 1985.

3. J. C. Doyle, "Analusis of Feedback Systems with Structured Uncertainty", *I.E.E. Proc.*, Part D, V129, p. 242, 1982.

4. J. C. Doyle et al., *Lecture Notes*, 1984 ONR/Honeywell Workshop on Advances on Multivariable Control.

5. J. C. Doyle, "Structured Uncertainty in Control System Design", 1985 CDC, Fort Lauderdale, FL.

6. C. E. Garcia and M.Morari, "Internal Model Control. A Review and Some New Results", *Ind. and Eng. Chem., Proc. Des. and Dev.*, 21, p.308, 1982.

7. T. Kailath, *Linear Systems*, Englewood Cliffs, NJ: Prentice Hall, 1980.

8. M. Morari, E. Zafiriou and C. G. Economou, *An Introduction to Internal Model Control*, in preparation, 1987.

9. M. Vidyasagar, *Control System Synthesis*, MIT Press, Cambridge, MA, 1985.

10. E. Zafiriou and M. Morari, "Synthesis of the IMC Filter by Using the Structured Singular Value Approach", 1986 ACC, Seattle, WA, p.1-6.

11. G. Zames, "Feedback and Optimal Sensitivity: model reference transformations, Multiplicative semi-norms and approximate inverses", *IEEE Trans. Aut. Control*, AC-26, p. 301, 1981.

D Acknowledgements

The author would like to thank Prof. V. Manousiouthakis for useful discussions and remarks.

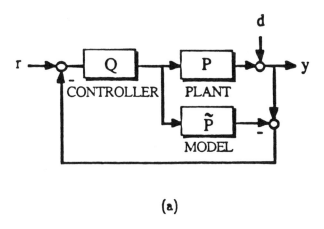

(a)

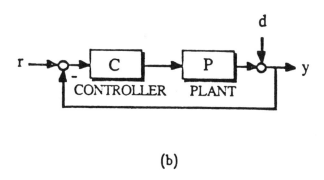

(b)

Figure 1: (a) Internal Model Control Structure. (b) Feedback Control Structure.

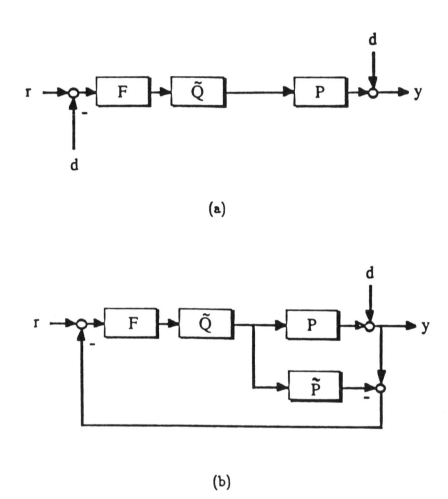

Figure 2: IMC Structure with the Filter $\mathbf{F}$. (a) $\mathbf{P} = \tilde{\mathbf{P}}$. (b) $\mathbf{P} \neq \tilde{\mathbf{P}}$.

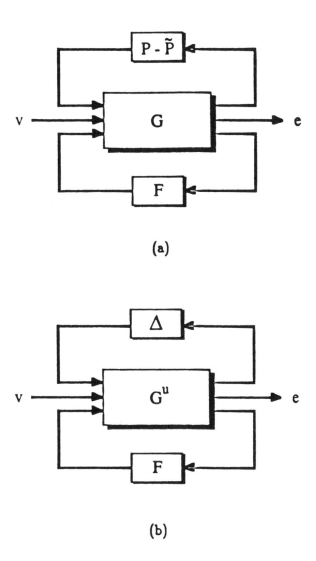

(a)

(b)

Figure 3: Model Uncertainty Block Diagrams.

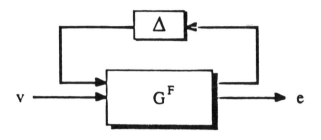

Figure 4: SSV Block Diagram.

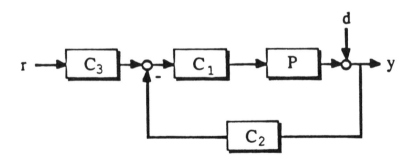

Figure 5: Two-Degree-of-Freedom Feedback Structure.

Characterization of Distillation Nonlinearity for Control System Design and Analysis

Karen A. McDonald
Department of Chemical Engineering
Univerisity of California
Davis, CA 95616

December 15-18, 1986

Abstract

This paper presents a summary of work done to quantitatively characterize the nature of nonlinearity in moderate and high purity distillation towers. Physically realistic representations of process nonlinearity are essential to obtaining practical answers to robustness issues and for designing controllers which attempt to compensate for process nonlinearity. This paper discusses the various ways in which distillation nonlinearity can be represented and presents techniques for using our knowledge about the structure of the nonlinearity to improve control performance and answer questions concerning control system robustness.

I Introduction

Many recent articles have attempted to narrow the "gap" between chemical process control theory and chemical process control practice by addressing the important issues of robust stability and robust performance. Recognizing the fact that linear, time-invariant process models used to design and analyze control system performance only approximate the steady state and dynamic characteristics of real nonlinear chemical processes, these articles provide a general framework in which the implications of plant-model mismatch can be analyzed. In particular, Doyle (1982, 1984) has developed a framework based on the Structured Singular Value (SSV) for incorporating model uncertainty which provides necessary and sufficient conditions for robust stability (i.e. any plant which is a member of a predefined "uncertainty family" will be stable) and robust performance (i.e. all plants in the uncertainty family will satisfy the predefined performance specifications). Skogestad and Morari (1985) have used this framework to evaluate the effects of input uncertainty (i.e. uncertainty in the manipulated variable), output uncertainty (i.e. due to measurement errors) and nonlinearity in distillation. Zafiriou and Morari (1986) have used the SSV to design the filter for an IMC controller. Thus,

279

the SSV approach appears to be useful for both control system analysis and design.

It is important, however, to understand the assumptions inherent in the description of the uncertainty set and to recognize that the usefulness of answers obtained using this framework will only be as good as the information on which it is based, the nominal model and the uncertainty description. For example, Figure 1 shows the norm bounded uncertainty family used in the SVD approach. It is critical, therefore, that we understand the structure and characteristics of sources of uncertainty in real chemical processes so that we can quantitatively specify realistic uncertainty sets and evaluate the validity and/or conservativeness of assumptions made in the SSV approach. Now that a framework exists for analyzing the effects of model uncertainty, systematic methods are needed to define realistic uncertainty sets for real processes.

There are many sources of potential model inaccuracy in real chemical processes. These include the selection of low order model structures to describe high order processes, inaccurate identification of model parameters due to poor measurements, uncertainty in the manipulated variables and time-varying phenomena. One of the major sources of uncertainty which is present in almost all chemical processes is process nonlinearity. Many nominal process models are determined by linearization around a particular operating point and the model will become more and more inaccurate as the process is perturbed from its steady state operating point. To make matters worse, control systems are now expected to perform over much wider operating regimes as directed by optimization routines. In many cases though we have some understanding, at least qualitatively, of the nature of the nonlinearity present in the process as it is perturbed over some operating regime. In fact, although nonlinearity contributes to inaccuracies in the process model and can be treated as an "uncertainty," there is definitely a structure (although not exactly known) associated with it. For example, in MIMO systems, when the process is perturbed the gain associated with one element might increase when another element decreases. Nonlinearity cannot be considered a random or uncorrelated uncertainty. If we can somehow include the nonlinearity in our model then we can determine the uncertainty family which arises due to nonlinearity. To get a complete picture of the uncertainty family, we would have to include the uncertainties due to all other sources as well, but as a first step we will analyze the "uncertainty" due to nonlinearity.

A quantitative understanding of the "uncertainty" that nonlinearity introduces is not only important for robustness analysis but is also a prerequisite for designing controllers which attempt to compensate for process nonlinearity. In order to design scheduling controllers, or controllers based on linearizing transformations, it is first necessary to mathematically characterize the nonlinearity. Although it may be possible to develop detailed mathematical models based on first principles, it should be emphasized that these models may not be available at the design stage when simple models which contain the essential features of the nonlinearities would be more useful.

In this paper, moderate and high purity binary distillation processes will be used as an examples of processes which exhibit nonlinear gain and dynamic characteristics. The first part of the paper discusses the characterization of nonlinearity for these towers, and the implications in terms of robustness analysis. The second part of the paper summarizes work which uses a particular characterization of the nonlinearity to implement a gain/time constant scheduled version of Dynamic Matrix Control on the nonlinear towers.

II   Characterization of Distillation Nonlinearity

To begin with I would like to make some qualitative observations concerning the nature of nonlinearity in moderate and high purity distillation towers. Based on numerous open and closed loop nonlinear simulations of moderate and high purity towers, I have found the following general characteristics (McDonald, 1985):

i)   Process Order:

Open loop responses to step changes in standard manipulated variables (L, V or D, V) and typical disturbances (x_f or F) can be fairly well represented as first order, deadtime systems (deadtimes are usually small however compared with time constants). These observations agree with other investigators (Wong and Seborg, 1986, Kapoor, 1986) and result from the fact that one time constant is dominating (Moczek, 1965). Thus, one way of analyzing the nonlinear behavior of towers is by evaluating process gain, time constant and dead time variations for an approximate first order system as the process is perturbed from its design point.

ii)   Gain Characteristics:

In all cases, when the mole fraction of the impurity in one of the product streams approaches zero and the impurity in the other product stream increases (as is often the case for most inputs), the gain corresponding to the pure stream decreases and the gain for the more impure stream increases. These process gain variations arise because the input changes which increase distillate product purity, for example, cause the composition profile to move up the tower towards the pinch zone, whereas changes which decrease distillate purity move the composition profile away from the pinch zone. Process gains must go to zero as the stream purity approaches unity since the purity cannot exceed unity. Of course the magnitude of the gain changes will depend on how far the process has been perturbed, but for the high purity towers order of magnitude changes process gains were observed when the distillate product was changed over a range typically encountered during transients. There are many simple models which can predict this kind of behavior *even* if the models do not accurately predict the gains. For example, the simple analytical expressions developed by McAvoy (1983) which require specification of only five parameters (x_D, x_B, α, N, x_F, F) do a fairly good job of representing the gain variations obtained using a detailed steady state package for the moderate and high purity towers (McDonald, 1985). Wong and Seborg (1986) have also presented a linear model in which the gains, and time constants are predicted empirically by fitting open loop response data to particular functions of product composition.

iii) Time Constants:

While process gains can be predicted using steady state design packages, process time constants are much more difficult to predict. One method which has been used to estimate time constants for high purity towers is to numerically generate the frequency response (using the "stepping" technique of Lamb and Rippin, 1969) of a set of linearized differential equations which represent a simplified dynamic distillation model (Luyben, 1973). This method invariably leads to extremely large time constants if the linearization point is one in which both product purity specifications are high. Work done by Kapoor et al. (1986) has shown that distillation towers contain positive recycle systems and the loop gain of the positive recycle system approaches unity at the high purity state. As the tower is perturbed away from the high purity design state (i.e. as one of the product streams becomes more impure) the gain of the positive feedback loop drops and tower time constants are greatly reduced. This behavior has also been observed in nonlinear simulations where time constants for high purity towers can change by an order of magnitude depending on the size of the perturbation (McDonald, 1985).

Kapoor (1986) has presented analytical expressions to predict time constants using a minimum amount of information: N, x_F, R, F/D, H and α. These expressions agree well with the numerical stepping technique. Other investigators have used various model reduction techniques to predict time constants but these methods usually require a substantial amount of information (i.e. such as tray compositions).

iv) Dead Times:

The numerical stepping technique can also be used to evaluate dead times (Kapoor, 1986). In most cases the dead times are small compared with the process time constants and do not change much over typical operating regimes.

Given these numerical and analytical methods to predict the changes in the process due to nonlinearity, can the uncertainty due to nonlinearity be adequately modeled using the uncertainty descriptions implicit in the formulation of the SSV approach? To answer this question an analysis of a high purity ($x_D = 0.994$, $x_B = 0.0062$) binary tower is carried out. This tower, referred to a Tower D, has been studied previously by several investigators (Kapoor, 1986, McDonald, 1987).

To analyze the uncertainty due to nonlinearity, Tower D was perturbed from its steady state design point over an arbitrarily chosen range: $0.988 < x_D < 0.998$, $0.002 < x_B < .012$. Reflux and vapor boilup are chosen as manipulated variables. At each new perturbed operating point, the frequency response was determined using the stepping technique described by Luyben (1973). The frequency response was carried out by linearizing the nonlinear equations representing the simplified (i.e. assuming equimolar overflows, constant relative volatility etc.) dynamic model about the perturbed operating point. Figure 2 shows the frequency response for the x_D–L response at the design point (solid line) and also over the perturbed states (squares) for two frequencies. The first thing one notices is that the "uncertainty" families are not easily represented by a disk centered about the nominal point. By doing so, one introduces a substantial amount of conservatism (i.e. there are many addition members of the uncertainty set). It is important to remember, however, that other uncertainties have veen neglected and that the overall uncertainty may contain the relationship between phase and gain uncertainty that is implicit in a disk shaped region. However, if nonlinearities are considered to be the major source of inaccuracies, it may be worthwhile to introduce a more realistic uncertainty description. There are other methods which are less structured in their uncertainty description (Owens, 1984) and these may prove to be more useful in these cases.

Another point which should be mentioned in connection with MIMO systems such as distillation, is the relationship between the uncertainty sets for each element of the process transfer function matrix. The Nyquist plot for x_B–V looks very similar to the x_D–L plot. It is important to recognize that a particular perturbed condition in the x_D–L uncertainty set corresponds to a particular point on the x_B–V uncertainty set. Thus in the SSV analysis, an uncertainty family in which the actual plant can be represented by *independently* selecting a member from each of the elements' uncertainty sets is physically unrealistic and introduces addition conservatism. Given our understanding of the relationship between the process gains, time constants and dead times in the various elements it should be possible to include these dependencies in the analysis.

Based on the Nyquist plot shown in Figure 2, the multiplicative uncertainty

$$l_m(i\omega) = \frac{(p(i\omega) - \tilde{p}(i\omega))}{p(i\omega)}$$

was determined. Figure 3 shows a plot of $\bar{l}_m(\omega)$

$$\bar{l}_m(\omega) = \max_{\Pi(\omega)} l_m(i\omega)$$

where

$$\Pi(\omega) = \text{uncertainty set (see Fig. 2)}.$$

In this case $\bar{l}_m$ does not increase monotonically with frequency but reaches a maximum at some frequency. Changing the upper and lower limits on the ranges of x_D and x_B raised or lowered this curve and caused a shift in the peak frequency but did not change the shape substantially. It should be pointed out that the frequency response was generated using the entire set of ODEs and no assumption was made concerning the order of the process.

In using the SVD approach, an estimate of $\bar{l}_m$ (or equivalently the unstructured multiplicative input (L_i) or output uncertainty (L_o) for MIMO systems) is required. This preliminary analysis shows one way in which the uncertainty due to nonlinearity might be estimated and also indicates the conservatism associated with uncertainty description used in the SVD approach.

III   Compensating for Process Nonlinearity

Another way in which our knowledge of process nonlinearity can be used is to develop scheduling algorithms which update process models on-line. For example, multivariable gain and time constant scheduling has been evaluated using simple analytical models to update key process model parameters (process gains and time constants) (McDonald, 1987). In this study, multivariable gain and time constant scheduling was used in conjunction with Dynamic Matrix Control for dual composition control of a moderate ($x_D = 0.98$, $x_B = 0.02$) benzene/toluene tower (Tower B) and a high purity ($x_D = 0.994$, $x_B = 0.0062$) isobutane/n-butane tower (Tower D).

First, nominal open loop responses were determined by averaging the open loop step responses generated by the nonlinear dynamic simulation. These responses were then scaled so that the gains matched those calculated from the steady state design package at the design point. As the process moved away from it's initial design state, simple analytical expressions were used to predict the process gains (McAvoy, 1983) and the time constants were predicted empirically (McDonald, 1985) as a function of product compositions. Thus, the nominal open loop responses were scaled to adjust for changes in the process gains, or contracted to adjust for the reductions in the time constants at perturbed operating points. Scheduling in this way (i.e. based on a nominal process model) allows for a smooth transition in the process model parameters as the product compositions move away from the design state and insures that the process model parameters return to their original values as the control system bring the process back to the design state. For the 2 × 2 dual composition control problem, only eight parameters (the gains and time constants for each of the responses) needed to be estimated at each time step. Although, in general, updating the process model on-line requires on-line matrix inversion of the dynamic matrix, matrix inversion can be avoided if the nominal process model can be factored from the updated process model i.e.

$$\mathbf{A}_p = \mathbf{D}_1 \mathbf{A}_o \mathbf{D}_2$$

where $\mathbf{D}_1$ and $\mathbf{D}_2$ are diagonal matrices. For energy balance control, the a_{11} and a_{12} gains change simultaneously, as do the a_{21} and a_{22} gains, so on-line matrix inversion can be avoided.

Several gain and time constant scheduled DMC algorithms were compared to the standard (nonscheduled) DMC for regulatory control performance for several choices of manipulated variables. Tuning parameters were selected according to the guidelines proposed by Cutler (1983). The same tuning parameters were used in both the modified and standard DMC algorithms for a particular tower and choice of manipulated variables. The feedforward capability of the DMC algorithm was not used; the feed disturbances were assumed to be unmeasureable.

For both towers, the scheduled DMC controllers resulted in improved control performance compared with the standard DMC controller. As would be expected, the largest differences between the scheduling approaches and the standard DMC were observed for cases in which product compositions deviated the farthest from the design point (i.e. larger disturbances, tunings with larger move suppression factors, and larger analyzer deadtimes). Figure 4 presents results for Tower D for a 20% change in feed composition. The peak in the bottoms composition is reduced by about 25% and the closed loop response is much faster.

IV   Summary

In conclusion, understanding and modeling process nonlinearity is essential for addressing issues of robust performance and stability as well as designing controllers which compensate for nonlinearity in real processes. In designing compensating controllers, simple analytical or empirical models which contain the main features of the nonlinearity can be used to reduce plant-model mismatch during transients or when the setpoint is changed. It may also be possible to use our knowledge of process nonlinearity to automatically retune controllers (i.e. adjust the filter constant in an IMC controller, or the move suppression factors in a DMC controller) by calculating l_m on-line as the process is perturbed and recalculating the tuning parameter to satisfy the robust stability and performance criteria. Future research is needed to provide estimates of the nature and magnitude of other uncertainties in process models and to determine how important the uncertainty description is in determining robust stability and performance.

References

[1] Cutler, C. R., "Dynamic Matrix Control, An Optimal Multivariable Control Algorithm with Constraints," *Ph.D. Thesis,* University of Houston, August (1983).

[2] Doyle, J. C., *IEE Proc.,* Part D, V. 129 (1982).

[3] Doyle, J. C., Lecture Notes, 1984 ONR/Honeywell Workshop on Advance or Multivariable Control.

[4] Kapoor, N., T. J. McAvoy, and T. Marlin, *AIChE J.,* Vol. 32, No. 3 (1986).

[5] Luyben, W. L. *Process Modeling, Simulation, and Control for Chemical Engineers.* New York: McGraw-Hill, 1973.

[6] McAvoy, T. J., "Interaction Analysis, Principles, and Applications," *ISA Monograph,* 1983.

[7] McDonald, K., "Predictive Control od Distillation Processes," *Ph.D. Dissertation,* University of Maryland, December 1985.

[8] McDonald, K., "Application of Dynamic Matrix Control to Moderate and High Purity Distillation Towers," *I&EC Proc. Des. Dev.,* in press (1987).

[9] Moczek, J. S., R. E. Otto, and T. J. Williams, "Approximation Model for the Dynamic Response of Large Distillation Columns," *CEP Symposium Ser.,* 61, pp. 136-146 (1965).

[10] Owens, D. H., "The Numerical Range: A Tool for Robust Stability Studies," *Systems and Control Letters,* 5, pp. 153-158 (1984).

[11] Skogestad, S. and M. Morari, "Model Uncertainty, Process Design, and Process Control," AIChE Annual Meeting, Chicago, November 1985.

[12] Wong, S. and D. Seborg, " Low-Order, Nonlinear Dynamic Models for Distillation Columns," *ACC Proceedings,* Paper TP6, Seattle, WA (1986).

[13] Zafiriou, E. and M. Morari, "Design of the IMC Filter by Using the Structured Singular Value Approach," *ACC Proceedings,* Paper WA1, Seattle, WA (1986).

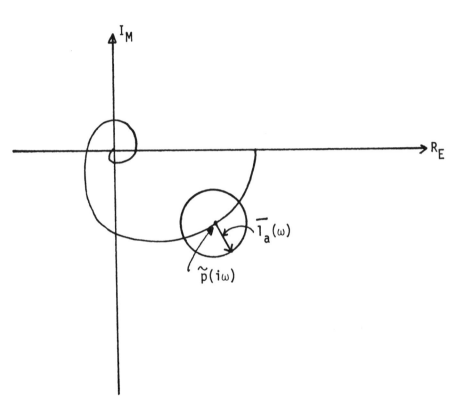

Figure 1: Norm Bounded Additive Uncertainty.

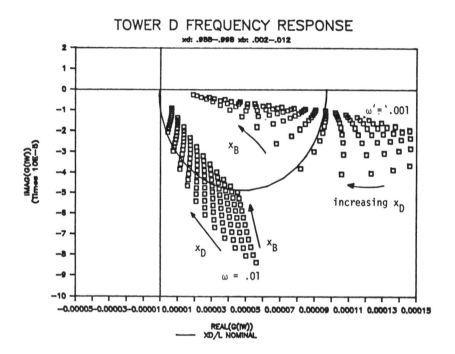

Figure 2: Tower D Nominal Frequency Response and Uncertainty Families at $\omega = 0.001, 0.01$.

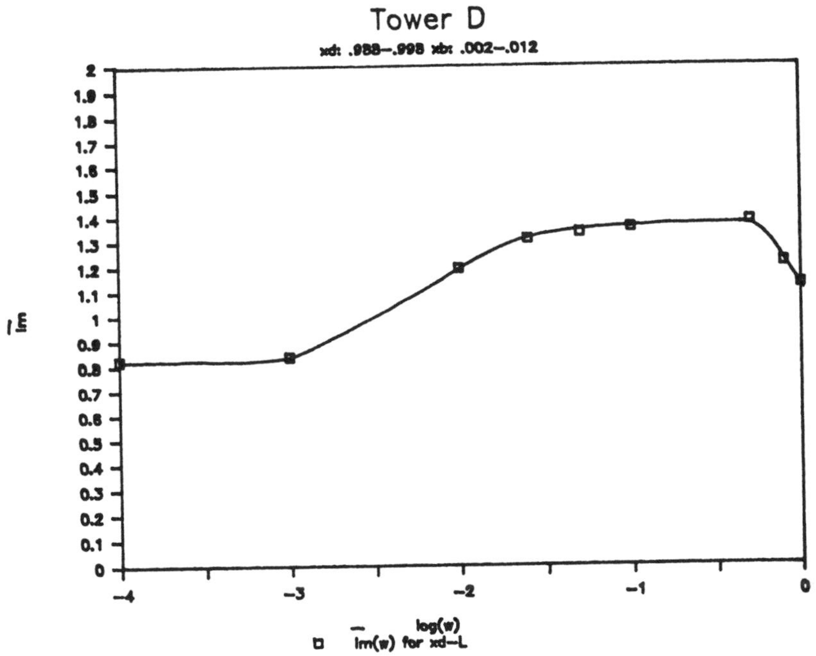

Figure 3: Multiplicative Uncertainty Bound for $x_D - L$ element, Tower D.

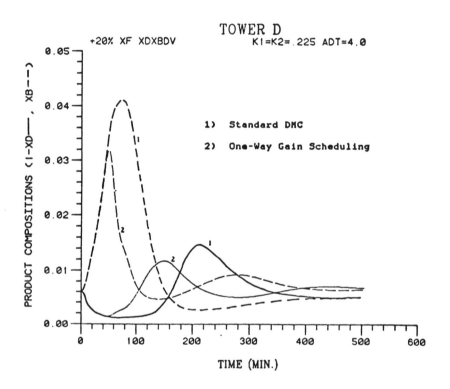

Figure 4: Tower D, Energy Balance Control, $+20\%$ Step in x_F.

Selecting Sensor Location and Type for Multivariable Processes

Charlie Moore, James Hackney, and David Canter
University of Tennessee
Department of Chemical Engineering
Knoxville, Tennessee 37996

December 15-18, 1986

Abstract

This paper presents several systematic procedures for selecting sensor type and locations for multivariable processes. The procedures are based on a singular value analysis of a process gain matrix and provide choices which are a practical compromise between independence and sensitivity.

Introduction

For many multivariable processes there is a decision to be made about sensor location and type which can be critical to the success or failure of a multivariable control system. It is important in multivariable control that the sensors selected be both sensitive and exhibit a low degree of interaction; otherwise, a control system may be doomed from the very beginning. Making a good judgement on the balance which must be made between these two typically conflicting objectives is a difficult problem. Sensor choices which are sensitive enough for good loop control are quite frequently very interactive in a multiloop control system. On the other hand, sensor which are relatively non-interacting can typically be too insensitive to be effectively used for feedback control.

Ethanol-Water Example

The question of proper sensor selection and placement is demonstrated clearly in the problem of the operation and control of distillation processes. Consider, for example, the ethanol-water distillation column shown in Figure 1 (ref. 4). Assume for this multivariable process that the first level objective is to stabilize the column by controlling two column temperatures. Note that with this 50 tray column there are 1225 possible combinations of sensor locations which

291

could be chosen. The basic concern is determining which is the best combination from the point-of-view of column control. It is not an easy problem to answer heuristically. It is simple enough to determine from the temperature profile (see Figure 2) that the two most sensitive trays in the column are trays 18 and 19. The problem with using these two locations is that they do not represent independent information about the column. Changes in the operation of the column will have about the same effect on these two adjacent trays. It is also reasonable to expect that the trays at either end of the column should be relatively independent, but the problem with this choice is that the temperature sensitivity is too low to be used for effective control.

For the ethanol-water column there is also the question of sensor types. Instead of using temperature sensors, analyzers could be used to more directly measure composition of some key component. The question includes the type of analyzer, what type of analysis is to be made, as well as whether the analysis should be made on the product streams or inside the column. This additional dimension adds greatly to the complexity of the sensor placement problem but it also adds to the possibility of improving the overall multivariable control problem.

Singular Value Decomposition

Several procedures will be developed in this paper for selecting sensor location and type, all of which are based on the principal of Singular Value Decomposition. SVD is a numerical algorithm developed to minimize computational errors involving large matrix operations but has also been used extensively in recent years to characterize and analyze various aspects of the multivariable control problem (1, 3, 5, 6, 10-13, 15-19). The singular value decomposition of a process gain matrix $\mathbf{K}$ results in three component matrices as follows:

$$\mathbf{K} = \mathbf{U}\mathbf{\Sigma}\mathbf{V}^{\mathrm{T}}$$

where

> $\mathbf{K}$ is an $n\mathrm{x}m$ matrix which describes the steady-state relationship between the manipulated variables and the various process sensors. The elements of $\mathbf{K}$ should be physically scaled to be representative of what the control system will actually see and should include the range and span of the valves and senors.

> $\mathbf{U} = \mathbf{U}_1 : \mathbf{U}_2 : \ldots \mathbf{U}_n$. $\mathbf{U}$ is an $n\mathrm{x}n$ orthonormal matrix, the columns of which are called the "left singular vectors." It is an appropriate coordinate system for viewing the sensors arranged such that the first column vector, $\mathbf{U}_1$, points in the direction of the most sensitive combination of sensors. The second column vector, $\mathbf{U}_2$, points in the direction of the next most sensitive combination of sensors, and so on.

$\mathbf{V} = \mathbf{V}_1 : \mathbf{V}_2 : \ldots \mathbf{V}_m$. $\mathbf{V}$ is an $m \times m$ orthonormal matrix, the columns of which are called the "right singular vectors." It is a coordinate system for viewing the manipulated variables arranged such that the first column vector, $\mathbf{V}_1$, points to the combination of control actions which have the most effect on the sensors. The second column vector, $\mathbf{V}_2$, points to the combination which has the next largest effect on the sensors, and so on.

Σ is an $n \times m$ diagonal matrix of scalars called the "singular values" that are organized in descending order such that $\sigma_1 \leq \sigma_2 \leq \sigma_3 \leq \ldots \leq \sigma_k \leq 0$. In terms of the multivariable control problem, the singular values represent the "decoupled" open loop gains of the multivariable process.

$CN = \sigma_1 / \sigma_k$. The "condition number" is another important parameter which can be determined from the singular value decomposition. It is a ration of the largest singular value to the smallest nonzero singular value and is an indication of the difficulty of the "$k \times k$" multivariable control problem. The larger the condition number, the more difficult it will be to control all "k" variables.

Algorithms for the determination of the SVD of a matrix have been studied by numerical analysts for some time and can be performed with great accuracy. It is discussed in many text books (8, 14, 20, 23) and software is readily available through a number of standard numerical packages (25, 26, 27, 28).

To demonstrate how SVD can be applied to a process gain matrix, consider the ethanol-water column operating with a DQ scheme (distillate rate and steam rate are the manipulated variables). At the normal operating conditions the gain matrix is shown in Figure 3 and was determined by a detailed steady-state column simulation package (6). Note that, with no specification on the location of the control sensors, the steady-state gain matrix is a 50x2 array of constraints which describes the temperature sensitivity on each tray with respect to each of the two manipulated variables. The decomposition of the gain matrix (shown in Figure 4) results in a 50x2 $\mathbf{U}$ matrix, a 2x2 $\mathbf{V}$ matrix, and it has 2 singular values.

Determining Sensor Locations

The SVD analysis of the gain matrix provides considerable insight into the problems of operation and control of a multivariable process. In terms of the decision of where to place sensors, it provides a very clear framework to evaluate each potential location in terms of both the sensitivity and the interaction aspects of the multivariable problem. Presented below are several SVD-based procedures for locating sensors which have been applied with considerable success to a number of industrial processes.

Principal Component Analysis

One straight forward way to select sensor locations using information from an SVD analysis is the principal component method. This method bases the choice on the location of the principal components in left singular vectors.

Consider again the properties and the physical significance of the column vectors in the U matrix. Each column in U is an orthonormal vector whose coordinate directions are described by each one of the process sensors. In the case of the SVD analysis of the 50 tray ethanol-water column, there are two vectors in U (U_1 and U_2), each has 50 elements, one for each of the 50 potential sensor locations in the column. U_1 is a vector which points in the most sensitive direction for the column, U_2 points in the next most sensitive direction which is perpendicular to the major direction of sensitivity. It stands to reason that the location of the principal component of each vector would be good choices for sensors which are relatively sensitive and also relatively independent.

The procedure for the principal component method of locating sensors in a multivariable system is very simple. Perform the SVD analysis on the gain matrix of the system and then choose the locations which correspond to the most sensitive element in each of the left singular vectors. For the ethanol-water example, the largest element in U_1(-0.61305) occurs on stage 18. The largest element in U_2 (0.41914) occurs on stage 13.

The principal component method is straight forward and works reliably in most cases. It is based on the assumption that the complete U vector can be reasonably well approximated by a much smaller vector. This may not always be true. Methods for testing this assumption as well as a more detailed method for sensor selection will be discussed later in this paper.

Modified Principal Component Method

Experience has shown that the location of sensors based on the principal component normally work well; however, there are cases where sensor sensitivity should be sacrificed for the sake of reduced sensor interaction. The trade-off between sensitivity and interaction is easier to see graphically by plotting the elements of each U vector versus sensor location. The "U vector plots" for the ethanol-water column are shown in Figure 5. Note, at the tray which corresponds to the peak of U_1 and also at the tray corresponding to the peak at U_2 that the coefficient of the other vector is not zero. This indicates that there will be, in fact, some level of interaction between the two sensors chosen by the principal component procedure.

The question is whether the overall balance between sensitivity and interaction might be better served by modifying the choice of the principal component method. Figure 6 provides a modified graphical presentation which more directly answers the question. This method plots the difference between the absolute value of the two vectors, and the selection procedure is based on the position of the positive and negative peaks (the positive peak indicates the position of the primary sensor and the negative peak indicates the position of the secondary sensor). Note that, for this example, the modified procedure suggests that trays 18 and 13 are good choices, even considering the interaction between the two "simplified" U vectors.

Overall vs. Partial SVD Analysis

It is important to realize, in the analysis of processes which have a choice of sensor locations, that there are at least two separate SVD analyses of significance. One is an overall SVD analysis based on a complete system with all possible sensor locations included. The other is the partial SVD analysis which is based only on those sensors which will be directly used in the control strategy. (There may also be other reduced systems which might need to be analyzed; for example, in distillation control it is not uncommon to base a control strategy on the temperature difference between two trays rather than actual temperature.)

The ethanol-water column provides a good example of the difference between an overall analysis and a partial analysis. In the table below the condition number and singular values are presented for the 50x2 overall analysis and for the 2x2 partial analysis using distillate rate and steam rate as manipulated variables. The 2x2 analysis is based on temperature sensors on tray 18 and 13 and is more representative of the multivariable problem which the actual control system would see. Note that the condition number is worse for the 2x2 partial analysis and that the open loop sensitivities for the smaller system are lower, but are still roughly the same magnitude.

		Overall Analysis (50x2)	Partial Analysis (2x2)
Condition Number	=	66.5	96.8
σ_1	=	0.08116	0.04984
σ_2	=	0.00122	0.00052

In general, the singular values of the partial analysis are less than the singular values for the full system. The condition number for the partial system, however, can be worse or better than predicted by the overall analysis, depending on the relative changes in each of the singular values. An improved condition number usually means a greater loss in sensitivity is in the first singular value than in the smallest. A condition number which is larger means that the greater loss in sensitivity is in the smallest singular value.

From the point-of-view of a control system design, the overall analysis is useful in gauging the magnitude of the problem and in determining appropriate sensor locations. The partial analysis, however, is the most appropriate to use for a more detailed phase of the control system design. In terms of the problem of selecting sensor locations, the difference between the condition numbers determined by the overall and the partial SVD analysis is a good validity test for the locations selected by the principal component method. If the condition number for the partial analysis is about the same as (or better than) the overall analysis, then the location of the sensors determined by the principal component analysis is probably a good choice. On the other hand, if the condition number from the partial analysis is considerably worse, then a more detailed method of sensor selection may need to be considered.

Global Method – The Intersivity Index

Location of sensors may, in some cases, require a more detailed study than a simple principal component analysis. One such method is a global search of all possible sensor combinations. A partial singular value analysis can be determined for each and every combination of sensors which makes physical sense. This procedure is more definitive but it is much more time-consuming and, in most cases, does not yield results which are significantly better than the principal component method.

Consider the global behavior of the ethanol-water column under a DQ first level scheme. A 2x2 SVD analysis of the 1225 possible combinations of two sensors is presented in Figures 7 and 8. These figures show in 3-D graphical form the two aspects of sensor selection which must always be considered: loop interaction and loop sensitivity. The loop interaction problem is presented as a reciprocal of the condition number. The higher the reciprocal of the local condition number, the better "conditioned" the multivariable control problem. The loop sensitivity is presented as a plot of the second (smallest) singular value. This represents the weakest part of the control system and must be at least greater than the noise level of the sensor if it is to be controlled. Note that from the condition number plot that there is a large number of sensor combinations which yield a much better "conditioned" multivariable system than does the combination selected by the principal component analysis. The problem with these choices, however, can be seen when the actual loop sensitivity is considered at these combinations. The sensitivity throughout most of the well conditioned region is far too low to be considered for good control. Note that the most sensitive region is around the sensor locations chosen by the principal component method. This choice, however, falls in a region where the condition is relatively poor compared to other choices.

When interaction and sensitivity are both considered, the sensor is somewhere between the two extremes. Consider Figure 9, which presents the "intersivity" one possible compromise between interaction and sensitivity. It is a plot of the product of these two measures (σ_2/CN) and predicts that a good location which balances the two problems is tray 25 and tray 12. The main difference between this choice and the choice offered by the principal component method is that the most sensitive temperature location has been avoided in order to improve the interaction problem. Although tray 25 is not as sensitive as tray 18, it is a healthy compromise between low interaction and strong sensitivity.

Selecting Sensor Type

The SVD analysis tools can also be used to help determine the type of sensor which should be used in a multivariable control system. Consider again the problem of the control of a distillation column. Temperature sensors may be used as an indirect indication of the composition or an on-line analyzer might be used to determine the composition directly. In deciding between using temperature or composition in the multivariable control system, there are many questions such as availability, reliability, cost, speed of response, and control effectiveness. It is the question of steady-state control effectiveness that SVD addresses. The SVD analysis can be used to quantitatively rank various choices with respect to control effectiveness in a specific multivariable application.

Consider, for example, a detailed analysis of the ethanol-water column operating under a DQ first level strategy. The table below compares a choice of temperature with a possible choice of six analyzers. The six analyzers include one for each component (ethanol, water, fusel oil) in each of the two phases (liquid, gas).

SVD ANALYSIS

	Overall CN	2x2 CN	Sensor Locations
Temperature:	66.5	96.84	18, 13
Vapor Phase: Analyzer for:			
Ethanol	14733.00	4874.47	7, 1
Water	*	*	50,21
Ethanol	*	*	50,15

(* extremely large numbers)

This analysis clearly shows that an analyzer which measures the composition of the vapor phase for ethanol at trays 1 and 18 would yield a very well-conditioned multivariable control problem. In this example, measuring, at well-placed locations, the composition of either the ethanol or the water gives a better-conditioned multivariable system than using temperature. It is interesting to note, however, that analysis of the liquid phase for any of the three components yields very poor results.

Summary

This paper presents two approximately and one detailed procedure for determining the proper sensor location and type for use in multivariable process control. The procedures are all based on the singular value decomposition of a steady-state gain matrix description of the process. The procedures are relatively straight forward and have been applied with very good success to a number of industrial unit operations.

References

[1] Arkun, Y. and S. Ramakrishnan, "Structural Sensitivity Analysis in the Synthesis of Process Control Systems," *Chemical Engineering Science,* 39, 7&8, 1167 (1984).

[2] Bristol, E. H., "On a New Measure of Interaction for Multivariable Process Control," *IEEE Trans. Auto Contro.,* AC-11, 133 (1966).

[3] Bruns, D. D. and C. R. Smith, "Singular Value Analysis: A Geometrical Structure for Multivariable Process," Presented at AIChE Winter Meeting, Orlando, Florida.

[4] Canter, D. L., "A Detailed Steady-State Control Analysis of an Ethanol-Water Distillation Column," M. S. Thesis, The University of Tennessee (1987).

[5] Downs, J. J. and D. F. Moore, "Steady-State Gain Analysis for Azeotropic Distillation," *Proc. JACC,* paper WP-7C (1981).

[6] Downs, J. J., "The Control of Azeotropic Distillation Columns," Ph.D. Dissertation, University of Tennessee (1982).

[7] Edgar, T. F., "Status of Design Methods for Multivariable Control," *Chemical Process Control, AIChE Symposium Series,* 72(159), 99 (1976).

[8] Forsythe, G. E., M. A. Malcolm and C. B. Moler, *Computer Methods for Mathematical Computations,* Prentice-Hall, Englewood Cliffs, New Jersey (1977).

[9] Hackney, J. E., "Control of a Xylene Distillation Column," Term Project, Ch.E. 6900, University of Tennessee, March 1985.

[10] Keeton, J. M., "A General-Purpose Multivariable Controller Using a Singular Value Decomposition Structure," Ph.D. Dissertation, The University of Tennessee (1982).

[11] Klema, V. C. and A. J. Laub, "Singular Value Decomposition: Its Computation and Some Applications," *IEEE Trans. Auto. Contro.,* AC-25, 164 (1980).

[12] Lau, H., J. Alvarez and K. F. Jensen, "Synthesis of Control Structures by Singular Value Analysis: Dynamic Measures of Sensitivity and Interaction," *AIChE J.,* 31, 3, 427 (1985).

[13] Lau, H. and K. F. Jensen, "Evaluation of Changeover Control Policies by Singular Value Analysis - Effects of Scaling," *AIChE J.*, 31, 135 (1985).

[14] Lawson, C. L. and R. J. Hanson, *Solving Least Squares Problems,* Prentice-Hall, Englewood Cliffs, New Jersey (1974).

[15] Moore, B. C., "The Singular Value Analysis of Linear Systems, Parts I and II," Systems Control Report No. 7801-7802, University of Toronto, Toronto, Canada (1981).

[16] Moore, B. C., "Principal Component Analysis in Linear Systems: Controllability, Observability, and Model Reduction," *IEEE Trans. Auto. Contro.*, AC-26, 17 (1981).

[17] Moore, B. C., "A Theory of Computer Process Control," Systems Control Report No. 8112, University of Toronto, Toronto, Canada (1981).

[18] Moore, C. F., "Application of Singular Value Decomposition to the Design, Analysis, and Control of Industrial Processes," *Proc. ACC,* paper TA1-8:30, pp. 643-650 (1986).

[19] Morari, M., W. Grimm, M. J. Oglesby, I. D. Prosser, "Design of Resilient Processing Plants - VII: Design of Energy Management System for Unstable Reactors - New Insights," *Chem. Engr. Sci.*, 40, 2, 187 (1985).

[20] Noble, B. and J. W. Daniel, *Applied Linear Algebra,* Prentice-Hall, Englewood Cliffs, New Jersey (1977).

[21] Smith, C. R., "Multivariable Process Control Using Singular Value Decomposition," Ph.D. Dissertation, University of Tennessee (1981).

[22] Smith, C. R., C. F. Moore and D. D. Bruns, "A Structural Framework for Multivariable Control Applications," JACC paper TA-7, Charlottesville, Virginia (June 1981).

[23] Stewart, G. W., *Introduction to Matrix Computations,* Academic Press, New York, New York (1973).

[24] ———, ASPEN, Mass Institute of Tech., Cambridge, Mass.

[25] ———, SYCOPACK, Department of Electrical Engineering, University of Tennessee.

[26] ———, EISPSCK, Applied Mathematics Division, Argonne National Labs.

[27] ———, MATLAB, Department of Computer Science, University of New Mexico.

[28] ———, IMSL, International Mathematical & Statistical Libraries, Inc., NBC Building, 7500 Bellaire Boulevard, Houston, Texas.

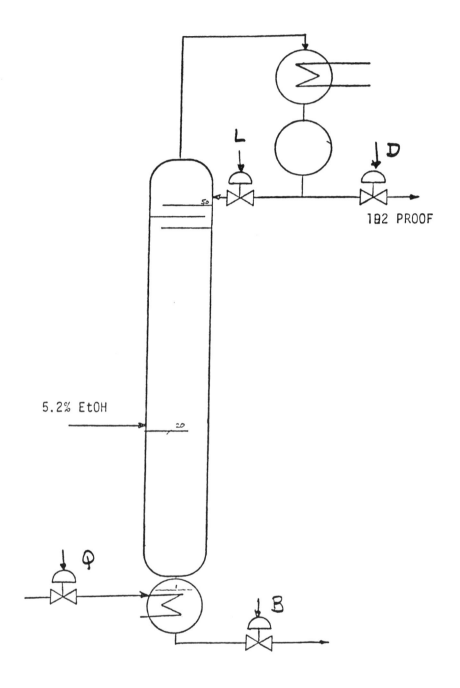

Figure 1: Ethanol-Water Distillation Column.

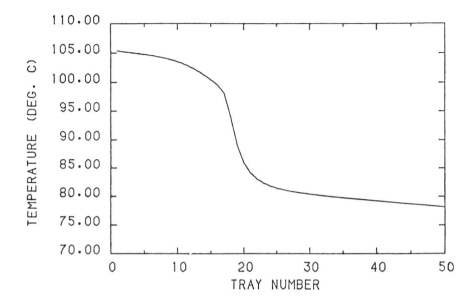

Figure 2: Ethanol-Water Temperature Profile.

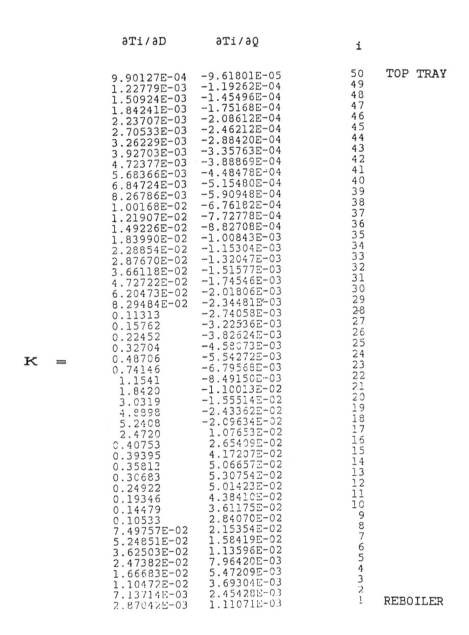

	$\partial T_i / \partial D$	$\partial T_i / \partial Q$	i	
	9.90127E-04	-9.61801E-05	50	TOP TRAY
	1.22779E-03	-1.19262E-04	49	
	1.50924E-03	-1.45496E-04	48	
	1.84241E-03	-1.75168E-04	47	
	2.23707E-03	-2.08612E-04	46	
	2.70533E-03	-2.46212E-04	45	
	3.26229E-03	-2.88420E-04	44	
	3.92703E-03	-3.35763E-04	43	
	4.72377E-03	-3.88869E-04	42	
	5.68366E-03	-4.48478E-04	41	
	6.84724E-03	-5.15480E-04	40	
	8.26786E-03	-5.90948E-04	39	
	1.00168E-02	-6.76182E-04	38	
	1.21907E-02	-7.72778E-04	37	
	1.49226E-02	-8.82708E-04	36	
	1.83990E-02	-1.00843E-03	35	
	2.28854E-02	-1.15304E-03	34	
	2.87670E-02	-1.32047E-03	33	
	3.66118E-02	-1.51577E-03	32	
	4.72722E-02	-1.74546E-03	31	
	6.20473E-02	-2.01806E-03	30	
	8.29484E-02	-2.34481E-03	29	
	0.11313	-2.74058E-03	28	
	0.15762	-3.22536E-03	27	
	0.22452	-3.82624E-03	26	
	0.32704	-4.58073E-03	25	
$K \ =$	0.48706	-5.54272E-03	24	
	0.74146	-6.79568E-03	23	
	1.1541	-8.49150E-03	22	
	1.8420	-1.10013E-02	21	
	3.0319	-1.55514E-02	20	
	4.8998	-2.43362E-02	19	
	5.2408	-2.09634E-02	18	
	2.4720	1.07653E-02	17	
	0.40753	2.65409E-02	16	
	0.39395	4.17207E-02	15	
	0.35813	5.06657E-02	14	
	0.30683	5.30754E-02	13	
	0.24922	5.01423E-02	12	
	0.19346	4.38410E-02	11	
	0.14479	3.61175E-02	10	
	0.10533	2.84070E-02	9	
	7.49757E-02	2.15354E-02	8	
	5.24851E-02	1.58419E-02	7	
	3.62503E-02	1.13596E-02	6	
	2.47382E-02	7.96420E-03	5	
	1.66683E-02	5.47209E-03	4	
	1.10472E-02	3.69304E-03	3	
	7.13714E-03	2.45428E-03	2	
	2.87042E-03	1.11071E-03	1	REBOILER

Figure 3: Ethanol-Water Gain Matrix.

$$U =$$
```
-1.15845E-04   -1.43652E-04
-7.27120E-04   -9.01615E-04
-1.77177E-04   -1.09961E-03
-2.16279E-04   -1.32333E-03
-2.62589E-04   -1.57508E-03
-3.17525E-04   -1.85761E-03
-3.82859E-04   -2.17410E-03
-4.60820E-04   -2.52826E-03
-5.54245E-04   -2.92444E-03
-6.66783E-04   -3.36777E-03
-8.03176E-04   -3.86431E-03
-9.69670E-04   -4.40126E-03
-1.17461E-03   -5.04726E-03
-1.42930E-03   -5.75268E-03
-1.74932E-03   -6.55005E-03
-2.15648E-03   -7.45457E-03
-2.68186E-03   -8.48474E-03
-3.37052E-03   -9.66303E-03
-4.28891E-03   -1.10167E-02
-5.53677E-03   -1.25787E-02
-7.26608E-03   -1.43879E-02
-9.71211E-03   -1.64894E-02
-1.32442E-02   -1.89325E-02
-1.84501E-02   -2.17670E-02
-2.62769E-02   -2.50335E-02
-3.82708E-02   -2.87459E-02
-5.69912E-02   -3.28632E-02
-8.67503E-02   -3.72601E-02
-0.13502       -4.17745E-02
-0.21549       -4.68344E-02
-0.35468       -5.72102E-02
-0.57201       -8.64782E-02
-0.61305       -5.28765E-02
-0.28907        0.13565
-4.75439E-02    0.21494
-4.58861E-02    0.33268
-4.16561E-02    0.40148
-3.56458E-02    0.41914
-2.89204E-02    0.39513
-2.24278E-02    0.34496
-1.67708E-02    0.28388
-1.21903E-02    0.22310
-8.67139E-03    0.16903
-6.06674E-03    0.12428
-4.18827E-03    8.90898E-02
-2.85722E-03    6.24467E-02
-1.92469E-03    4.28993E-02
-1.27531E-03    2.89479E-02
-8.23620E-04    1.92336E-02
-3.30684E-04    8.69678E-03
```

$$V = \begin{matrix} -1.0000 & 2.69431E-03 \\ 2.69431E-03 & 1.0000 \end{matrix}$$

$\sigma 1 = 8.5492$

$\sigma 2 = 0.12861$

$CN = 66.47593$

Figure 4: Ethanol-Water SVD Analysis.

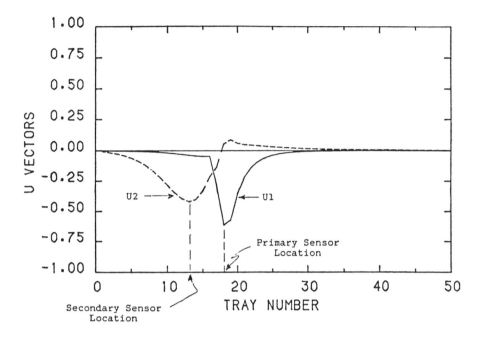

Figure 5: Principal Component Analysis: **U** – Vector Plots.

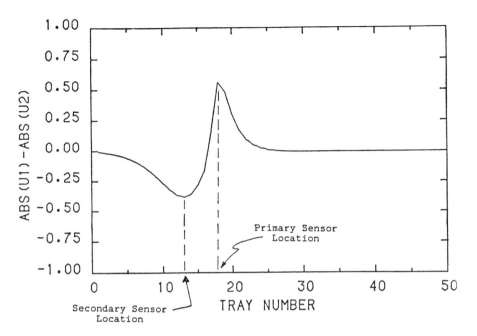

Figure 6: Modified Principal Component Analysis: $Abs(U_1) - Abs(U_2)$ vs. Sensor Location.

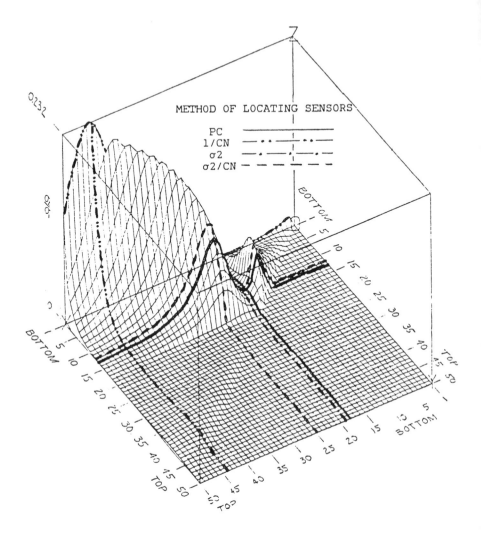

Figure 7: Sensor Interaction (1/CN) vs. Sensor Location.

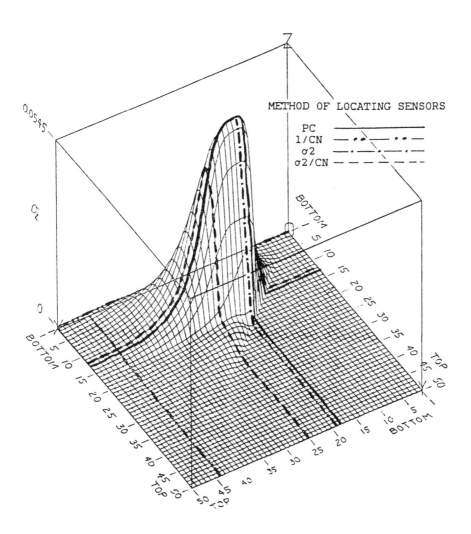

Figure 8: Sensor Sensitivity (σ_2) vs. Sensor Location.

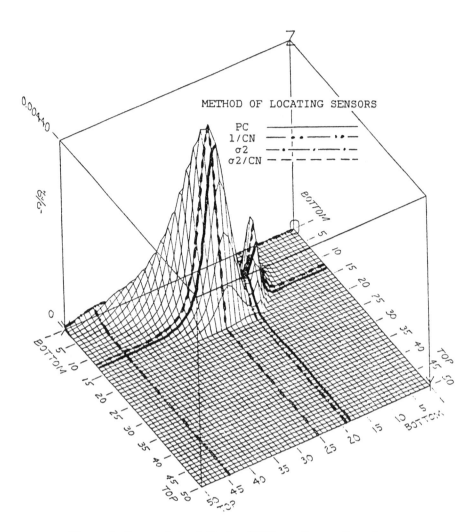

Figure 9: Sensor Intersivity (σ_2/CN) vs. Sensor Location.

Three Critiques of Process Control Revisited a Decade Later

Manfred Morari
Chemical Engineering, 206-41
California Institute of Technology
Pasadena, CA 91125

December 15-18, 1986

Abstract

During a period of disillusionment with process control research in the 1970's, three critiques were published. In this paper, these critiques are analyzed from today's perspective and conclusions for future research are drawn.

Introduction

After the enthusiasm about the developments in optimal control theory in the 1960's, disappointment set in in the early 1970's when it became evident that the newly developed techniques addressed the most pressing practical issues inadequately. A number of papers appeared criticizing the state-of- the-art, three of which have received wide attention. Foss (1973) discusses at length several practical examples and criticizes that the "modern" theory provides little help in generating a solution. Lee and Weekman (1976) argue from an industrial viewpoint. They cite a number of case studies to demonstrate that "constrained control" is one of the most pressing issues because on-line optimization tends to push the optimal operating conditions toward the intersection of different process constraints. They felt that the "current" theory has little to offer to guide the control engineer in the design of these control systems. Furthermore they stated that the importance of the optimization layer and the requirements it imposes on the feedback control layer was inadequately addressed in the literature. Kestenbaum, Shinnar and Thau (1976) finally define more specifically the general objectives a feedback control system should satisfy and make some strides to bring the modern theory in line with classical proven analysis techniques.

The focus points of these critiques were:

1. **Modern (Linear) Multivariable Control Theory.** The theory seemed to promise much but deliver little and had virtually no impact on industrial practice.

2. **Model Requirements/Inadequacies.** All modern control techniques need a model in the form of a set of linear differential and algebraic equations. In a practical environment, a model is rarely available or often very inaccurate. It is unclear how to deal with the model inaccuracies when designing the control system.

3. **Effect of Process Design on Process Control.** Control can help to obtain the best performance for an existing plant but if the plant has been designed inadequately, then control has little to offer to remedy the situation. Therefore, as had been suggested earlier by Ziegler and Nichols (1943) and Rosenbrock (1974) the control engineer should focus on the *design* process and influence it such that the *control* performance is improved.

4. **Choice of Control Structure.** In a similar manner as design decisions, the selection of the manipulated variables, measurements and controller structure usually has a much more significant impact on the resulting closed loop performance than the design of the controller itself. Few systematic tools for control structure selection are available.

5. **Process Constraints.** Process optimization forces the operating conditions to lie at the intersection of constraints. Ad hoc logic schemes are designed to switch between different constraints and manipulated variables when saturation or constraint violation are imminent.

6. **Operator Acceptance.** In order to be acceptable to the operator, control schemes have to be transparent and allow on line adjustments. While simple classic schemes have this feature, it is not present in modern control techniques.

7. **Use and Misuse of "Optimal Control".** While the idea of optimal control was viewed quite negatively in most of the critiques, it is our perception now that this criticism was only partly justified.

8. **Nonlinear/Adaptive Systems.** A theory for nonlinear control does not exist and adaptive controllers appear far away from a practical implementation stage.

We will now analyze each of these issues in more detail, point out what progress has been made in the last decade and what issues remain to be resolved.

Modern Linear (Multivariable) Control Theory

Four techniques were particularly popular in the academic literature at that time.

Modal/Pole Placement Control

As was pointed out eloquently by Foss (1973) and has been realized by almost everybody today, poles are a highly inadequate measure of closed loop system performance. Thus, while there are plenty of techniques available to place poles at prespecified positions, there is very little insight available to guide the designer on where to place these poles. The closed loop performance depends also strongly on the zero locations which are invariant under feedback. In particular, right half plane zeros (nonminimum phase behavior) limit the pole locations for which acceptable closed loop behavior can be obtained in a non-obvious fashion. Therefore, the pole placement techniques have been abandoned by almost everybody today.

Decoupling

Decoupling was the earliest method suggested for multivariable control (Boksenbom and Hood, 1949). With all the knowledge on single-input-single-output design available it seemed natural to convert a multivariable problem into a set of independent single-input-single-output problems. In the critiques it was pointed out that the "interactions" might be beneficial and that a removal might not be necessary.

In the last decade we have learned much about the benefits and pitfalls of decoupling. The general conclusion is not if one should utilize the interactions but if one can remove them without serious performance deterioration. The deterioration can be a consequence of multivariable nonminimum phase behavior (Holt and Morari, 1985 a,b) or can be due to robustness problems. If the open loop system is ill-conditioned (Morari and Doyle, 1986), then an inverse based (decoupling) controller can give rise to serious performance problems in the presence of model uncertainty.

Multivariable Frequency Domain Design Techniques

The Nyquist Array method developed by Rosenbrock (1969) and the Characteristic Loci technique developed by MacFarlane (1970) can be viewed as generalizations of decoupling. The class of systems to which they can be applied successfully, i.e., multivariable systems which can be treated as a set of almost independent single-input-single-output systems, is very limited. The most important consequence was to redirect attention from the state space to the frequency domain. Today, these insights are still valuable but the techniques have been virtually abandoned as a practical design tool.

Linear Quadratic Optimal Control

Because of its special importance then as well as now, the technique is discussed in more detail.

It is my perception that much confusion arose in the minds of the application engineers because of the Kalman state space interpretation of the linear quadratic optimal control problem. Suddenly, people started to suggest to measure all the physical states of a distillation column (temperature and concentrations on each tray) or to build large filters to estimate all these states. The concept of state was elevated to a position which was not justified. Also the state space turns out to be quite unsuitable for stating design specifications like robustness.

Because of this confusion, it is often overlooked that the theory is able to solve a number of problems which appear to puzzle some researchers even today. It is quite widely recognized that the tradeoff between control quality and control effort, and disturbance rejection and measurement noise amplification is addressed very well in that framework. It is less known that the whole issue of time delay compensation can be handled easily and optimally in this framework. For continuous systems a design based on a first or second order Páde approximation leads to performance which is, for practical purposes, indistinguishable from that obtainable through a Smith Predictor (Galloway and Holt, 1986). For sampled data systems delays do not offer any difficulties at all and can be dealt with in the optimal control framework in a straightforward fashion. Also, the problem of control system design for setpoint following versus disturbance rejection can be addressed easily by designing a two degree of freedom controller using standard optimal control theory.

From a tutorial viewpoint, the basic idea of feedback control appears to have been missed. In the usual derivation of the Kalman filter/feedback controller, the appearance of a feedback controller strikes one almost as accidental. Also, until recently no connections were drawn between the state space oriented optimal control techniques and the frequency domain oriented classical design tools which had a proven track record. Without this connection, the experience gained in the past appeared worthless and the efforts to gain familiarity with the new techniques too large to be justified by the promised gains. Finally and most importantly, while classical techniques address the issue of model uncertainty very explicitly as one of the key reasons for feedback design, this problem was completely forgotten as soon as state space methods appeared in the literature.

Model Requirements/Inadequacies

It is intuitively obvious that the more we know about a system, the better we can influence its performance, i.e., control it. It is not so clear how to express this implicit tradeoff between model quality and performance mathematically. The challenge is to find a model uncertainty description and a measure of performance which makes sense from a physical and an engineering viewpoint and leads to a mathematically tractable problem.

In classical control theory, model uncertainty is expressed as gain and phase variations in the crossover region and robustness is measured in terms of gain margin and phase margin. For today's design techniques and particularly in the context of any "optimal control" method, these robustness measures are inadequate. For example, Palmor and Shinnar (1981) show that for an optimal time delay compensator design, the gain margin is always larger than two and the phase margin always ∞. Nevertheless, for all practical purposes, the performance can be arbitrarily bad. Also, the concepts of gain and phase margin cannot be extended to multivariable systems.

The current approach is to model the expected uncertainty and to design the control system to optimize the worst expected performance in the range of possible plants. The uncertainty is described in the frequency domain and the performance specification is expressed as a bound on the sensitivity operator (H_∞ performance specification).

Example. Consider the control system design for a process with time delay which is inaccurately known (it can vary up to an amount $\pm\delta$). The possible variations can be captured by the multiplicative uncertainty description depicted in Figure 1. This form of norm bounded uncertainty description allows more than just time delay uncertainty and is therefore inherently conservative. From some simple arguments, it can be shown that for robust stability the complementary sensitivity function should always lie below the inverse of the model uncertainty bound. Figure 2 shows clearly how the time delay uncertainty restricts the achievable bandwidth of the closed loop system. It is illustrative and suggests a very simple robust control system design procedure.

If the control system has been designed for disturbance rejection (a step entering through a lag at the system output) rather than setpoint following then the complementary sensitivity function usually has the shape shown in Figure 3. From our graphical arguments, it becomes evident that the special effort of modelling the disturbance and designing specifically for disturbance rejection is only justified if the disturbance spectrum falls inside the limits imposed by the time delay uncertainty. This will often not be the case and therefore involved modelling of the disturbance can rarely be justified for the purpose of control system design.

For multi-input-multi-output systems there is much freedom to model uncertainty. A multiplicative norm bounded description can be used as we discussed for single-input-single-output systems. This description is very crude however, lumping all the uncertainty which might arise from different parameter and element changes into one big block. The most general uncertainty description we can deal with in the current mathematical framework is shown in Figure 4. Here M is a transfer matrix which includes information on the system model, the uncertainty and the performance specifications and Δ is a block diagonal matrix where the norm of each block is bounded by unity. Within the H_∞ framework the Structured Singular Value (SSV) (Doyle, 1982) can then be used to assess robust stability and robust performance.

The SSV holds much promise for chemical engineering systems. As an example, consider the high purity distillation column studied by Skogestad and Morari (1986). On the basis of a very simple model and quite straightforward assumptions on the model uncertainty, they were able to make very fundamental conclusions. In particular, they showed that a two point composition control scheme involving the reflux (L) and the boilup (V) as manipulated variables will never perform as well as a control scheme involving the distillate flow (D) and the boilup (V). The LV scheme has inherent robustness problems. For all practical purposes the required model accuracy makes it impossible to achieve acceptable performance.

It should be noted that for this example very accurate modelling of the physical process is neither necessary nor in any way beneficial for control purposes. It is critical, however, to include uncertainty considerations in the assessment of the performance of the different schemes. Even in a nonlinear simulation, both control schemes can appear to perform very well. In the presence of slight uncertainty in the valve positions, however, the LV scheme performance deteriorates drastically while the DV scheme remains largely unaffected.

Open Problems

Though the H_∞ framework appears quite useful for addressing the robust stability and performance problems, it is not clear how to take advantage of this powerful tool when it comes to solving practical problems. In particular, a number of "weights" have to be selected which reflect the class of input signals to be considered, the performance specifications on the outputs and the model uncertainty. These weights are related to more natural engineering specifications in an indirect and non-obvious manner. The advantage of the H_2 (linear quadratic optimal) problem formulation is that the weights (power density spectra) can be obtained in principle from well defined experiments on the real process. In the H_∞ framework, we have to deal with sets of power bounded signals and it is not clear how to define the appropriate set from experiments or a study of the physical system to be controlled. It is conceivable that the uncertainty description which can be utilized in the H_∞ framework can be obtained from a study of the fundamental equations describing the physical system or from well defined experiments or both, but a systematic procedure is not available to date.

The choice of weights would be greatly simplified if we understood how they influence the final solution of the problem. At present, we are largely lacking a qualitative understanding of how a robust controller should look like in a specific situation. Thus, a number of detailed case studies and experiments are needed to help us gain qualitative insight and understanding regarding the appropriage weight selection.

Finally, efficient algorithms for finding the controller which optimizes robust performance are not available. Work in this area, however, is better carried out by numerical analysts and mathematicians.

Effect of Process Design on Process Control

As mentioned in the Introduction, design modifications usually have a more drastic effect on control system performance than the design of the control system itself. We distinguish two tasks – analysis and synthesis. By *analysis* we mean finding the best achievable performance (dynamic resilience) in a certain situation *independent* of controller complexity. *Synthesis* refers to the task of finding the process design configuration which leads to the optimal dynamic resilience. In the synthesis area, essentially no progress has been made. At present, synthesis is carried out by a sequence of trial and error steps succeeded by analysis.

Until recently only the H_2 framework was available for analysis. It can provide information on the best achievable output variance given information on the measurement noise and the disturbance spectrum. It points at the problems caused by nonminimum phase elements but fails to address the question of performance deterioration in the presence of model uncertainty. For the distillation column mentioned above this is the dominating factor affecting the achievable closed loop performance.

Within the H_∞ framework simple criteria have been developed during the last few years to assess the sensitivity of closed loop performance to model uncertainty (Skogestad and Morari, 1987). Depending on the available information on the process, more or less involved tools can be employed. The condition number and the Relative Gain Array play a key role.

The main open problems are synthesis and a refinement of the available analysis tools.

Choice of Control Structure

We consider the following problem:

Given the control objectives, the measurements, manipulated variables and controller structure (interconnection between measurements and manipulated variables) are to be selected for optimal performance. The measurement selection question refers to the case when the variable to be controlled cannot be measured, as well as to the problem of defining additional secondary measurements like, for example, in a cascade arrangement to improve the dynamic performance.

Little progress has been made on the measurement selection problem since the paper by Weber and Brosilow (1972) The problem is again that most available measurement selection criteria do not account for model uncertainty but only for measurement noise and disturbance spectra. It is clear that with our improved understanding of robust control, the development of a superior measurement selection procedure should be possible.

The choice of the appropriate set of manipulated variables is basically a design decision and the same remarks as in the previous section apply to this problem.

As a controller structure single control loops are preferred because of the low modelling, maintenance and tuning requirements. The criterion for choosing which measurement should be connected to which manipulated variable is usually that if one of the single loops fails, the remaining control system should remain stable without readjustment of the tuning parameters. In the same spirit, it should be possible to design each control loop separately and to obtain satisfactory performance when all the loops are operating together. A necessary condition to satisfy the fault tolerance/independent design criteria is that the diagonal elements of the Relative Gain Array should be positive (Grosdidier et al., 1985). A sufficient condition is provided by the μ Interaction Measure (Grosdidier and Morari, 1986): If each one of the control loops is designed to satisfy the bound imposed by the μ criterion, then fault tolerance is automatically guaranteed.

Open Problems

The most serious issue is that the gap between the necessary condition provided by the RGA and the sufficient condition provided by the μ Interaction Measure can be arbitrarily large. With other words, the μ Interaction Measure which is based on norm-type arguments can be very conservative. A second problem is that these conditions address stability only but not the performance achievable with a decentralized control system design.

Process Constraints

Algorithms for the optimal operation in the presence of constraints have been developed by industry and have received much attention during the last decade. The two most popular examples of so-called "model predictive control" techniques are Dynamic Matrix Control (DMC) developed by Shell (Cutler and Ramaker, 1979; Prett and Gillette, 1979) and Model Algorithm Control (MAC) developed by Richalet et al. (1978).

Despite all the claims to the contrary, model predictive control is very similar to linear quadratic optimal control with a few very important differences. MPC can deal with general linear constraints involving both process inputs, outputs and inferred variables. It is the only example of optimal control applied routinely and on a large scale in real time in an industrial environment. By employing a finite moving horizon and by estimating the effect of the disturbance on the controlled output, closed loop disturbance rejection and model error compensation are possible.

The on-line implementation has been made possible through an efficient formulation of the control problem as a quadratic program or with a slightly modified objective function as a linear program which is solved at each time step on-line. This type of control scheme offers a very clean alternative for the often confusing override/split range/logic schemes which have been used traditionally.

The main open problem in this context is how to correct the algorithm in the presence of uncertainty. If the model is inaccurate, the performance might deteriorate significantly and it cannot be guaranteed that the constraints will be observed by the real system.

Operator Acceptance

It is our claim that a control scheme is acceptable to the operating personnel if its actions can be understood and influenced rationally without a switch to "manual". This does not imply in any way that the scheme has to be restricted to a PID type controller.

Based on this premise, LQG was a failure because any on-line adjustment is impossible nor is it clear how the different controller parameters affect the control performance. On the other hand, the model predictive control techniques, DMC and MAC were clearly a success. This might be attributable to their predictive nature which allows the communication of the projected result of the control action to the operating personnel. Thus it helps to explain to the operators why a particular control action which was taken is reasonable in the context of the overall control objectives.

SISO-IMC is also attractive from an operator point of view because it yields essentially "optimal" control and has a single tuning parameter which is directly related to the closed loop bandwidth or closed loop speed of response. Moreover, in many cases the controller simplifies to a PID controller whose parameters are related directly to the parameters of the process model and the adjustable closed loop bandwidth parameter.

For a majority of systems, the ideas of SISO-IMC extend in a straight-forward fashion to multivariable systems. For some systems, in particular ill-conditioned systems, the IMC technology does not carry over in a transparent manner to the multivariable case without possibly significant sacrifices in robustness and/or performance.

Use and Misuse of "Optimal Control"

Though criticized heavily in the 1970's, optimal control is now applied routinely in a slightly modified form as MAC and DMC in industry. Apart from the direct application, the main benefit of optimal control is that it allows to determine the achievable performance under precisely stated assumptions but no constraint on the controller complexity. Thus, it provides an absolute target against which other simple controllers can be measured.

In the case of LQG, we obtain information on the optimal achievable variance of the outputs after specifying the noise and disturbance characteristics and the input penalty weight. In the SSV/H_∞ problem, we can test if the performance specification can be met for given disturbance characteristics, input penalty weights and a model uncertainty description. Thus, these techniques provide the ultimate dynamic resilience assessment.

As reported in the critiques, the experience with optimal control was often bad, mainly because the techniques were applied in situations when the necessary assumptions were not satisfied. Needless to say, there is no physical problem where all the mathematical assumptions are satisfied exactly. The main contribution of the new robust control technqiues is that they removed one of the basic assumptions (the assumpton of a perfect model) from the problem formulation, thus significantly broadening the applicability of these techniques.

Nonlinear/Adaptive Systems

Some successful attempts to crack the general nonlinear control system synthesis problem have been published recently. The work by Hunt, Su and Meyer (1983) has received much attention by chemical engineers (Kravaris and Chang, 1986; Alsop and Edgar, 1986, Kantor, 1986). The surface has been barely scratched and a number of important issues like robustness remain virtually unaddressed.

Several companies are now selling adaptive controllers (ASEA, Foxboro, Leeds and Northrup, etc.) despite many unresolved theoretical issues. It appears that adaptive control works much better than it should from a theoretical viewpoint. The robustness issue has received much attention but no generally applicable algorithm with desirable characteristics has emerged yet. The multivariable problem has not been studied seriously yet except in an ad hoc manner and through applications.

Conclusions

Overall, much progress has been made during the last decade in narrowing the often criticized theory-practice gap because the theoreticians have started to address more practical problems and issues. The main accomplishments were a better understanding of the robustness and the development of a comprehensive technology to address it. Another breakthrough were the model predictive control techniques which allowed to replace the ad hoc constraint handling policies which had been transferred from analog hardware to process control computers. Apart from a significant number of minor issues which have been mentioned previously, the most promising and widely open areas are the control of nonlinear systems and adaptive control.

Acknowledgements

Partial financial support from the National Science Foundation and the Department of Energy are gratefully acknowledged.

References

1. Alsop, A. W. and T. F. Edgar. "Nonlinear Heat Exchanger Control Through the Use of Partially Linearized Control Variables", *Proc. American Control Conf.*, 1006-1013 (1986).

2. Boksenbom, A. S. and R. Hood. "General Algebraic Method Applied to Control Analysis of Complex Engine Types", Report NCA–TR–980, Washington, D. C. (1949).

3. Cutler, C. R. and B. L. Ramaker. "Dynamic Matrix Control – A Computer Control Algorithm". AIChE National Mtg., Houston, TX (1979); also *Proc. Joint Automatic Control Conf.*, San Francisco, CA (1980).

4. Doyle, J. C. "Analysis of Feedback Systems with Structured Uncertainties" *Proc. Inst. Elec. Eng.*, 129, Pt. D, 242-250 (1982).

5. Foss, A. S. "Critique of Chemical Process Control", *AIChE J.*, 19, 209-214 (1973).

6. Galloway, P. J. and B. R. Holt. "Multivariable Time Delay Approximations for Analysis and Control". AIChE Annual Meeting, Miami Beach, FL (1986).

7. Grosdidier, P. M., M. Morari and B. R. Holt. "Closed Loop Properties from Steady State Gain Information". *Ind. Eng. Chem. Fundam.*, 24, 221-234 (1986); also *Proc. American Control Conf.*, San Diego, CA 1290-1295 (1984).

8. Grosdidier, P. and M. Morari. "Interaction Measures for Systems Under Decentralized Control". *Automatica*, 22, 309-319 (1986).

9. Holt, B. R. and M. Morari. "Design of Resilient Processing Plants–VI. The Effect of Right–Hand–Plant Zeros on Dynamic Resilience". *Chem. Eng. Sci.*, 40, 59-74 (1985).

10. Holt, B. R. and M. Morari. "Design of Resilient Processing Plants–V. The Effect of Deadtime on Dynamic Resilience". *Chem. Eng. Sci.*, 40, 1229-1237 (1985).

11. Hunt, L. R., R. Su and G. Meyer. "Global Transformations of Nonlinear Systems". *IEEE Trans. Autom. Control*, AC-28, 24-31 (1983).

12. Kantor, J. C. "Stability of State Feedback Transformations for Nonlinear Systems – Some Practical Considerations". *Proc. American Control Conf.*, 1014-1016 (1986).

13. Kestenbaum, A., R. Shinnar and F. E. Thau. "Design Concepts for Process Control". *Ind. Eng. Chem. Process Des. Dev.*, 15, 2 (1976).

14. Kravaris, C. and C.-B. Chung. "Nonlinear State Feedback Synthesis by Global Input/Output Linearization". *Proc. American Control Conf.*, 997-1005 (1986).

15. Lee, W. and W. V. Weekman, Jr. "Advanced Control Practice in the Chemical Process Industry: A View from Industry", *AIChE J.*, 22, 27-38 (1976).

16. MacFarlane, A. G. J. "Return Difference and Return–Ratio Matrices and Their Use in the Analysis and Design of Multivariable Feedback Control Systems". *Proc. Inst. Elec. Eng.*, 117, 2037-2049 (1970).

17. Morari, M. and J. C. Doyle. "A Unifying Framework for Control System Design Under Uncertainty and its Implications for Chemical Process Control" in *Chemical Process Control – CPC III* (M. Morari and T. J. McAvoy, eds.), CACHE and Elsevier, Amsterdam, 5-52 (1986).

18. Palmor, Z. J. and R. Shinnar. "Design of Advanced Process Controllers", *AIChE J.*, 27, 793-805 (1981).

19. Prett, D. M. and R. D. Gillette. "Optimization and Constrained Multi-variable Control of a Catalytic Cracking Unit." AIChE National Mtg., Houston, TX (1979); also *Proc. Joint Automatic Control Conf.*, San Francisco, CA (1980).

20. Richalet et al. (1978).

21. Rosenbrock, H. H. "Design of Multivariable Control Systems Using the Inverse Nyquist Array". *Proc. Inst. Elec. Eng.*, 116, 1924-1936 (1969).

22. Rosenbrock, H. H. *Computer–Aided Control System Design*, Academic Press, London (1974).

23. Skogestad, S. and M. Morari. "Control of Ill–Conditioned Plants: High Purity Distillation." AIChE Annual Mtg., Miami Beach, FL (1986).

24. Skogestad, S. and M. Morari. "Design of Resilient Processing Systems. Effect of Model Uncertainty on Dynamic Resilience. *Chem. Eng. Sci.* (1987).

25. Weber, R. and C. B. Brosilow. "The Use of Secondary Measurements to Improve Control," *AIChE J.*, 18, 641 (1972).

26. Ziegler, J. G. and N. B. Nichols. "Process Lags in Automatic-Control Circuits." *ASME Trans.*, <u>65</u>, 433-444 (1943).

Chapter 2

Discussion Sessions

Introduction

Each of the papers presented at the Shell Process Control Workshop are encapsulated here in the form of a summary of the oral presentation and the main topics of the discussion session which followed each paper.

The summaries of the presentations attempt to capture the varied styles of the academic and industrial speakers. In some instances, questions during a presentation prompted discussion before the general question and answer session. For all presentations, we also attempt to capture the essence of the ensuing discussion.

2.1 "Design Methodology Based on the Fundamental Control Problem Formulation," C. E. Garcia (Speaker) and D. M. Prett

Summary of Presentation

The first research presentation began with a brief introduction relating the current paper to previous papers.[1] [2] The earlier papers addressed the issue of model predictive control and introduced the concept of the fundamental control problem. Here the speaker expounded on these ideas giving a clear statement of the directions for control research as proposed by Shell Development Company. The presentation closely followed the content of the paper.

A significant point emphasized during the presentation was that Shell is advocating a top down design procedure which begins with the general problem and then reduces it through intelligent assumptions. The advantage of such an approach was illustrated by the example of a simple PI controller being proposed and implemented for a process. Later, the process model gain changes and the controller is inadequate, thus necessitating the addition of a "trick" to the controller. If the system had been analyzed properly, the change in the gain would have been anticipated, a controller capable of handling the gain change would have been implemented, and there would have been no need to "fix" the controller. It is quite possible that a PI controller would still have been chosen. However, the limitations would have been recognized and dealt with more efficiently.

Summary of Discussion

The discussion following the presentation addressed the formulation of the fundamental control problem, as well as several practical issues concerning

---

[1]Garcia, C. E. and D. M. Prett, "New Developments in Model-Predictive Control," *11th InterAmerican Conference on Chemical Engineering,* San Juan, Puerto Rico, December, 1985.

[2]Garcia, C. E. and D. M. Prett, "Advances in Industrial Model Predictive Control," *Chemical Process Control Conference – III,* Asilomar, CA, January, 1986.

the implementation and support of control systems in chemical plants. The following two paragraphs are a summary of points raised primarily by B. L. Ramaker concerning the implementation and support issue.

There are several factors driving industry toward a single standardized approach. The first and most obvious is that there is a large cost associated with maintenance of a series of specialized control algorithms. The economics of a single universal algorithm cannot be refuted. Second, since the personnel generally responsible for maintaining control loops are engineers having either a Bachelor or Master of Science degree and little specialized training in control, a single straightforward algorithm is desirable.

Another, more complicated factor is the interaction between operators and the control systems of processes that are "being pushed to their limits" and operated close to constraints. An acceptable control algorithm must instill enough confidence in the "average" operator so that it is left on-line some high percentage of the time (98 percent), and equally as important, handle time variant process constraints. The current profile for this "average" operator is a person with 1-2 years of a non-technical college education and who is responsible for 60 to 300 control loops as well as some plant maintenance. It is important to note that gaining the confidence of the operator does not mean that the operator must understand the algorithm, but rather that the control system must operate in a consistent and reliable manner. Therefore, it may be necessary to find a method of incorporating human preference into control design. Alternately, this problem may be alleviated in control rooms of the future by dividing plant personnel into dedicated control room engineers and maintenance operators. The long term trend in industry is to "get the operator out of the loop" for various reasons. However, it is important to realize that operators have "intangible" information which control systems do not know how to handle (e.g., the maintenance record of an error prone pump, and start-up procedures). Therefore, the operator's role will be changing, but it is unlikely that he will be eliminated.

During the discussion of the formulation of the fundamental control problem it was pointed out that there are parallels in the formulation of control problems in space systems where the approach must be general enough to encompass all aspects of the problem. However, it was also pointed out that the presented formulation does not currently include the design of a process with the specific intention of giving it desirable control characteristics. This concept is illustrated in the research paper "Control in Autoclave Processing of Polymeric Composites" by B. R. Holt. Industrial participants showed that the overwhelming need in process control today and in the foreseeable future is to control existing processes. They agreed, of course, that Holt's message is valid.

Two significant issues raised during the discussion dealt with the modeling of the process and the structure of the process. Several participants felt that modeling the complete process is much more difficult than breaking the problem into parts and modeling them individually. In addition, the uncertainty in the control problem (e.g., noise, errors in measurements, etc.) should be addressed. Another disadvantage of a unified approach is that it will not lead to an examination of the structure of the control problem. This single point has major implications with regards to academic research, because there are constraints on the research that academics may undertake. In general, academics are looking for the underlying principles of control and their relation to structure of the processes. While this can be addressed in part by the fundamental control problem, it is not in the spirit of the formulation. There was a consensus that Shell is addressing a very important part of the problem and that Shell has had a definite effect on the academic community, but there are other items that are of equal importance. These thoughts were captured well in a statement by J. C. Kantor:

> "What has been presented is the modern view of process control, namely that the quality of good control is something that a process or model admits, not something that the controller provides. The key issue still to be resolved is what control objective will a certain model together with its uncertainty description admit. This is a vital part of the current research and if anything characterizes what is interesting about process control, this is it."

2.2   "Control of Autoclave Processing of Polymeric Composites," B. R. Holt (Speaker)

Summary of Presentation

The presentation was substantially the same as the paper and questions were held until after the presentation was complete. The ensuing discussion was broadly based upon three themes. The first involved the idea that the design of a process should be directly influenced by control considerations, the second considered the fact that model development is a significant hurdle to be overcome, and the third questioned how process control and dynamics should be incorporated into chemical engineering education.

Summary of Discussion

Again, as in the discussion following the opening research presentation, it was pointed out that the petrochemical industry is progressing toward less stable operating conditions with an emphasis on control systems to maintain stability and optimization systems to enhance profitability. There are many forces driving us to these less stable operating regimes, most notable of which are energy conservation and tighter process designs. The first results in greater interaction among process streams and the second in decreased capacitance of the process. Operators are uncomfortable making decisions in these operating regimes, and therefore, our plants are relying more heavily on control systems.

Philosophically, there was little disagreement that control considerations should influence process design. However, given the tremendous investments in capital and design methodologies, the problem must be viewed from an economic perspective. As an example in a related area, theoretically, there is a tremendous opportunity for improving the efficiency of chemical processes, some of which use 10 to 100 thousand times the minimum free energy. However, the lack of improvements in this area must be viewed as a lack of economic incentive. Therefore, control considerations will influence process design only when the benefits significantly outweigh the disadvantages of the status quo, or when the process can only be effected through the use of advance control techniques.

An issue in control is the trade-off between models and the quality of control. At one extreme is no model and a "recipe", while at the other is an ultra-complex nonlinear model which may be used for on-line optimization. Occupying the middle ground is an approach which may involve the use of a statistician to gradually adapt parameters to describe the process and control it. As the chemical industry moves more toward the latter extreme, process models assume an increasingly important role. Unfortunately, it is possible to spend a considerable amount of time developing a model for a process only to find that it is inadequate. While it may eventually be possible to generate models quickly and cheaply, there will always be risks associated with the systems approach to control design that need to be anticipated and evaluated in the design stage.

An important point made with regard to Chemical Engineering education is that the omission of process dynamics from the curriculum cannot be addressed by adding one or two process control courses. In many cases, field engineers have difficulty communicating priorities to design people, because the process designers are versed primarily in the steady state.

With regard to the specifics of control courses, a new course that introduces general concepts such as the effects of constraints and the use of models to predict process behavior is needed. Such a course, preferably taught in conjunction with a control laboratory, would greatly enhance the student's intuition for process dynamics. A caveat was made at this point that too strong an emphasis on intuition may lead to "experts" who do not have the prerequisite theoretical understanding. A course and textbook which strikes a balance between theory and intuition and which channels intuition along organized lines would be very beneficial.

The structure of the process control courses in undergraduate education greatly influences the people who continue on to perform graduate work in control. If the undergraduate education emphasizes aspects of electrical and aerospace engineering, then students identifying with this approach will continue on to graduate work, and the resulting research will be closely aligned with electrical and aerospace control systems.

Finally, on a more general note, the point was made that two of the primary purposes of an undergraduate education are the development of critical reading skills and the teaching of a logical, systematic approach to problem solving. If one moves too far toward the qualitative side, neither of these skills will be developed.

Closing Discussion for the Monday Morning Session

The discussion following the opening presentation, "Design Methodology Based on the Fundamental Control Problem Formulation," closed with the statement that "Shell has underestimated the impact of its presentation of DMC." There was a general desire on the part of the Shell participants to expand on this statement during the closing discussion of the morning session. As such, the question "What do you feel has been the impact of DMC?" was posed to the participants. The response is best summarized by the following individual comments.

- The introduction of DMC[3] in 1979 and its success has heightened the interest in model based control and is responsible for much of the research analysis in this area. It has also encouraged a number of researchers to recognize the importance of constraints. Unfortunately, much academic work is "following" rather than "leading" the field. Therefore, the presentation of DMC has been very important.

- The lack of influence of DMC on industrial companies negatively impacts the direction of academic research by diluting the role of DMC in overall industrial applications. Paradoxically, this is harmful to Shell.

- The impact of DMC has been twofold. On the one hand it has shown the academic community an exciting area of research and has spawned an interest in a broad set of approaches to process control (e.g., optimization and constrained control). It has also played a role in demonstrating to the academic community that advanced control is a viable research area. On the other hand, it has encouraged people to take the attitude that here is a good algorithm; let's tune the dickens out of it and see how it compares to other methods. This distracts researchers from other areas of potential value.

- DMC has provided a spark which has enticed younger people to select control as an area of research.

- The major contribution of Shell's presentation of DMC is that it has demonstrated a serious commitment by a major oil company to a large

---

[3]Prett, D. M. and R. D. Gillette, "Optimization and Constrained Multivariable Control of a Catalytic Cracking Unit," *AIChE 86th National Meeting,* Houston, TX, April, 1979.

research effort in process control. This statement along with the publications and continuing interest is the major impact. The details may be misunderstood, but if you look at the amount of work done in process control in chemical engineering, you can see the effects. Shell has had a significant impact on the content of academic research.

2.3   "Expert Systems in Process Control and Optimization: A Perspective," D. B. Garrison (Speaker), D. M. Prett, and P. E. Steacy

Summary of Presentation

In keeping with the general outline of the paper, the speaker presented a general critique of expert systems based on several areas of investigation: numerous assessments of existing systems, extensive interaction with researchers and practitioners within the field of A.I. and those interested in this technology, and the development of a prototype intelligent interface to an in-house control system design package.

While great expectations have recently been voiced, D. B. Garrison cautioned that this enthusiasm is best confined at present to the specific application domains which have proven amenable to the basic tenets of existing knowledge handling techniques. The majority of problems being addressed, however, particularly in the control and optimization areas, will require extensive research and development in such fundamental issues as knowledge representation and program control before satisfactory performances can reasonably be expected. The state of the technology can best be viewed as immature.

The presentation closely followed the major points of the paper. Of special interest was the viewpoint which questioned the usefulness of the knowledge engineer. Reasons were given in support of this view, and the prospect of eliminating this function by the application of machine learning technology was discussed.

Garrison ended the presentation by reiterating the role of expert systems in control applications as a means of disseminating a formalized technology, not as a means of facilitating the implementation of ad hoc solutions of limited applicability. While this role represents significant advancements in fundamental A.I. techniques, several problems can be addressed in the interim with the available technology. Foremost among these is the transfer of technology through the use of intelligent interfaces to control software.

A further benefit noted for adopting the formalisms provided within the A.I. framework is the insight gained from a systematic review of one's expertise. While it was pointed out that this is not unique to A.I., it was agreed that the ultimate potential of (and challenge for) A.I. is to facilitate the externalization of this mental exercise. The ultimate man-machine interface will capture the essence of one's expertise in a facile and timely manner.

Summary of Discussion

It was asked whether the concerns of A.I. were not merely a subset of the more general issues addressed by information technology, and as such might be better handled within the latter framework. As evidence for the shockingly low efficiency with which information is retrieved, O. A. Asbjørnsen offered that many companies do not even try to retrieve useful information, because the time and money spent in the retrieval of this information is greater than simply regenerating the information. The extent that research in A.I. does not attempt to address the information handling problem casts doubt upon its status as a viable and independent field of study. While it is probably true that efforts toward better knowledge representations will have a direct and beneficial impact on information processing in general, the fundamental problem remains how best to achieve the efficient and effective handling of information. As computing speeds increase (as represented by novel computer architectures), fundamental database management issues will be more adequately addressed, allowing for our sloppiness or incompetence in characterizing information. For a company to devote extensive resources to the study of A.I., when progress in related (more fundamental) areas will probably solve the knowledge processing problem in due time, must be viewed as a risk and should be evaluated as such.

A greater technological risk was seen, however, in not pursuing the developments in A.I. D. M. Prett declared "we do not wish to be ignorant in this area; this represents a greater risk to us. We must be in a position to be able to evaluate others' claims intelligently and make informed decisions concerning the use of this technology."

It was generally agreed that while it is relatively easy to take issue with the particulars of this technology, the A.I. framework offers a very real and substantial benefit toward the effective transfer of technology within an organization. Prett offered the following in support of this view. "When we (as Shell Development) 'sell' a product to Shell Oil, we sell the algorithm. We find it extremely difficult to transmit the expertise gained during development. Expert systems provide a wonderful philosophy for embedding in and around an algorithm the expertise which is really needed to make that algorithm work."

Prett also asked of the group, what is an expert? He offered the following definition and discussed some of the consequences of such a definition: An expert is an individual who, as an expert, performs an act outside of his or her awareness. Within this definition there is not necessarily an attribute of intellectual depth to this activity. It was noted that some of the more successful expert systems use precisely this form of rote information. However, when this type of information is either lacking or does not support an application well, it is very difficult to translate the current expert system methodologies. In some cases it may be difficult to identify an expert. In cases where expertise does demand true intellectual prowess, the externalization of this knowledge to another individual takes the form of a tutorial and the true power of the knowledge is lost because of this incomplete transmission. The result is a system which does not do anything wrong per se, but to an expert in the area, what the system does do is very simple, almost childlike, and the original expert ultimately wants nothing to do with it. The process of transmitting the expertise from the expert to the system via the knowledge engineer is very precariously balanced between "teaching" and "learning." Much improvement to this process is a prerequisite to progress in the field.

While it was generally realized that expert systems differ from simply "plugging in the appropriate formula," the following example was offered to demonstrate that the benefits can be just as real. Prett described a scenario whereby an intelligent interface is able to extract from an operator a process model containing enough detail for our control technology to support the design of a control system. The system might possibly prevail upon the operator, based on the presentation of a very poor closed-loop response, to go to the control room and return with more information. The interaction closely follows how an experienced control engineer might interact with the operator and relies upon the operator's general perception of the behavior of the process. Since it involves expertise, it is declared an expert approach. Since it is automated, it is declared an expert system.

2.4 "An Artificial Intelligence Perspective in the Design of Control Systems for Complete Chemical Processes," G. Stephanopoulos, J. Johnston (Speaker), and R. Lakshmanan (Speaker)

Summary of Presentation

The paper was presented jointly by R. Lakshmanan and J. Johnston. A general overview of the area of knowledge-based systems was first presented by Lakshmanan. Knowledge representation in process analysis, planning, and design was then addressed. A description of the "human-aided machine design" paradigm followed in the second half of the presentation delivered by Johnston. The final segment of the presentation, which was shortened due to the length of the presentation, was a description of a prototype design package under development at the Laboratory for Intelligent Systems in Process Engineering (LISPE) of the Department of Chemical Engineering at MIT.

Summary of Discussion

The discussion focused on what should be the role of academia in the development of knowledge-based systems and to what extent, if any, this development constitutes research. Industry, it was agreed, should develop a first-hand understanding of the issues in order to make informed decisions about the claims of this technology. The immediate benefits to some applications were also recognized, providing more support for industry involvement. It was less clear to what extent, if any, academia should involve itself with the exercise of integrating techniques that have possibly been around for years. The research content and ultimate contribution of this exercise were questioned by the academics present.

While all agreed that research can be an investigation of the underlying principles governing a phenomenon, D. M. Prett, speaking on behalf of the presenters, claimed that it can also be the investigation into some aspect of the validity of the combination of known guiding principles in such a way as to represent some package of value to society. However, Prett also emphasized that this particular area, Expert Systems, is so poorly understood that it is difficult to reach any sound conclusions regarding its value as yet. It thus seems reasonable to applaud any organized research effort, such as that proposed, which enlightens the field.

2.5 "Process Identification - Past, Present, Future," D. M. Prett, T. A. Skrovanek, and J. F. Pollard (Speaker)

Summary of Presentation

The presentation was given by J. F. Pollard, and was substantially the same as that in the submitted paper. He first outlined the definition of the model identification problem as practiced at Shell, that of linear parameter estimation. A general overview of the state-of-the-art of Shell's identification technology was presented, with a discussion of the advantages and disadvantages of increasing levels of complexity and sophistication.

One of the primary points emphasized during the presentation was that the choice of identification technique is not nearly as important as the quality of experimental data. That is, if the data quality is very high (correct test signal design, low noise, no disturbances, etc.) then any identification algorithm will give good results. The true test of a "sophisticated" technique is its performance on a poor set of data. One important aspect of acquiring good experimental data is that of proper design of the test. Future research in the area of identification will be concerned with the end use of the model (process control), rather than blindly striving for the "best" fit to the data.

Summary of Discussion

The discussion began with J. J. Haydel asking what kind of improvement in control system performance could be expected using Shell's best identification technology. It was emphasized that the bottom line test of a model is how well it works in a controller. In this regard we have no direct comparisons between a model developed using step tests and one derived using more sophisticated techniques. D. M. Prett indicated that until the development of the more sophisticated tools we had gone into a control house, developed a model using some simple technique, put it in a QDMC controller, and tuned it. When the loop later failed to meet specifications and was taken off control, there was no way of knowing whether it was due to a poor model or to poor tuning. The major advantage of these latest techniques is that they give some indication of the validity of the model.

Pollard suggested that if the system has low noise levels and the operators will tolerate large step tests, then this is often the best we can do. From a practical perspective, however, these conditions are very rare. For most systems, a large increase in model accuracy has been found using the advanced tools. Jim stressed that we do not want to completely replace step testing, as there is obviously still a place for it.

The point was made that the cost of conducting experiments is very high, suggesting the need for optimal design of experiments. This is the reason that control requirements should be stated before experimental design. The experiment should be designed so that the model will be accurate over the frequency range of interest. There was some opinion expressed that there is more emphasis on test signal design than is justified. The correct identification of steady state gain is not important to the control system. Rather, correct identification around the crossover frequency is of more importance. Also, correct identification does not depend on having a perfect test signal.

The suggestion was made that a large corporation such as Shell could benefit from developing detailed nonlinear models for classes of systems (e.g. distillation columns). The integrated effort of doing this will be less than that for identifying models using field experiments alone for all the units of a similar class. Shell personnel commented that similar pieces of equipment exhibit remarkably different behavior, both dynamic and steady state. An added problem with this approach is the extensive manpower requirement. First principles modeling is very difficult, and often bears little resemblance to the actual unit. American oil companies have spent thousands of man years developing first principles models, and few would feel comfortable using the results for any control house activity involving closed-loop end uses.

2.6 "Adaptive Control Software for Processes with Significant Deadtime," C. Brosilow (Speaker), W. Belias, and C. Cheng

Summary of Presentation

The presentation was given by C. Brosilow and was very similar to material in the submitted paper describing the INTUNE^{TM} controller by Controlsoft, Inc. The internal block diagram structure of the controller was shown and discussed. The three modes: *Autotune-1, Autotune-2,* and *Learn* mode, were described, along with the operator's interface and the engineer's interface. A detailed example detailed a test run on a laboratory heat exchanger.

Summary of Discussion

The question and answer session began with a comment that incorporating the nonlinear structure of a specific problem would improve performance. Brosilow agreed, indicating that most applications of adaptive control are more truly of a nonlinear control nature.

There was some discussion concerning the proof of convergence for general adaptive control systems. The standard assumptions used in adaptive controller convergence proofs do not hold for real systems. These assumptions include a linear process, linear model, and bounded zero mean noise. Brosilow responded that he has an approach which makes sense and has a reasonable opportunity of stabilizing any system. A filter time constant can always be chosen long enough that the system will be stable. Others made the comment that convergence proofs are extremely difficult. Why these methods work in practice is a complete mystery at times. It is unreasonable to ask for a proof here.

D. M. Prett made the point that the industry currently has 10 to 20 percent of its control loops on manual. This is because people in the control house don't understand automatic control action sufficiently well to tune and maintain the loops. Effort should first be focused on simply tuning these properly and then observing which, if any, need the added complexity of on-line adaptive control technology. These adaptive control systems may increase the tendency of the operators to put the loops on manual, because they are not as "understandable" as PI controllers. Furthermore, we have millions of dollars of equipment already in place that has "nothing wrong" with it except that it is poorly tuned. Brosilow responded that the problem of loops being off control is what this product seeks to address. He stressed the point that one should not use an adaptive controller as a PI tuner, that is, tuning the PI controller at the local conditions and then going away and leaving the tuning fixed as the conditions change. Prett commented that we could come back ten years from now and count the number of adaptive controllers running without operator intervention, and it will be as small as it is today, with those controllers generally applied in specific instances to difficult and expensive problems. Brosilow responded that adaptive control is not something that will go away. Finally, Prett reiterated his point that there is a significant need to tune the controllers already in place before it can be determined whether advanced techniques such as adaptive control are required. However, he generally concurred that adaptive control will not "go away."

2.7 "A Systems Engineering Approach to Process Modeling," O. A. Asbjørnsen (Speaker)

The Tuesday evening session of the workshop was devoted primarily to process modeling in process control. The first presentation was given by O. Asbjørnsen of the Systems and Research Center and the Department of Nuclear and Chemical Engineering at the University of Maryland entitled "A Systems Engineering Approach to Process Modeling."

Summary of Presentation

Asbjørnsen's presentation began with a working definition of "Systems Engineering." It is the analysis and design of units that function together in a total system to perform specified tasks under specified tolerances. Systems Engineering is a synergy of many important areas that include optimization, control, identification, reliability, safety, and communication. A unit is any technical, financial, social, or abstract mathematical representation including another subsystem. A system is a general network or connection of cause and effects, and the system tools are assumed to be independent of the applications. The definition of modeling used in the context of Systems Engineering was that of any collection of abstract cause and effect relationships. The thrust of the presentation was that modeling within a Systems Engineering framework as it occurs in industry should be a learning process with a feedback path identified as connecting research and development, plant design and plant operation.

To indicate how results are obtained in an industrial setting, Asbjørnsen deviated from the content of his paper to present examples of pressure control in an ethylene plant and the determination of safe operating limits for a propane storage facility. Additionally, many examples of combining first principles with information from the learning process were given.

Summary of Discussion

The discussion began with a comment from D. B. Garrison indicating that he disagreed with Asbjørnsen's statement that general principles were a good foundation for rules in an Expert Systems. It has been his experience that it is difficult to build a system in any discipline simply because of the nature of building the system and the current inadequacy of representations. The representations available to a programmer through the Expert Systems methodology for building a system are very limited. The feeling was that AI is still in it's infancy as compared with mental expertise for tasks of the complexity that humans routinely perform. This may change as we gain more insight into Expert Systems.

The remainder of the discussion focused primarily upon the question of developing first principle models versus obtaining a simple step response model for the purpose of process control. J. J. Haydel and D. M. Prett emphasized that process control engineers at Shell generate a process model that is just capable of accomplishing the process control objective and no more. This model will obviously obey all first principle concepts such as mass and energy balances, but will contain various levels of empiricism in expression. The level of empiricism is where the adequacy issue is focused. We generally focus on the minimum level of first principle concepts that are able to provide a usable model. The underlying reason for this policy is that process modeling is a very costly and time intensive task for which the recovery of extensive modeling costs are questionable. Often, a simple step response model (or model generated for a plant by a number of other techniques as discussed by J. F. Pollard in the Tuesday morning session) is preferred to first principles modeling, unless a more sophisticated model is readily available from another source. First principles modeling is aesthetically pleasing to Chemical Engineers, because it is the method by which most are educated. In practice though, it is difficult to justify modeling from first principles because of the complexity one is faced with in today's plants and the high costs associated with the additional modeling effort. Asbjørnsen suggested his opinion was not significantly different, but he still saw merit in modeling from first principles.

2.8 "Modeling and Control of Dispersed Phase Systems," J. B. Rawlings (Speaker)

Summary of Presentation

The presentation focused on the modeling, behavior, and control of two dispersed phase industrial processes, a continuous crystallizer and an emulsion polymerization system. The goal of the work was to model a class of dispersed phase systems and to use the new models as a vehicle for developing control algorithms able to address reactor stability, increased product yields, and lowered energy costs in the presence of disturbances. Following the organization of the paper, the presentation began with a brief introduction to the population density function and continued with the development of a generalized population balance over an arbitrary volume in a Lagrangian formulation. This function would be used in the examples.

The first industrial example was that of a continuous crystallizer designed to grow crystals of a specific size. Four equations were introduced for modeling the crystallizer: a growth rate expression as a function of supersaturation, a nucleation rate expression as a function of supersaturation, a population balance, and a solute balance. The equations indicated explicitly that the boundary condition is coupled to the integral of the population density function due to the supersaturation driving force for nucleation. This coupling is a source of internal feedback and means that uncontrolled continuous crystallizers are often unstable, as confirmed by experimental data. The second industrial example was that of a continuous emulsion polymerization reactor. As with the crystallizer, the boundary conditions for the model equations are coupled to the integral of the population density function. This coupling explains the unstable steady states that can also be exhibited by this system.

After a brief review of control strategies used previously with dispersed phase systems, Rawlings deviated from the content of his paper to introduce the development of an optimal controller for batch crystallization based upon a population balance model. The goal was to optimize the crystallizer temperature as a function of time subject to constraints and the model nonlinearity.

Summary of Discussion

Much of the discussion was related to the formulation and solution of the batch crystallizer problem introduced near the conclusion of the presentation. Rawlings explained that the formulation of his controller was nonlinear, because the reactor temperature affects all the rate constants in the model. Getting the solution to the partial differential equation (PDE), then solving the optimization problem, is unacceptable. The two problems must be solved together. The difficulty in obtaining a solution is mostly due to finding the proper numerical technique. One particular problem that arises is that the PDE is defined on a semi-infinite interval of zero to plus infinity. Collocation methods for the solution of PDE's on unbounded intervals are not well developed at present and may not be useful. Some suggestions were offered to solve the problem by using various numerical methods and coordinate transforms. Rawlings then asked if there was existing technology to solve his problem as a QDMC problem.

After discussing numerical techniques, Rawlings pointed out that his interest is not to get a powerful, fast PDE solver. "I'll do that if I have to, because I am limited by the hardware I have and I want to get the solution to work. It's not the goal of the work though." He also didn't necessarily think that using a bigger computer is the right approach either. The idea is to get the solution to a class of PDE's characterized by a set of properties using existing hardware and some refinements in technology. This solution can then be used whenever similar problems occur, without any additional effort tuning the underlying numerical technique.

A suggestion was offered for solving the stability problem in the continuous flow systems. It was mentioned that stability might be achieved in the process design phase by using a CSTR and a plug flow reactor in series. The proper combination may solve the control and design problem at the same time. Rawlings agreed. It is one of the reasons he is not looking at the emulsion polymerization reaction control problem, even though he does have a better understanding of the process. It is known that the dynamics of the reactants are too fast to perform control in a reasonable way. The solution is to try and correct the problem in the design phase and not approach it as a control problem. The batch crystallization problem is much more interesting problem from that point of view.

2.9 "Robust Control, An Overview and Some New Directions," V. Manousiouthakis (Speaker)

Summary of Presentation

The speaker presented some of the concepts of algebraic control theory to the audience. He also detailed some of the results for the design of robust controllers outlined in his paper, but the presentation was shortened somewhat due to time limitations.

Summary of Discussion

Comments from the audience concerning this paper were mainly on the motivation and practicality of this work and on technical details on some results presented. On the first issue, there was a concern that this same formulation has been around since the 60's and has had little impact so far. A question was asked as to what is the author's motivation behind employing this formulation. The author replied that the formulation used in this paper justifies everything being done in linear time invariant systems to date, gives characterizations for sets of stabilizing controllers for linear time invariant systems, and gives characterizations for sets of stabilizing controllers for nonlinear systems.

Specifically, a comment was made that characterizations for stabilizing controllers for nonlinear processes are almost impossible to construct, except in special cases. A main issue is the fact that the approach used by the author to find factors for the fractional representation (solving nonlinear Riccati equations) is not necessary if the objective is only to find a stabilizing controller for nonlinear processes. It was pointed out that according to the author's characterization, a coprime factorization is stabilizing, but coprimeness can be achieved by using a constant parameter as a controller without having to solve Riccati equations.

On the robustness issue, a theorem was presented by the author where an assumption about the plant uncertainty was made as follows: $P - P_0 + I$ has to be full rank over the entire right-half plane for every uncertain plant, where P is the plant and P_0 is its model. It was pointed out by the audience that this condition is very hard to verify in practice. Other robustness results only require that the number of unstable poles of the system remain constant for all possible plants given by the uncertainty description. It was claimed that this condition is easier to verify.

2.10 "An Overview of Nonlinear Geometrical Methods for Process Control," J. C. Kantor (Speaker)

Summary of Presentation

The author presented material intended to clarify concepts of geometrical methods. A number of examples were discussed. The author said that his motivation for looking at geometrical methods for analyzing nonlinear processes comes from the need to understand the whole idea behind building a control system. That is, what are the important combinations of variables to control and manipulate for nonlinear processes?

Summary of Discussion

Comments from the audience were focused along two lines. The first one was a concern on the physical significance of the resulting transformation of variables used for control. The author pointed out that in his experience he has found that the linearizing combinations of variables usually yields a practical physical quantity. This was demonstrated through the examples. Some people mentioned that, similar to modal control for linear processes, the "best" combination of variables, although possibly containing physical meaning, do not reflect the true performance of the control system. That is, the performance might be defined in terms of a variable that yields necessarily a nonlinear control system. The fact that the transformation is nonsingular only guarantees zero offset at steady state, but does not guarantee that the true performance variable will exhibit the desired dynamics. It was also pointed out that in finding the transformations the assumption of continuous derivatives might be too restrictive. For many models where correlations are used the derivatives may not be continuous. The author said that the purpose of this tool is mainly for analysis and, therefore, a simpler model that is indeed continuous might be sufficient. From the industrial side, a warning was made to the fact that the kinds of models used in Chemical Engineering are so imprecise and incomplete that this kind of analysis could not be applied. The author rebutted that, at least within the class of models presented, much can still be learned by using this method.

The second line of discussion was along the fact that the Hunt, Su and Meyers method relies on the nonlinear process having some desired (and restrictive) properties in order for it to follow linear closed-loop dynamics. In the absence of those properties and, in particular, in the presence of constraints (which would limit the invertibility of the plant) it was felt by some members of the audience that an approach like QDMC would have to be used. Since this approach would work for all nonlinear processes, then there was a question as to whether the geometric approach was useful at all. A member of the audience made the point that the issues of transformation into a linear system and whether the resulting system is invertible and unconstrained are separate. The author stressed that via the geometrical approach a lot can be learned about the process, then a method like QDMC could be applied for implementation in the face of constraints.

2.11 "Recent Advances in the use of the Internal Model Control Structure for the Synthesis of Robust Multivariable Controllers," E. Zafiriou (Speaker)

Summary of Presentation

This paper addressed the issue of designing an IMC controller for open-loop unstable systems. The author also presented a methodology for designing robust controller by calculating a set of optimal filter time constants within the IMC framework. This work represents a merger of the IMC structure and the Structured Singular Value approach. The design is suboptimal when compared with J. Doyle's μ–Synthesis method. However, the methodology is easier to use than the μ–Synthesis approach.

Summary of Discussion

There were several issues raised by Shell representatives. One issue was the method for the selection of the filter order. Although the minimum order is chosen to achieve realizability of the controller, this might not be enough for satisfying robust stability. At the present time, there is no standard method for selecting the order other than the realizability criterion. Another question concerned the implementation via a classical structure which is not suited for constraint handling. The academics pointed out that if the manipulated variable is constrained for open-loop unstable processes there is nothing a controller can do. However, if predictive control is used, constraint violation prevention (in particular for output variables) could still be achieved.

Another question was raised as to whether this method is a specialized tool with expert requirements for implementation. The speaker indicated that, compared to the requirements for implementation of μ–Synthesis, there is no need to know the theory as well. Also, the value of designing a controller (using two filters as proposed by the author) for decoupling an ill-conditioned system, rather than accepting interactions using a simpler design such as two PIDs, was questioned. Both the author and others indicated that the correct procedure should be to try the simplest design first and evaluate the performance.

The author also stated that this work is not the final answer, but is a first step in attacking the robustness issue within the IMC framework.

2.12 "Characterization of Distillation Nonlinearity for Control System Design and Analysis," K. McDonald (Speaker)

Summary of Presentation

This paper presented a method of estimating variations in gain, time constants and dead times in distillation columns based on simple correlations. These correlations are functions of physical parameters such as relative volatility. The simple correlations may not be accurate in predicting the absolute values of the model parameters. However, they do predict well the rate of change in the first order model parameters which is sufficient for control system analysis and design. The uncertainty generated by using these correlations is then used to determine uncertainties in the Nyquist plot and used for robust design. One theorem stated that, for stability, it was sufficient to look at vertices of the Nyquist region generated by the maximum and minimum bounds of the parameters. Another use for these correlations was to update on-line the simple model used in a predictive control method such as QDMC. The author presented examples in the paper in which such an update method provides better performance than the standard method.

Summary of Discussion

A point was brought up concerning the computational effort involved in verifying all the Nyquist vertices. Another point was that a significant problem with uncertainty description is in accounting for correlation among elements of multivariable processes. How one can address this issue with the approach proposed in this paper is still a research problem.

The discussion then turned to the issue of process representation. It was questioned whether these simple models should be used on-line to update a model predictive controller such as QDMC, rather than using them for determining uncertainty in linear models. Shell representatives indicated that direct use of the simplified nonlinear dynamic model would provide a better representation and consequently better performance. Conceptually, it would be simple to implement the nonlinear dynamic model approach within the model predictive control framework. Technically, there are many issues, such as the degree of simplification versus the performance level which is gained, which need to be addressed. This is a research area which needs further investigation. Some academics were concerned with using these models in the control implementation without having a theory for analyzing the stability of the resulting controller.

2.13 "Selecting Sensor Location and Type for Multivariable Processes," C. Moore (Speaker), J. Hackney, and D. Canter

Summary of Presentation

This presentation was divided into two parts. In the first part, the author presented methods for selecting sensor location applied to distillation columns based on singular value decomposition. Since singular value decomposition is dependent on the scaling of the system it is very important that the system be scaled properly. The author recommends strongly that before the analysis is done the system be scaled physically. From the practical viewpoint, the selection of the correct measurement locations can have a much higher impact than the selection of the appropriate control algorithm. For a two input problem (distillate and boilup), it was shown that for minimum condition number the sensors must be close to the top and bottom of the column, respectively. Using a principal component analysis, the trays were close to the middle where the temperature profile was more sensitive. Therefore this problem requires a tradeoff between measurement locations which provide good sensitivity and measurement locations which provide independent information.

During the second part, a more general talk was given, proposing a change in emphasis in control research away from design and into development and maintenance of control systems. The author proposed the development of tools which analyze the information buried in process data to develop hypotheses concerning the current state of operation and control. These tools produce images of the plant at discrete times, which can be archived as the historical evolution of the plant and the control system. These images can be used for comparison later. This will help in a number of plant operations area such as failure detections, optimization, etc. The author used one of the slides used by R. Lakshmanan and J. Johnston in the context of Expert Systems indicating the importance of developing methods for analysis of operations. He proposed the use of Expert Systems to analyze and process graphical representations of system performance in order to detect failure and system change.

The author also said that there is an urgent need for the theoreticians to develop theories which can be used within this image processing framework. One example which he cited is the statistical process control technique. Traditionally, this technique has been applied to discrete processes. Certainly it would be quite challenging to develop techniques which would work for continuous processes controlled by a control system.

Summary of Discussion

The discussion concentrated almost exclusively on the first part of the talk. It was mentioned that the disturbance effects were important and that the selected measurements might be different if a disturbance enters the process.

The author pointed out that it was ridiculous to move the sensors and, even if the profile is controlled, the issue of placing the sensors remains. The author also said that, based on his experience, the sensor locations selected by the analysis with respect to the manipulated variables are sufficient for disturbance rejection. This is true because, in general, the disturbance directionality is similar to that of the manipulated variables. However, this is not true for setpoint changes. Then the issue of inferring composition with temperatures was brought up. If composition analyzers were available, why use temperatures at all? The author indicated that in general these analyzers are so slow that temperatures are necessary for disturbance rejection. Also, it was pointed out that the method presented does not guarantee that the selected temperatures will be correlated with the compositions. The author replied that it does not matter which temperature one looks at for disturbance rejection. If the appropriate analyzer measurement is available then a cascaded control structure can be implemented where the inner loop is designed for fast disturbance rejection and the outer loop meets the control objectives.

A comment was made from Shell that any "appropriate" temperature is sufficient for most operations and, if desired, that analyzers are generally not a problem to install and operate. Also, it was asked whether this analysis can indicate for which measurement locations any control strategy will not work. The author said that systems for which the largest singular value was much smaller than the valve resolution or for which the smallest singular value was much larger than sensor noise, will exhibit problems. The problem can be that one does not have the right instrument span. A warning was made by Shell that these are important issues, but one should not make common sense decisions into issues of theoretical significance.

The second part of the talk was discussed briefly at Shell's insistence. Some academics asked for Shell to comment on Statistical Process Control. It was pointed out that Shell supports C. Moore's comments and that there is an "extremely sad" percentage of advanced control loops in operation. A "fingerprint" of the process to use for diagnosis would be very helpful.

Due to time constraints, a section of time was allotted Thursday morning to provide another opportunity to discuss this new direction in research.

2.14   "Three Critiques of Process Control Revisited a Decade Later," M. Morari (Speaker)

Summary of Presentation

In the spirit of the Workshop, M. Morari presented what he loosely described as a "critique of critiques." The critiques were prepared by Foss (1973), Lee et al. (1976), and Kestenbaum et al. (1976). These well-known critiques of process control were timely and important. Morari said that in reviewing these critiques he was interested in reviewing progress to date and did not wish to appear to be in any way critical. In fact he said that each researcher could benefit from reviewing his own work over the years in a similar fashion. He also wished to point out that this presentation would not be a comprehensive critique, but would rather focus on key topics. D. M. Prett pointed out that, for historical perspective, it was interesting to note that the last of these reviews was completed in 1976, the same year that Shell first implemented its on-line, multivariable constrained form of DMC as documented in the landmark Cutler, Ramaker, Prett, Gillette series of papers of 1979.

With respect to "Modal/Pole Placement," Morari made the point that the technique was "all well and good," but the unanswered question was "where should we place the poles?" On "Decoupling" the forgotten issue was the required level of model accuracy. On "Frequency Domain Control Design," it is very interesting, has increased our understanding of control, but has not been a useful nor effective control design technique. On "LQOC" the forgotten issue is again the high level of accuracy required in the process models. The point being made was that many study areas have not resulted in effective tools for control design nor have they represented the basis for a control algorithm in practice, because significant "other" issues were ignored. This is not to say that the research efforts were without value. They have individually and in composite represented a significant step forward in understanding and comprehension of the control problem. However, let's advertise them as such and not as "products." A good example of a shortcoming in most individual studies is the degree to which the issue of the level of modeling and model uncertainty can affect the validity of application of a given control design. Sometimes the modeling step is the most critical, but this is often overlooked.

The Garcia/Prett paper and discussion during this workshop also brought out the value of such understanding. Their point is that accounting for performance objectives and process representation considerations up front may still generate a PI controller as a solution, but it will now be one that is appropriately considerate of model accuracy, etc. Their paper does not demand DMC or other sophisticated approaches as the solution to all control problems.

Morari then proceeded to review progress over the years in many areas of process control. This was completed without significant comment. Morari emphasized that great advances have been made in the area of constraints handling with the invention of Shell's DMC and MAC. Shell's DMC is the first application of on-line dynamic constrained optimization implemented (and documented in the open literature) in industry. He said it is very important for us all to realize that what DMC achieves in a unified methodology is what some have tried to achieve in isolated cases with ad hoc collections of software and hardware, including switching relays and logic overrides. However, even in the context of DMC, what do we have to say about model uncertainty? Morari said that his research group at CalTech is doing much work on this area, but that we must realize that the issue is by no means resolved.

This point is obviously of great concern to industrial researchers who must produce products as well as "understand" problems. This concern is currently being met by academic researchers by way of the recent increase in the effort studying model requirements and model uncertainty. Industry's concerns and at least one overview perspective is well illustrated in the Garcia/Prett paper presented at this workshop.

Morari then proceeded to discuss model uncertainty issues. The hope is that many alternative approaches to model uncertainty handling will be developed over the next few years. Morari did emphasize that we now understand uncertainty sufficiently well that we can say that maybe we will not need sophisticated techniques such as μ–Synthesis, for example, to deal with this problem. Morari felt that this was an excellent example of the value of academic research. I.e.,

> "We, in academia, have the opportunity to study, conceptualize, and contemplate problems and their solutions. As a consequence of this process, we may develop techniques or tools for problem solution. However, we can in many cases understand a given area so well that we learn to restructure our existing approaches to problem solution in such a way that the original difficulty is no longer a problem."

Morari went on to emphasize the value of a control scheme whose actions can be understood by observers such as process engineers or operators. Morari felt that it was important for an operator to feel that he can interact with a control scheme in a manner other than putting the loop "on manual." He felt that DMC was a good example of this high acceptance of interaction, while LQG had been a failure in this respect.

D. M. Prett responded to this by saying that he felt that this 'interaction value' was being overemphasized. Prett felt that the fact that DMC and other controllers "did the job and did it well" from a purely technical perspective greatly overrode any interaction attributes. In this sense, DMC minimizes the need for interaction in that "it works." Hence operators do not distinguish between PI and DMC controllers. Operators do not "understand" PI or DMC or most other control algorithms. They do understand that a given controller works or does not work. This is important to them. Operator interaction beyond this point is not a necessary component for a successful application.

Morari then went on to discuss how, as control system designers, we often ignore the implicit assumptions inherent to the technique being implemented. This often leads to ad hoc quick fixes or what Brian Ramaker referred to as "tricks" in on-line application. In this sense, once again, the approach raised in the Garcia/Prett paper forces the designer to recognize the assumptions and compromises necessary to achieve a feasible solution.

Summary of Discussion

This was the end of the presentation proper. There was much discussion of a general content. The most significant statement made in the discussion was Morari's comment that DMC would be limited only by the ability of its users to adequately model the process or processes encompassed in the design. Included in the model question was the model uncertainty aspect.

Chapter 3

Shell Control Problem

3.1 Introduction

During the preliminary planning of the workshop the coordinators recognized the importance of providing tutorial sessions during the conference. The objectives of these sessions were

- To illustrate the difference between solving a problem via application of a single methodology rather than using ad hoc "tricks" and "quick fixes;"

- To give the academic participants the opportunity to work on a problem which is representative of the kind of problems which an industrial control engineer has to solve;

- To provide an environment other than the standard lecture and discussion format in order to enhance the communication process between the industrial and academic participants.

It was decided that there would be two tutorial sessions. Shell personnel would develop a problem definition from their experiences and resources. Shell would also provide the facilities which would help the participants in analyzing and understanding the problem. During the first tutorial session, Shell personnel would provide descriptions of the available hardware and software facilities and the problem definition. The remainder of the time of the first session would be used by the participants to understand the problem definition and to conceptualize solution approaches. The second session would be used to analyze and implement these possible solution approaches.

The next section summarizes the results of the two tutorial sessions.

3.2 Tutorial Sessions Summary

Participants:

Academia

    O. Asbjørnsen
    C. Brosilow
    B. Holt
    J. Johnston
    J. Kantor
    R. Lakshmanan
    V. Manousiouthakis
    K. McDonald
    C. Moore
    M. Morari
    J. Rawlings
    E. Zafiriou

Shell Personnel

> J. M. Fox
> C. E. Garcia
> J. M. Keeton
> H. K. Lau
> P. Marquis
> W. B. Schmidt
> P. Sommelet

Summary of First Day of the Problem Solving Session

J. E. Wheeler, the Shell Process Control Training Center coordinator, wel-
comed the participants to the training facilities. J. M. Keeton introduced the
participants to the computer facility and the software packages available at the
Shell Process Control Training Center. The two software packages available
to the users were CONSYD and CSD. CONSYD is a software system developed jointly by M. Morari and his students at CalTech and W. H. Ray and
his students at the University of Wisconsin. CSD is Shell's in-house control
system design package. A summary describing the basic structure of the CSD
package was distributed to the participants.

C. E. Garcia discussed the problem definition with the participants. The
unit for which the group was asked to develop a control strategy is a heavy oil
fractionator. The feed into this column contain fractions with boiling point
ranging from 0 to 1000 degrees F. The column splits the feed into the desired
products. The three circulating refluxes in the column remove heat. These
refluxes reboil other columns in the plant. There are three product draws from
the column. A summary of the process description and the control design issues
was distributed to the participants.

Numerous questions regarding the problem and the software packages were
raised by the academic participants. Two common complaints made were that
the problem definition was vague and that they had to work with unfamiliar
software. Another interesting complaint was that some expressed doubts concerning the apparent difficulty of the problem.

After the first session many questions were raised concerning the value of
the tutorial sessions. The academics were interested in learning more about a
broad range of control problems which occur in industry. Some members of
academia were also interested in analyzing the tutorial problem further. To
accommodate this dual need it was decided that the group be separated into
two subgroups. One group, which consisted of J. Johnston, R. Lakshmanan,
K. McDonald, J. M. Keeton, H. K. Lau, and P. Marquis, returned to the
training center to work on the problem on the second day session. The rest of
the group remained at the hotel conference room. Shell personnel presented
practical examples illustrating the control issues with which our industry is
commonly faced.

Summary of Second Day of the Problem Solving Session

D. M. Prett described a furnace process control problem in detail to the participants who remained at the conference room. There was much interest in the process engineering aspects of the problem described. However, the problem was dismissed as being not of a high degree of difficulty from a control engineering viewpoint. Prett, after much discussion, again redirected the focus of the group study. He reviewed the objectives of the tutorial sessions with particular emphasis on the subject of the need for the development of methodologies for problem solution rather than so called quick fixes completed in an ad hoc fashion. Hence, when Shell describes a particular process and its particular control problems one can be sure that the solution will be found within the framework of the currently available methodologies. When a particular problem demands an increase in the basic technical sophistication of the so called "standard" solution, then this improvement will be completed and installed in standardized fashion into our existing methodology. We will in most cases have in our tools for problem solution more than is needed for a particular problem, but the solution will be of a standard type. In this way we find that we can maintain our existing facilities at the lowest manpower cost. Therefore, rather than exposing the academic participants to a wide range of typical industrial problems, Prett suggested that Shell create a composite "real" problem that would include, within a single problem definition, most of the features present in industrial problems. In this way we could all have a base on which to compare solution methodologies.

All the participants agreed to the idea that Shell will publish a generic control problem to illustrate the relevant design issues, such as multiple variables and constraints, as a standard example. The academic participants agreed to work on this problem. All participants hoped that this example would serve as a standard test for new control theories. At some future date the whole group will meet again to discuss the solution developed by each. This exchange of information was also felt to be beneficial to industrial researchers in the control skill area.

K. McDonald, one of the group who went to the process control training center to work on the workshop problem, reported that she felt that the second day at the training center was a beneficial experience for her. She also stated that in order to meet all the required objectives of the workshop problem the control system would have to be able to handle constraints. That is exactly the point which Shell personnel were trying to convey with this example. P. Marquis echoed her statements and stated that perhaps the majority of the group had given up too quickly after the first day.

Conclusion

Shell agreed to publish a generic control problem which will serve as a standard control problem for the academics to study. This problem will contain all the control issues which are considered important by Shell. The academic participants in the tutorial sessions agreed to study this problem and to develop solutions. We will gather some time in the future to discuss and compare results.

The Shell Control Problem is described in the next section.

3.3 Problem Description

Figure 3.1 shows a heavy oil fractionator with three product draws and three side circulating loops. The heat requirement of the column enters with the feed, which is a gaseous stream. Product specifications for the top and side draws are determined by economics and operating requirements. There is no product specification for the bottom draw, but there is an operating constraint on the temperature in the lower part of the column. The three circulating loops remove heat to achieve the desired product separation. The heat exchangers in these loops reboil columns in other parts of the plant. Therefore, they have varying heat duty requirements. The bottom loop has an enthalpy controller which regulates heat removal in the loop by adjusting steam make. Its heat duty can be used as a manipulated variable to control the column. The heat duties of the other two loops act as disturbances to the column.

The relevant information regarding the Shell Control Problem is stated in these five subsections:

1. Control Objectives

2. Control Constraints

3. Process Model

4. Uncertainties in the Gains of the Model

5. Prototype Test Cases

We have tried to encapsulate the relevant control issues in this one problem while staying as realistic as possible. The problem is stated such that an infinite number of scenarios can occur in controlling the unit. We would encourage the development of solution methodologies that are flexible enough to deal with varying (and possibly conflicting) problem requirements, and can be readily automated such that control designs can be carried out by plant personnel with only a modest knowledge of control concepts.

A complete solution to the problem should describe, in detail, the analysis and synthesis procedures that indicate that the proposed controller satisfies the control objectives for all the plants in the uncertainty set. However, because of possible discrepancies between investigators on analysis techniques, we have formulated a number of prototype test cases which form a common frame of reference for evaluating different designs.

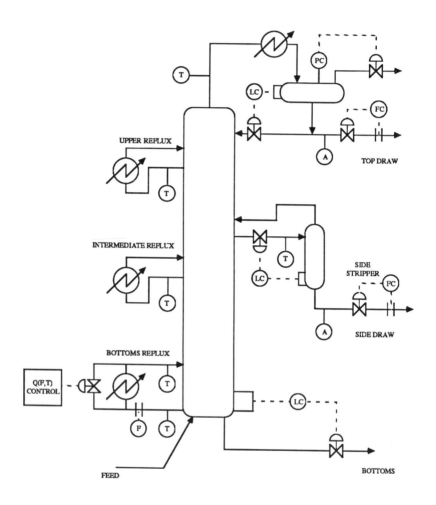

Figure 3.1: Diagram of a Heavy Oil Fractionator (Shell Control Problem).

Control Objectives

1. Maintain the top and side draw product end points at specification (0.0 ± 0.005 at steady state).

2. Maximize steam make in the steam generators (i.e., maximize heat removal) in the bottom circulating reflux (Important note: heat duties are expressed in terms of heat input to the column. Decreasing heat duty implies increasing the amount of heat removed).

3. Reject the unmeasured disturbances entering the column from the upper and intermediate refluxes due to change in heat duty requirements from other columns. (Upper and intermediate reflux duties range between −0.5 and 0.5.) Reject disturbances even when one or both end point analyzers fail.

4. Keep the closed-loop speed of response between 0.8 and 1.25 of the open-loop process bandwidth.

Control Constraints

1. All draws must be within hard maximum and minimum bounds of 0.5 and −0.5.

2. The bottom reflux heat duty is constrained within the hard bounds of 0.5 and −0.5.

3. All manipulated variables have maximum move size limitations of magnitude 0.05 per minute.

4. Fastest sampling time is 1 minute.

5. The bottom reflux draw temperature has a minimum value of −0.5.

6. The top endpoint must be maintained within the maximum and minimum values of 0.5 and −0.5.

Process Model (First Order Deadtime)

$$\frac{Ke^{-\theta s}}{\tau s + 1}$$

Units for θ, τ are in minutes.

	TOP DRAW	SIDE DRAW	BOTTOMS REFLUX DUTY	INTER. REFLUX DUTY	UPPER REFLUX DUTY
TOP END POINT	$K = 4.05$ $\tau = 50$ $\theta = 27$	$K = 1.77$ $\tau = 60$ $\theta = 28$	$K = 5.88$ $\tau = 50$ $\theta = 27$	$K = 1.20$ $\tau = 45$ $\theta = 27$	$K = 1.44$ $\tau = 40$ $\theta = 27$
SIDE END POINT	$K = 5.39$ $\tau = 50$ $\theta = 18$	$K = 5.72$ $\tau = 60$ $\theta = 14$	$K = 6.90$ $\tau = 40$ $\theta = 15$	$K = 1.52$ $\tau = 25$ $\theta = 15$	$K = 1.83$ $\tau = 20$ $\theta = 15$
TOP TEMP	$K = 3.66$ $\tau = 9$ $\theta = 2$	$K = 1.65$ $\tau = 30$ $\theta = 20$	$K = 5.53$ $\tau = 40$ $\theta = 2$	$K = 1.16$ $\tau = 11$ $\theta = 0$	$K = 1.27$ $\tau = 6$ $\theta = 0$
UPPER REFLUX TEMP	$K = 5.92$ $\tau = 12$ $\theta = 11$	$K = 2.54$ $\tau = 27$ $\theta = 12$	$K = 8.10$ $\tau = 20$ $\theta = 2$	$K = 1.73$ $\tau = 5$ $\theta = 0$	$K = 1.79$ $\tau = 19$ $\theta = 0$
SIDE DRAW TEMP	$K = 4.13$ $\tau = 8$ $\theta = 5$	$K = 2.38$ $\tau = 19$ $\theta = 7$	$K = 6.23$ $\tau = 10$ $\theta = 2$	$K = 1.31$ $\tau = 2$ $\theta = 0$	$K = 1.26$ $\tau = 22$ $\theta = 0$
INTER. REFLUX TEMP	$K = 4.06$ $\tau = 13$ $\theta = 8$	$K = 4.18$ $\tau = 33$ $\theta = 4$	$K = 6.53$ $\tau = 9$ $\theta = 1$	$K = 1.19$ $\tau = 19$ $\theta = 0$	$K = 1.17$ $\tau = 24$ $\theta = 0$
BOTTOMS REFLUX TEMP	$K = 4.38$ $\tau = 33$ $\theta = 20$	$K = 4.42$ $\tau = 44$ $\theta = 22$	$K = 7.20$ $\tau = 19$ $\theta = 0$	$K = 1.14$ $\tau = 27$ $\theta = 0$	$K = 1.26$ $\tau = 32$ $\theta = 0$

Uncertainties in the Gains of the Model

	Top Draw	Side Draw	Bottoms Reflux Duty	Inter. Reflux Duty	Upper Reflux Duty
Top End Point	$4.05 + 2.11\epsilon_1$	$1.77 + 0.39\epsilon_2$	$5.88 + 0.59\epsilon_3$	$1.20 + 0.12\epsilon_4$	$1.44 + 0.16\epsilon_5$
Side End Point	$5.39 + 3.29\epsilon_1$	$5.72 + 0.57\epsilon_2$	$6.90 + 0.89\epsilon_3$	$1.52 + 0.13\epsilon_4$	$1.83 + 0.13\epsilon_5$
Top Temp	$3.66 + 2.29\epsilon_1$	$1.65 + 0.35\epsilon_2$	$5.53 + 0.67\epsilon_3$	$1.16 + 0.08\epsilon_4$	$1.27 + 0.08\epsilon_5$
Upper Reflux Temp	$5.92 + 2.34\epsilon_1$	$2.54 + 0.24\epsilon_2$	$8.10 + 0.32\epsilon_3$	$1.73 + 0.02\epsilon_4$	$1.79 + 0.04\epsilon_5$
Side Draw Temp	$4.13 + 1.71\epsilon_1$	$2.38 + 0.93\epsilon_2$	$6.23 + 0.30\epsilon_3$	$1.31 + 0.03\epsilon_4$	$1.26 + 0.02\epsilon_5$
Inter. Reflux Temp	$4.06 + 2.39\epsilon_1$	$4.18 + 0.35\epsilon_2$	$6.53 + 0.72\epsilon_3$	$1.19 + 0.08\epsilon_4$	$1.17 + 0.01\epsilon_5$
Bottoms Reflux Temp	$4.38 + 3.11\epsilon_1$	$4.42 + 0.73\epsilon_2$	$7.20 + 1.33\epsilon_3$	$1.14 + 0.18\epsilon_4$	$1.26 + 0.18\epsilon_5$

$$-1 \leq \epsilon_i \leq 1 \quad ; \quad i = 1, 2, 3, 4, 5$$

Prototype Test Cases

Demonstrate, through simulation, that the proposed controller satisfies the control objectives without violating the control constraints for the following plants within the uncertainty set (assume all inputs and outputs are initially at zero; magnitudes for the upper and intermediate reflux duty step changes are indicated below):

1. $\epsilon_1 = \epsilon_2 = \epsilon_3 = \epsilon_4 = \epsilon_5 = 0$. Upper reflux duty $= 0.5$, intermediate reflux duty $= 0.5$.

2. $\epsilon_1 = \epsilon_2 = \epsilon_3 = -1$; $\epsilon_4 = \epsilon_5 = 1$. Upper reflux duty $= -0.5$, intermediate reflux duty $= -0.5$.

3. $\epsilon_1 = \epsilon_3 = \epsilon_4 = \epsilon_5 = 1$, $\epsilon_2 = -1$. Upper reflux duty $= -0.5$, intermediate reflux duty $= -0.5$.

4. $\epsilon_1 = \epsilon_2 = \epsilon_3 = \epsilon_4 = \epsilon_5 = 1$. Upper reflux duty $= 0.5$, intermediate reflux duty $= -0.5$.

5. $\epsilon_1 = -1, \epsilon_2 = 1, \epsilon_3 = \epsilon_4 = \epsilon_5 = 0$. Upper reflux duty $= -0.5$, intermediate reflux duty $= -0.5$.

3.4   Schedule of Events

SHELL PROCESS CONTROL WORKSHOP

TUTORIAL ON CONTROL DESIGN – I

Monday, December 15, 1986

Shell Process Control Training Center

2:00	–	*2:15*	Introduction to PCTC Facilities – J. Wheeler
2:15	–	*2:45*	Introduction to Computing Facilities – J. M. Keeton (DEC, CONSYD, CSD)
2:45	–	*3:45*	Control Strategy/Structure Workshop
3:45	–	*4:30*	Individual Group Discussions
4:30	–	*4:45*	Wrap-Up Discussion
4:45	–	*5:00*	Return to Hotel

GROUP MEMBERS

1. C. Moore
   E. Zafiriou

2. B. Holt
   J. Johnston

3. J. Kantor
   K. McDonald

4. V. Manousiouthakis
   O. Asbjørnsen

5. C. Brosilow
   R. Lakshmanan
   J. Rawlings

SHELL PROCESS CONTROL WORKSHOP

TUTORIAL ON CONTROL DESIGN – II

Tuesday, December 16, 1986

Shell Process Control Training Center

1:15	–	*1:45*	Review of First Day's Workshop
			Control Objectives
			Control Structure
1:45	–	*2:45*	Unconstrained Control Strategies Workshop
2:45	–	*3:15*	Discussion
3:15	–	*3:30*	Introduction to QDMC – H. K. Lau
3:30	–	*4:15*	Constrained Control Strategies Workshop
4:15	–	*4:45*	Discussion/Wrap-Up
4:45	–	*5:00*	Return to Hotel

GROUP MEMBERS

1. C. Moore
   E. Zafiriou
   J. Kantor
   K. McDonald

2. B. Holt
   J. Johnston
   V. Manousiouthakis
   O. Asbjørnsen

3. C. Brosilow
   R. Lakshmanan
   J. Rawlings

Chapter 4

Summary

Preamble

Following M. Morari's presentation on the last day of the Workshop, ninety minutes were set aside for general discussion along the lines of summarizing the week's activities. D. M. Prett introduced and chaired this discussion and is the author of the following report.

Summary Discussion

D. M. Prett opened the summary discussion with a statement encouraging all present to fully participate in this summary so that the final workshop report could capture the essential points raised during the week. Recognizing that most people were very tired basis the long work sessions over the previous few days, he requested that the group attempt to focus on a small number of points that would represent the "message" we wish to capture as a result of our efforts.

To facilitate this process, attention was directed to the opening pages of the workshop notes to the extract from Sindermann's *The Joy of Science,* reproduced below for emphasis:

> "Science is an art form. Just as with painting or sculpture there are those who develop its commercial aspects, using 'formulas' to produce a salable product. Some professionals stumble on a formula early and grind out successive papers which elaborate on a single theme. Then there are those who search for elegant portrayals of concepts–of visions–and who become bored quickly with the routine and the mundane. Such individuals may spend an entire career searching for but never finding the right experiment or never making the unique synthesis. It is from the ranks of those searchers, though, that brilliant new insights can be expected."

It was pointed out that, in all likelihood, all present had, at one time or another, experienced the various kinds of activity noted in this interesting statement. Many have, for example, developed the commercial aspects of science "using 'formulas' to produce a salable product." Equally, many have been fortunate enough to "stumble on a formula early" while others have searched long and hard to develop "elegant portrayals of concepts." As Sindermann points out, "It is from the ranks of those searchers, though, that brilliant new insights can be expected." To this, Prett added the thought that there is tremendous value in all these different endeavors, but most especially when they can be synergized in such a fashion that the "visionary" researchers are working in a complementary fashion with the "single theme" researchers. This is how really significant unified progress can be made.

In this context Prett reviewed the development of Shell's Dynamic Matrix Control algorithm which has had such a significant impact on both industrial and academic research in the process control field. He pointed out that through its evolution there were periods of brilliant insight of short duration followed by long periods of laboring towards producing an industrially viable product. It was most important to have a research organization capable of executing both activities simultaneously. Brilliant insights are of intellectual interest only if no basis for generating an industrially significant product exists. On the other hand, many products do not stand the test of time without a series of creative insights resulting in innovations that can improve product performance. Throughout this discussion these thoughts should be kept in mind.

Another thought raised in this discussion was the concept of the value of the individual researcher. This point was discussed at length during the Wednesday evening dinner. Dr. Ron Jensen (Director, Corporate Research & Development-Engineering, Shell Development) had made this the central focus of his speech after dinner. There are many research groups, teams, etc., but we should never lose sight of the individual researcher and his individual contribution as an innovator and inventor. It is the innovator as an individual, albeit while acting as a member of a team, that we should all be primarily interested in promoting.

At this point Prett reviewed the workshop objectives as presented in the front of this book. They are reproduced here for emphasis.

Objectives

"The objective of this process control research workshop is to improve the communication process between academic researchers, industrial researchers, and the engineering community in the field of process control. From this communication process it is intended for us all to improve our understanding of the nature of the control problems facing us. Where does the theory apply and where does it not apply? What is the optimal role for academic researchers, and what is the correspondingly appropriate role for industrial researchers? Are these roles different? In what ways must we modify our respective approaches to problem definition and solution so as to improve the complementary aspect of our efforts? These are some of the issues we explore in the workshop."

It was pointed out that these objectives were "lofty" and "noble" in expression. However, they are also in the nature of being very difficult to measure in the sense of improvement or progression towards achievement. From the communication standpoint it can be said that representatives of academic research, industrial research and the engineering community were present at this workshop and did indeed communicate with each other. As will be seen from this discussion summary, some important progress was made as a result of this communication.

To initiate open discussion Prett made the following observation:

"Academic researchers should become more comfortable with the concept of studying and understanding problems without necessarily trying to produce a product that solves the problem. Rather, let us recognize that very often the result of the study can actually allow us in industry to solve the problem with existing techniques together with this new understanding.

"Academic researchers are particularly well placed to perform this role if industrial researchers and the engineering community will put increased effort into communicating the nature of the industrial problem to the academic community.

"Thought leaders in industry and academia must nourish and promote this 'understanding' attitude through promotion of the individuals who are interested in this approach. This can be done by encouraging the publishing of papers and presentations along these themes. It can also be achieved by promoting increased efforts to organize tutorial sessions and focused seminars on these themes to run concurrently with major national conferences such as operated by AIChE and ACC.

"It should be emphasized, for instance, that within Shell the control engineering community as represented by B. L. Ramaker is at least as interested in the control research group performing a 'studying and understanding' role as he is interested in their producing so-called new algorithms, etc. There is a lesson in this for the academic community.

"An excellent example of this relationship between industrial and academic research groups is evidenced by that existing informally between Shell and M. Morari's academic research group. The understanding we in Shell have gained as to the nature of the control problem due to the work along the lines of IMC and model uncertainty is significant. However, IMC will not represent a product or tool for us, but rather a vehicle we will use to better understand and tackle our control problems. We would even claim that the Shell developed DMC approach with its sophisticated constraints handling scheme is a superior product. This in no way minimizes the value of the outstanding research associated with many individual efforts based on IMC. This is true of many similar research efforts, and I single out IMC only because it is well known to all present."

Having made this point, Prett invited C. Moore to dialogue with the group along the lines of one of his research presentation's suggestions for future areas of endeavor for process control engineers. Moore summarized his thoughts with the following statement:

"There is an urgent need for a new area of research which is separate and distinguishable from the traditional *process control system design* which receives so much theoretical focus today. This new area could perhaps be called *process control system cultivation* and would address the monitoring evaluation and evolution of an operating control policy."

With this suggested idea in mind, Prett summarized his comments as noted earlier, directed attention to Moore's statement, and opened the floor for general discussion and comment.

This general discussion, to which all workshop participants contributed very actively, resulted in significant progress in understanding the control problem and its solution. It would not be possible to document each and every comment made, so this summary will only highlight those points that were raised and upon which there was a consensus. These were:

- The innovator, the inventor, the individual researcher is an invaluable resource that must be nourished, supported, and promoted.

- Process control must establish itself as an independent pursuit working with, but not dominated by, the control theorists in fields other than chemical processes. Continuing to follow rather than lead these other control theorists can only result in our experiencing great difficulty in generating methodologies for problem solution in our field. Chemical process control engineers need methodologies that can handle constraints, time varying performance objectives, and many other difficult problems. Following in the Electrical Engineering Linear System Theory vein will not lead to this problem solution.

- It became apparent to us all that there is a great gap between the theory and the practice of process control and that increased emphasis must be placed on theoretical developments to facilitate technology transfer from academia to industry and from industrial research labs to the engineering community. Closing of this technology gap is at least as important as the invention of new technology.

- The value of the study of process control problems in order to gain an increased understanding of the nature of the problem must be recognized and such efforts rewarded. Products may or may not result from such efforts (i.e., products, tools, or techniques in the traditional sense), but the increased understanding is of inherent value and can of itself make the problem a non-issue.

- Although it is difficult to design a perfect or even optimal relationship between academia and industry, there is one thing that can be said with certainty. It is that industry needs to open up its communication process with academia and expose them to the real world of control problems. This is not intended to promote problem solution, per se, in the individual sense, but to promote the overall problem "understanding" issue raised by Moore.

- To meet the needs of academic researchers it was decided to include in this final report a complete specification of a complex control problem that could represent a basis for academic research on so-called "real" problems.

With these points in mind as representative of our collective views resulting from the workshop, Prett closed the discussion.

If there was one point on which all participants agreed, it was that the workshop had been a positive and enlightening experience for all. There was a general feeling of having accomplished a difficult task, and for this all participants are to be congratulated. All present agreed that their perspective had been changed by the experience, and when this was compared to the noble objectives of the workshop, the conclusion must be that we were successful.

J. J. Haydel closed the workshop with the statement that given the success of this first workshop, we will meet again for a second workshop in the future to review progress. A suggested frequency of these workshops was agreed to by all present.

The last word went to Manfred Morari, who said in summary:

> "As the planning for the workshop proceeded, I had my doubts whether such an effort as proposed would be beneficial and whether significant subjects for discussion would arise. I am happy to say I was proven wrong on both counts. I must congratulate Shell Development on executing a well planned and organized workshop that has resulted in significant discussion of some very important topics. I hope this discussion will continue in the future. Again, I congratulate Shell Development on a job well done."

On this positive statement we end this report.